Y0-BYB-345

CRC SERIES IN BIOCOMPATIBILITY

Series Editor-in-Chief

David F. Williams, Ph.D.
Senior Lecturer in Dental and Medical Materials
Department of Dental Sciences
University of Liverpool
Liverpool, ENGLAND

FUNDAMENTAL ASPECTS OF BIOCOMPATIBILITY
Editor: David F. Williams, Ph.D.

BIOCOMPATIBILITY OF ORTHOPEDIC IMPLANTS
Editor: David F. Williams, Ph.D.

BLOOD COMPATIBILITY
Editors: David F. Williams, Ph.D. and Donald J. Lyman, Ph.D.

BIOCOMPATIBILITY OF DENTAL MATERIALS
Editors: D. C. Smith, Ph.D. and David F. Williams, Ph.D.

SYSTEMIC ASPECTS OF BIOCOMPATIBILITY
Editor: David F. Williams, Ph.D.

BIOCOMPATIBILITY IN CLINICAL PRACTICE
Editor: David F. Williams, Ph.D.

BIOCOMPATIBILITY OF CLINICAL IMPLANT MATERIALS
Editor: David F. Williams, Ph.D.

Systemic Aspects of Biocompatibility

Volume I

Editor

David F. Williams, Ph.D.

Senior Lecturer in Dental and Medical Materials
Department of Dental Sciences
University of Liverpool
Liverpool, England

CRC Series in Biocompatibility

Series Editor-in-Chief

David F. Williams, Ph.D.

Senior Lecturer in Dental and Medical Materials
Department of Dental Sciences
University of Liverpool
Liverpool, England

CRC Press, Inc.
Boca Raton, Florida

Library of Congress Cataloging in Publication Data
Main entry under title:

Systemic aspects of biocompatibility.

(CRC uniscience series on biocompatibility)
Bibliography: p.
Includes index.
1. Metals—Toxicology. 2. Metals in the body.
3. Metals in medicine. 4. Polymers and polymerization
—Toxicology. 5. Polymers in medicine. I. Williams,
David Franklyn. II. Series. [DNLM: 1. Biocompat-
ible materials. QT34 S996]
RA1231.M52S97 615.9'02 80-17499
ISBN 0-8493-6621-6 (v. 1.)
ISBN 0-8493-6622-4 (v. 2.)

Direct all inquiries to CRC Press, Inc., 2000 N.W. 24th Street, Boca Raton, Florida 33431.

International Standard Book Number 0-8493-6621-6 (Volume I)
International Standard Book Number 0-8493-6622-4 (Volume II)

Library of Congress Card Number 80-17499
Printed in the United States

SERIES PREFACE

One of the most noticeable and beneficial aspects of the recent developments in medicine has been the exploitation of technological advances and innovations. Modern hospitals and clinics contain many pieces of hardware that are used for diagnostic and treatment purposes and which testify to this progress. While it is difficult to single out any one particular branch of biomedical engineering, as this subject is now called, as giving the most benefit to the patient, the tremendous advances that have been made in surgery through the use of implanted devices must certainly number among the more significant. Anyone who has witnessed the transformation that is produced in an arthritic patient by treatment with a total joint replacement prosthesis can be left with little doubt as to the clinical importance of this type of development.

Although a wide spectrum of implanted devices now exists, fulfilling needs in such diverse surgical disciplines as ophthalmology, cardiology, neurosurgery, dentistry, and orthopedics, they all have one thing in common; they must have intimate contact with the patient's tissues, providing a real, physical interface. It is the existence of this interface that poses the most intriguing questions, provides the most interesting challenges, and restricts developments most seriously. This series of volumes is about this interface, the reactions which occur there, and the effects these reactions have on prostheses and, especially, on patients. The term biocompatibility has been introduced with reference to these interactions. It is the term used to describe the state of affairs when a material exists within a physiological environment without either the material adversely and significantly affecting the body, or the environment of the body adversely and significantly affecting the material. Absolute biocompatibility may be regarded as utopia, while in a more realistic way we have to consider the various degrees of biocompatibility that we find in practice. As discussed in the opening chapter to *Fundamental Aspects of Biocompatibility* there are many ways of looking at biocompatibility and the literature is replete with observation, comment and argument over what constitutes "adverse" and "significant" interactions and what "biocompatibility" really means.

It is partly because of the somewhat confused state that exists at this moment and partly because of the tremendous importance of this subject that this series of volumes on biocompatibility has been written and compiled. The planning of the series was based on two premises, first that a better understanding of biocompatibility could be achieved by considering both the fundamental aspects of these interactions and their clinical effects, and secondly that a collection of papers within one series derived from all branches of surgery would, through a process of cross-fertilization of ideas, serve to enhance this understanding.

The series starts, therefore, in the first volume, with a discussion of the fundamental aspects of biocompatibility, covering the principles of materials degradation in the physiological environment and the basic features of the tissue response to implanted foreign bodies. *Biocompatibility of Orthopedic Implants*; *Blood Compatibility*; and *Biocompatibility of Dental Materials* are volumes which deal with the major clinical areas that utilize implants: orthopaedic, cardiovascular, and dental surgery. In each of these volumes the clinical features of biocompatibility are emphasized, using descriptions of the devices, materials, and clinical procedures as a basis for discussion where necessary and appropriate.

Systemic Aspects of Biocompatibility returns to the theme of the fundamental interactions and explores the question of the systemic effects that may arise after biomaterial-tissue contact and especially reviews the relationship between biocompatibility and toxicology. *Biocompatibility in Clinical Practice* then brings together a wide range of surgical disciplines, the clinical features of biocompatibility associated with diverse procedures such as the treatment of hydrocephalus, the use of contact lenses, bladder replacements, and cardiac pacemakers being reviewed and compared. In *Biocompatibility of Clinical Implant Materials,* biocompatibility is discussed from the materials point of view, each major material being reviewed in terms of its properties, clinical use, and biocompatibility.

These volumes are not merely independent publications on biocompatibility, but rather provide a series of complementary sources of instruction and reference, providing what is believed to be the first comprehensive review of biocompatibility.

The series could not, of course, have been completed without the assistance and patience of the many contributing authors, to whom I am deeply grateful.

David Williams

PREFACE

Systemic Aspects of Biocompatibility is basically concerned with the effects that biomaterials may have on tissues remote from the site of their application. An examination of the literature on this subject will reveal that very little factual data exists for, in the main, attention has been primarily focused at the tissue-material interface and the local effects of any interaction. Since many substances are readily transported in the tissues, it must be expected that products of any interaction may be dispersed systemically and it is highly relevant to consider their fate and their potential for tissue damage at these distant sites. An attempt is made in these two volumes, therefore, to correlate systemic biocompatibility with the known toxicology of the materials and, indeed, many of the chapters have been contributed by acknowledged authorities on the toxicology of the specific biomaterials considered.

As with the other parts in the series, the rationale and planning are discussed in the first introductory chapter, and there is little need to preempt that discussion in a preface. It is hoped that, as a sequel to *Fundamental Aspects of Biocompatibility,* this work will provide a further basis for the understanding of the fundamentals of biocompatibility.

These volumes could not have been completed without the expertise of either the contributing authors or the editing staff at CRC Press. To all these I wish to express my gratitude.

David F. Williams

THE EDITOR

David F. Williams, Ph.D. is Senior Lecturer in Dental and Medical Materials in the School of Dental Surgery, University of Liverpool, England.

Dr. Williams received a B.Sc. with 1st class Honors in 1965 and a Ph.D. in 1969, both in Physical Metallurgy at the University of Birmingham. He has been on the staff at Liverpool University since 1968, apart from a year as Visiting Associate Professor in Bioengineering at Clemson University, South Carolina in 1975.

Dr. Williams is a Chartered Engineer and a Fellow of the Institution of Metallurgists. He has been a council member of the Biological Engineering Society and is currently chairman of the Biomaterials Group of that Society.

Dr. Williams has major research interests in the interactions between metals and tissues, polymer degradation and the clinical performance of biomaterials. He has authored many research papers and reviews in these areas and has previously written two books and edited two volumes.

CONTRIBUTORS

R. D. Bagnall, Ph.D.
Senior Lecturer
Dental Materials Science
Edinburgh University Dental School
Edinburgh, Scotland

C. Allen Bradley, Ph.D.
Professor and Chairman
Department of Biopharmaceutical Sciences
University of Arkansas for Medical Sciences
Little Rock, Arkansas

Stanley A. Brown, D. Eng.
Associate Professor of Bioengineering in Orthopedic Surgery
University of California
Davis, California

Ulrich Borchard, Ph.D.
Professor of Pharmacology
University of Dusseldorf
Dusseldorf, West Germany

Thomas Ming Swi Chang, M.D., F.R.C.P.(c)
Director, Artificial Cells and Organs Research Center
Professor of Medicine and Physiology
McGill University
Montreal, Quebec
Canada

G. C. Clark
Professor of Dental Sciences
University of Liverpool
Liverpool, England

Michael G. Crews, Ph.D.
Assistant Professor of Food and Nutrition
Texas Tech University
Lubbock, Texas

Jeanine Grisvard, Ph.D.
Assistant Professor of Cellular Biology
University of Paris XI
Orsay, France

Etienne Guille, Ph.D.
Assistant Professor of Molecular Plant Biology
University of Paris XI
Orsay, France

Thomas J. Haley, Ph.D.
Science Advisor to the Director
National Center for Toxicological Research
Jefferson, Arkansas

Stephen P. Halloran, B.Sc., M.Sc.
Senior Biochemist
Department of Clinical Biochemistry
St. Peter's Hospital
Chertsey
Surrey, England

Arne Hensten-Pettersen, Dr. Odont.
Research Associate
NIOM, Scandinavian Institute of Dental Materials
Oslo, Norway

Hogne Hofsøy, D.D.S.
Assistant Professor of Microbiology
University of Oslo
Oslo, Norway

Leon L. Hopkins, Ph.D.
Professor and Chairman
Department of Food and Nutrition
Texas Tech University
Lubbock, Texas

Robert A. Jacob, Ph.D.
Chief Clinical Nutrition Laboratory
Jamaica Plain, Massachusetts

Nils Jacobsen, Dr. Odont.
Associate Professor of Dental Technology
University of Oslo
Oslo, Norway

Jindřich Kopeček, Ph.D.
Head, Laboratory of Medical Polymers
Institute of Macromolecular Chemistry
Prague, Czechoslavakia

Sverre Langård, M.D.
Head, Department of Occupational Medicine
Telemark Sentralsjukehus
Porsgrunn, Norway

Laszlo Magos, M.D.
Toxicologist
Medical Research Council Toxicology Unit
Carshalton
Surrey, United Kingdom

Katharine Merritt, Ph.D.
Associate Professor of Microbiology in Orthopedic Surgery
University of California
Davis, California

Robert J. Pariser, M.D.
Assistant Professor of Medicine (Dermatology)
Eastern Virginia Medical School
Norfolk, Virginia

Trevor Rae, Ph.D.
Research Associate
University of Cambridge Clinical School
Cambridge, England

Igor Sissoeff, Ph.D.
Assistant Professor of Molecular Biology
University of Paris XI
Orsay, France

Gail K. Smith, Ph.D., V.M.D.
Assistant Professor of Orthopedic Surgery
School of Veterinary Medicine
University of Pennsylvania
Philadelphia, Pennsylvania

Michael K. Ward, M.D.
Consultant Physician
Royal Victoria Hospital
Newcastle, England

David F. Williams, Ph.D.
Senior Lecturer in Dental and Medical Materials
University of Liverpool
Liverpool, England

TABLE OF CONTENTS

Volume I

METALLIC MATERIALS

Volume II

Metallic Materials

Chapter 1

INTRODUCTION

D. F. Williams

In the volume *Fundamental Aspects of Biocompatibility* in this series the characteristics of the interactions between biomaterials and tissues that take place at their interface and the consequences of those interactions in the immediate vicinity were discussed. Thus, consideration was given to the mechanisms of metallic corrosion and polymer and ceramic degradation in the physiological environment and the resulting histological and biochemical effects in the tissue. Even within this context, however, the association between biocompatibility and general toxicology was introduced and discussed, since the contact between foreign substances and tissues arising from biomaterials applications (i.e., implantation and extracorporeal circulation) merely represent one of the several methods of administration of such substances to the body. A comparison of their fate following implantation to that following oral, inhalation, dermal, and intraperitoneal routes is, therefore, extremely useful for the understanding of the basic parameters of biocompatibility. This concept is taken one stage further in this part of the series, which is largely concerned with the fate of biomaterial-derived substances in the tissues and which calls extensively upon data from general toxicology and biochemistry. The effects described may be manifest in the tissue around an implanted device but this need not be the case and, indeed, this part has been arranged such that the phenomena are largely discussed independent of the site of the biomaterial; the volumes are therefore entitled *Systemic Aspects of Biocompatibility.*

There is very little common ground between the toxicity of metals and polymers and so the contributions are divided into two parts to deal with these two categories of material. The third class of material, ceramics, is not specifically included here since, as discussed in *Fundamental Aspects of Biocompatibility,* their toxicology can be related to that of metals.

The second part on polymeric materials is introduced separately in Chapter 4 of *Systemic Aspects of Biocompatibility,* Volume 2. This first part, dealing with metals, is effectively divided into three subsections. Chapters 2 to 6 cover some general aspects of metals in the tissue and the basic characteristics of their distribution and physiological effects. It will be recalled that all but the transuranic metals are found in nature, usually in the combined state as an ore and, indeed, many of the metals are ubiquitous and of widespread distribution. Many, therefore, are found in quite high concentrations in animals and plant tissues and a significant proportion, the so-called essential trace elements, actually have a positive role to play in physiological processes. In normal, healthy man there will be an optimum concentration of metals in the body fluids, which may range from zero with those which are highly toxic, to large amounts in the case of an essential metal such as iron. Normally the body, through homeostatic mechanisms, controls the amount of metal that is absorbed and stored. Problems arise when the body is presented with metals in large amounts, or in abnormal forms such that the normal equilibrium cannot be maintained, giving rise to the possibility of metal ion toxicity.

There are many factors that are important in this context and several have been selected for close examination in this volume. First, there is the general question of the types of biochemical reactions in which the metal ions can participate and in particular the nature of the complexes that they can form in biological fluids. This subject is addressed by Clark in Chapter 2. The most significant way in which metals exert an influence in physiological systems is via their relationship with enzymes. Many enzymes require the presence of certain metal ions for their activity, xanthine oxidase requiring molybdenum, carboxypeptidase requiring zinc, pyruvate carboxylase needing manganese, and so on. At the same time, however, this does mean that metal ions may be able to interfere with certain enzyme catalyzed processes if one metal which is present in unusually large amounts can bind to the site on an enzyme that is normally occupied by another metal, thereby producing inhibition. These interactions between metals and enzymes are discussed by Rae in Chapter 3.

Metal-enzyme interactions are of great significance in metal toxicology but represent only one of a series of types of interactions which may take place between metals and biological structures. The whole spectrum of interactions is largely dependent on the affinity that metal ions have for DNA and this particular subject is covered by Guille et al. in Chapter 4.

When dealing with metal ion toxicology it is tempting to consider each metal in isolation. This can be misleading, however, and in Chapter 5 Magos introduces the concepts of synergism and antagonism in metal toxicology, illustrating the discussion with examples. The final chapter in the first section by Jacob, deals with the problem that faces any research worker who is attempting to monitor trace metal levels in groups of people. One possible noninvasive technique of sampling is to use hair as a biopsy material and it is this which is discussed in Chapter 6.

Chapters 7 to 16 each deal with the biological properties of a specific metal. Included here are nickel (Jacobsen et al. Chapter 7), cobalt (Clark, Chapter 8), chromium (Langard and Hensten-Pettersen, Chapter 9), molybdenum (Williams, Chapter 10), titanium (Williams, Chapter 11), vanadium (Crews and Hopkins, Chapter 12), aluminum (Bradley, Chapter 13), copper (Halloran, Chapter 14), mercury (Williams, Chapter 15) and silver (Pariser, Chapter 16). These have all been chosen for their relevance to biomaterials. Nickel is a major alloying element in stainless steel, in some cobalt-based alloys, and also in some nickel-titanium alloys. Cobalt is the parent metal of numerous surgical alloys, most of which also contain chromium. Chromium is also found in stainless steel. Titanium is used either as a pure metal or in an alloy with small quantities of both aluminum and vanadium. Aluminum is also the metallic element bound to oxygen in the major surgical ceramic, alumina. Copper is utilized in certain intrauterine contraceptive devices. Mercury is a major component of dental amalgam and silver is widely used in dental alloys. The only metallic material used to any significant extent in surgical alloys, but not included here, is iron. This has been omitted because of the difficulty of extracting relevant information from the mass of data that has accumulated on the biological properties of the metal. A few aspects of this are, however, discussed later by Smith in Chapter 1 in Volume 2.

With all these metals the normal tissue levels, the metabolism and storage, and the toxicity are discussed, particular attention being paid, where appropriate, to the mechanisms of the reactions between the metal ions and the tissues. This section, therefore, provides a firm basis for the study of the subtle aspect of the biocompatibility of the metals.

At this stage, detailed clinical evidence of metal ion toxicity arising from the use of biomaterials is not readily available. However, some interesting observations are made in Chapters 1 to 3 of Volume 2 on this subject. In Chapter 1, Volume 2 Smith reviews the possibilities of systemic effects arising from the implantation of metals and uses his own experimental work on iron as the basis for discussion. A slightly different mode of entry of metal ions into the human body is the subject of Chapter 2 of Volume 2 in which Ward describes metal ion toxicity resulting from hemodialysis. In this case, metal ions are contained within the dialysate and are able to pass through the membrane directly into the blood stream. The very significant effects associated with small amounts of aluminum in this context testify to the relevance of metal ion toxicology. Finally, in Chapter 3, Volume 2 Merritt and Brown review the role of biomaterial-derived metal ions in the development of hypersensitivity.

This volume is meant to provoke thought amongst scientists and clinicians involved in the use of biomaterials on the possibilities of systemic effects arising from tissue-biomaterial interactions. The fact that overt systemic effects are not clinically obvious is a good sign, but it must not deceive us into thinking that no such effects can arise. The factual basis provided in this volume on the general toxicity of these materials should prove useful in the study of these more subtle aspects of biocompatibility.

Chapter 2

LOW MOLECULAR WEIGHT COMPLEXES OF TRANSITION METAL IONS IN BIOLOGICAL FLUIDS

G. C. F. Clark

TABLE OF CONTENTS

I. INTRODUCTION

The transition metals are those elements with a partially filled electron shell, either in their elemental or in one of their common oxidation states. They occupy the central block of the periodic table, filling the three rows from scandium to copper, zirconium to silver, and hafnium to gold. Although copper is not always considered a transition metal, it is included here because of the partially filled shell in the oxidized state. In the biological context it is often useful to include zinc, as it shares some of their characteristics, biologically if not chemically.

They are very much typical metals, with high melting points and high strength, and have been exploited by man as his major tool-making materials. The various surgical prostheses now in use for implantation[1] are all, with the exception of the ceramic, polymeric, and glassy types, alloys of various transition metals. The stainless steels are alloys of iron, nickel, and chromium. The cobalt-chromium alloys are, as their name suggests, made up of cobalt and chromium, plus molybdenum and sometimes nickel and tungsten. Titanium implants are either of pure titanium or of an alloy containing vanadium and aluminum, while electrodes for such devices as pacemakers are made from alloys of the various platinum metals.

Apart from such artificial situations, transition metals never occur in very large amounts in any organism. The majority of them have no biological function, and indeed the soluble salts of all of them are toxic if taken in large enough amounts. However, several of them are absolutely essential for the central metabolism of all organisms. Although several metals are known to be essential for mammals alone, iron, copper, cobalt, zinc, and manganese may be regarded as essential for all life.

Life, as we know it, is thermodynamically unstable. It depends above all on a continuous flow of energy to maintain it, and this energy flow derives from an electron flow from reducing agents, such as glucose, to oxidizing agents, such as oxygen itself. The range of reduction potentials encountered in biological systems is large, from −450 mV in sulfate-reducing bacteria to + 850 mV in the iron bacteria.[2] The range of values found in mammalian systems is less extreme, but it is still much wider than that found for most oxidation-reduction reactions of organic materials. To transport electrons from glucose to oxygen, retaining the energy released for use by the cell, requires a string of concatenated reactions, coupled to each other and to energy trapping reactions, such as the synthesis of high energy phosphate bonds in adenosine triphosphate.[3] The compounds used are nearly all iron complexes, either with sulfur, in the ferredoxins,[4] or with porphyrin rings, in the cytochromes.[5] It would be very difficult to imagine a series of purely organic compounds which could ever replace these metal complexes.

In addition to their role in electron transport, iron and copper proteins are an essential part of the oxygen transporting systems of animals[6], and metalloproteins play an essential role in the binding of oxygen and nitrogen prior to their reduction.[7] Although both of these gases are, on the basis of energetics alone, very easy to reduce, in practice the reactions are very slow. Only metal complexes can bind them in such a way as to lower their great activation energies and allow the reduction to take place at low temperatures and in aqueous systems. The reduction of oxygen takes place using cytochrome oxidase, which contains iron and copper, while nitrogen reduction uses nitrogenase, a molybdenum-iron protein found only in bacteria and blue green algae.[8]

Cobalt is of great interest in the context of the interrelationship of chemistry and biology. Biochemically the free metal is not of any great importance, but in the form of vitamin B_{12} it is essential for all higher animals. The vitamin acts as a form of handle by means of which cells are able to manipulate single carbon units, such as methyl and formyl groups.[9] The intermediates consist of complexes containing metal-

Table 1
OCCURRENCE OF TRANSITION METALS IN MAN AND THE ENVIRONMENT

Metal	Man(ppm)	Earth's crust (ppm)	Biochemical functions
Cobalt	0.01—0.7	25	All reactions involving vitamin B_{12}
Chromium	0.02—0.04	100	Glucose tolerance factor
Copper	4	55	Cytochrome oxidase, peroxidases
Iron	50	50,000	Redox reactions involving cytochromes or iron sulfur proteins
Manganese	1	950	Superoxide dismutase, mitochondrial membrane structure
Molybdenum	0.2	1.5	Xanthine oxidase, nitrogenase in prokaryotic organisms
Nickel	0.04	75	No known function

Note: Numerical data from Reference 10. The list of functions is not exhaustive, but indicates the relative importance of some of the metals.

carbon bonds, a form otherwise unknown in biological systems, and until recently a relatively unexplored area of chemistry.

The other essential metals are dealt with in more detail elsewhere. It should be noted that all of the transition metals, essential or not, are toxic if supplied in a high enough dose. At concentrations above those found naturally they will bind to the active sites of enzymes, disrupt membrane functions, and disturb the equilibrium controlling other metals. Table 1 shows the amounts of some transition metals in man and in the environment with their biological functions.

On the whole, the body only absorbs a certain proportion of any metal in the diet. For example, about 10% of the iron present in the diet may be regarded as absorbable with an average intake of between 0.5 and 2 mg/day. The actual quantity of an essential metal in the body may be in fact very much larger than the daily intake, which implies that there will be a constant recycling of the body's stores, as cells die and are replaced. The total quantity of iron in the body is 3 to 4 g, present as hemoglobin in the red cells, transferrin (an iron transport protein) in the blood, and hemosiderrin and ferritin in the liver. Due to the short life of red cells, there is a very great exchange between these pools. Although 99.9% of the iron in the body is protein bound, rapid exchange can only come about via its dissociation from one protein to form a low molecular weight complex, which is freely diffusible, and the reaction of this with another protein.[12] It is the properties of these complexes, their kinetic and thermodynamic stability, and their charge, which largely governs the exchange of iron and other metals.

The properties of these complexes are a function of both the metal and its ligands. It is not possible to fully understand the ways in which the properties change from metal to metal without properly understanding the chemical forces responsible for the formation of the complexes.

II. INORGANIC CHEMISTRY

A. General

The transition metals are shown in Table 2 which is in the form of an extract from the periodic table. As mentioned in the introduction, the transition metals are defined by the properties of their electron shell, the number of electrons for each element being given in Table 2. The properties which distinguish these elements from others are their

Table 2
TRANSITION ELEMENTS SHOWING
d ELECTRON DISTRIBUTION

Sc	Ti	V	Cr	Mn	Fe	Co	Ni	Cu
1	2	3	5	5	6	7	8	10
	Zr	N6	Mo	To	Ru	Rh	Pd	Ag
	2	4	5	6	7	8	10	10
	Hf	Ta	W	Re	Os	Ir	Pf	Au
	2	3	4	5	6	7	9	10

Note: Beneath each element are the number of d electrons in the elemental state. This does not increase in steps of 1 since other shells may be filling at the same time. Although Cu, Pd, Ag, and Au have filled shells in their elemental states their common ions have unfilled shells. Although y (preceding Zr) and Lu (preceding Hf) have an unfilled d shell, for practical purposes they can be excluded from this list.[16]

variable oxidation states, their very great propensity to form strong complexes with both inorganic and organic ligands, and the tendency of these complexes to be colored.

All of the transition elements have at least two oxidation states. For some, such as silver, only one is commonly found, and indeed in aqueous solutions there are several which commonly exist in only one state. However, all of the biologically important transition metals can exist in two or more states in the body. This is both a function of, and a major determinant of, the complexes formed. For example, cobalt (III) is an unstable state when the metal is complexed by water, its oxidation potential being so high that it oxidizes water. However, in the presence of ammonia the situation is changed so much that cobalt (II) is oxidized by oxygen to cobalt (III) forming the cobalt (III) ammonia complex which is extremely stable.[13]

The oxidation states of the first row range from Ti (IV) to Cu (I), with a continuous change along the row. In general the further to the right the more stable the lower oxidation states become. The second and third rows are very alike. Their oxidation states are higher than those for the first row. For molybdenum and tungsten the chemistry is dominated by the (VI) state. A major result of this is that they form large multicentered complexes, such as the so-called hetero-polyacids with phosphorus and arsenic e.g.,

$$[P_2\ Mo_{18}\ O_{62}]^{6-}$$

For this reason, and because they provide fewer elements of biological importance, the second and third rows will not receive as full a coverage here as the elements of the first row.

B. Ligand Field Theory

The properties of the complexes formed by the transition metals can be explained by a group of theories, covering a wide range of sophistication, which are collectively known as ligand field theory. The name derives from the electric field in which a metal ion finds itself when it is surrounded by the ligands which make up a complex. The electronic interactions between the metal and its ligands determine the stability, geometry, stoichiometry, and all the other properties of the complex.

A metal ion suspended in a vacuum is surrounded by an intense electric field. On

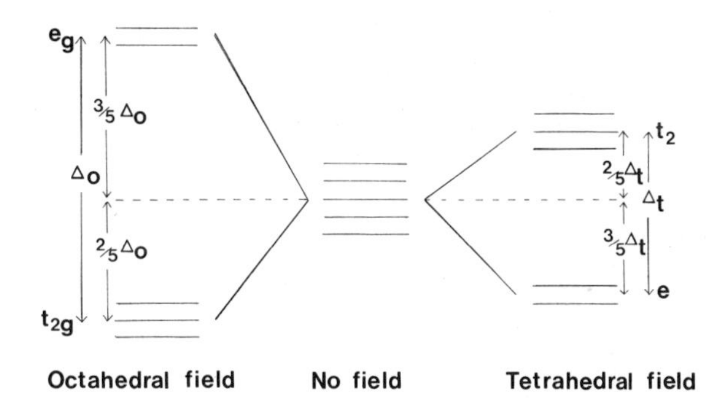

FIGURE 1. Splitting of orbital energies in an electrostatic field.

immersing it in a solvent such as water, it immediately loses at least 79/80 of the energy in the field.[14] This energy is used to form a shell of water molecules around the ion, with their dipole moments aligned towards the metal. This process occurs for all metals and accounts for the outstanding properties of water as a solvent. The energy loss is known as the hydration energy, and, as an example, is about 267 k cal mol^{-1} for manganese.

If the solution contains anions such as fluoride, then there will be a tendency for these to gather in the sphere surrounding the metal due to electrostatic forces. This process does not in general produce strong complexes unless the anion carries several independent negative changes, i.e., is multidentate. Thus, calcium acetate is almost totally dissociated in solution, while calcium citrate is extremely stable.

Metals with d electrons are capable of further interactions with their potential ligands. A metal with one d electron can be analyzed into five separate states of equal energy. In three of these the orbitals lie between the three cartesian axes, and in the other two they lie along them. If such a system is placed at the center of an octahedral array of negative charges, the orbitals will split into two groups, as shown in Figure 1.

The orbitals along the axes will be increased in energy, due to electrostatic repulsion, and the other three orbitals decreased in energy. The two high energy states are known as e_g orbitals, and the low energy ones t_{2g}. The difference between the states is denoted by Δ. It is not only charged ligands which are capable of splitting the orbitals like this, as any dipolar molecule will have this effect, even water. The magnitude of the splitting is primarily a function of the ligand, and these can be arranged to form the spectrochemical series in ascending order of magnitude, $I^- < Br^- < Cl^- < F^- < OH^- < C_2O_4^{2-} = H_2O < -NCS^- < NH_3 <$ ethylene diamine $< CH^-$.[15]

For a multi-electron atom the orbitals will be filled according to the normal rules. Electrons have a spin of ± 1/2, and each orbital can hold two electrons of opposite spin, so the net result is zero spin. The electrons are distributed so as to minimize their energy and maximize their spin. If Δ is large, the t_{2g} orbitals are all filled before any electrons are placed in the high energy e_g orbitals. There is then a net lowering of the energy of the complex of 2Δ/5 per electron, which is in the region of 100 kJ mol^{-1}.

This lowering of the energy of the complex due to the splitting of the d orbitals is known as the ligand field stabilization energy (LFSE). The energy per ligand is comparable with that for covalent bonds, and it accounts for the very great stability of the complexes formed. As mentioned, the assignment of electrons to orbitals is a function of Δ and of the spin distribution. For manganese (II) the most common arrangement is a high spin state in which there is one electron in each energy orbital. The high energy orbitals balance the low energy ones and there is no LFSE. Thus manganese is notable among the transition metals for the low stability of its complexes. The spectrochemical series can normally be used as a guide to the stability of complexes, as the LFSE is proportional to Δ. For manganese, with no LFSE at all, complexes are formed to a great extent according to electrostatic interactions and the law of mass action.

The LFSE alters for different metals according to the number of electrons. The metals can be arranged in order of the stability of their complexes in a similar way to the ligands. This is known as the Irving Williams order, $Mn^{2+} < Fe^{2+} < Co^{2+} < Ni^{2+} < Cu^{2+} > Zn^{2+}$.[16] There are exceptions to this rule, but for the ligands found in biological fluids it is generally true. The complexes of the higher oxidation states are generally stronger, because of the greater electrostatic interaction.

C. Kinetics

In the discussion so far, it is the thermodynamic stability that has been involved. However, there is always a possibility in chemistry of substances that are thermodynamically unstable but yet kinetically very stable, due to the high activation energies of the reactions by which they decompose.

For most metal complexes the reactions they undergo are almost immeasurably fast. Most reactions involve the displacement of a ligand by water and the subsequent displacement of this by another ligand. The rate constant for the exchange of water in copper complexes about 10^8 S^{-1}.[17] For some metals, especially cobalt (III) and chromium (III), this is not the case. Cobalt (III) complexes have reaction half-times of the order of weeks. They can be prepared by oxidation of ammoniacal solutions of cobalt (II), when they are thermodynamically stable.[13] On changing the conditions to dilute acid they become thermodynamically unstable, but do not decompose because of their kinetic stability.

Behavior of this sort is seen in the polymeric hydroxides formed on hydrolyzing some metal salts, expecially iron (III). These can be formed if the concentration of the free metal or the pH is suddenly raised, and form stable colloidal particles which only slowly dissolve when the conditions are returned to normal.

D. Hydrolysis

While it is convenient to talk of the free metal when considering metal ions in solution, one can rarely ignore the fact that the metal is complexed by water, normally forming an octahedral complex.

Water can be considered as a weak acid, with a pk of about 14. If it is complexed to a metal ion, electrons will be withdrawn from the protons, and the pk of the water will fall. Thus a complex of the type $M(H_2O)_6^{2+}$ will behave as a stronger acid than water.

As with other polyprotic acids there will be a series of ionizations with increasing pk's. Thus for iron (III) the following reactions occur[18]

$$Fe(H_2O)_6^{3+} = Fe(H_2O)_5(OH^-)^{2+} + H^+, \text{pk } 2.19 \quad (1)$$

$$Fe(H_2O)_5(OH^-)^{2+} = Fe(H_2O)_4(OH^-)_2^{+} + H^+, \text{pk } 5.67 \quad (2)$$

$$Fe(H_2O)_4(OH^-)_2^{+} = Fe(H_2O)_3(OH^-)_3 + H^+, \text{pk} > 12 \quad (3)$$

For divalent metal ions the first pk values are generally greater than seven, and hydrolysis does not play a very important role. For those trivalent ions like chromium and iron, the first pk is always less than seven and hydrolysis plays a dominant role in their complex chemistry in biological systems.

In addition to the simple hydrolysis outlined above, hydroxide can act as a bridge between two metal ions making possible these polymeric species mentioned in the previous section. Although two-center oligomeric forms are known for divalent cations, this is again a major phenomenon for ions of oxidation state equal to or greater than three. It is of major importance for the transition metals of the second and third row. A very thorough survey of the hydrolysis of all the metallic elements may be found in Reference 19.

E. Oxidation-Reduction

The reduction potential of a metal ion with variable oxidation state is obviously a function of its state of complexation. The potential of the iron (II)-iron (III) couple varies very greatly as shown by the following forms.[20]

$$Fe(CN)_6{}^{3-} + e^- = Fe(CN)_6{}^{4-} \quad E° = 0.36 \text{ v} \tag{4}$$

$$Fe(H_2O)_6{}^{3+} + e^- = Fe(H_2O)_6{}^{2+} \quad E° = 0.77 \text{ v} \tag{5}$$

$$Fe(phen)_6{}^{3+} + e^- = Fe(phen)_6{}^{2=} \quad E° = 1.12 \text{ v} \tag{6}$$

The relative stability of the two states is determined by the value of the change in the electrostatic interaction between the metal and ligands, the possible change of the spin state, and the relative magnitudes of the LFSE in the two states. The rate of the redox reaction may also be affected by the complexation of the metal. This phenomenon has been exploited by biological systems to its greatest extent in the iron-porphydrin complexes in cytochromes. Oxidation and reduction can be of very great importance during the absorption of iron from the intestine. Iron (III) in the diet can be reduced to iron (II) in the gut lumen, rendering it soluble.[21] Absorption still requires that the iron (II) be complexed by, for example, ascorbate, which also stabilizes the oxidation state. In the body fluids each metal may be regarded as retaining only one oxidation state. Changes may occur, but only inside cells or upon binding to protein. The common oxidation state of cobalt is II in aqueous solution, but it appears very likely that its complexes inside cells contain either cobalt (II) or cobalt (III).[22]

III. STABILITY CONSTANTS

In an aqueous system containing a metal ion and a potential ligand a series of reactions can be described,[23] each with its own equilibrium constant.

$$M + L = ML \quad , \quad k_1 = \frac{[ML]}{[M]\,[L]} \tag{7}$$

$$ML + L = ML_2 \quad , \quad k_2 = \frac{[ML_2]}{[ML]\,[L]} \tag{8}$$

$$ML_2 + L = ML_3 \quad , \quad k_3 = \frac{[ML_3]}{[ML_2]\,[L]} \tag{9}$$

The constants k are sometimes replaced by the overall stability constants β, e.g.,

$$M + 3L = ML_3 \quad \beta_3 = \frac{ML_3}{[M]\,[L]^3} \tag{10}$$

For the majority of metals and ligands forming labile complexes, these equilibria are attained very rapidly. The only way in which the actual species present in solution can be found is by calculation of their distribution using known stability constants, which is dealt with in a later section, or by careful measurement of some parameter of the system, such as pH and comparison of this with that predicted for various possible distributions. In practical situations one is forced to rely on the previously tabulated values. There are two major collections of these by Sillen and Martell[24] and Smith and Martell.[25] Although the principles by which these data are obtained have been known since the time of Tannik Bjerrum, in 1941,[26] the accurate determination of very large numbers of constants has depended upon advances in electronics rather than changes in understanding.

Modern methods of determination rely primarily on very accurate pH determination,[27] the use of calorimetric measurements to measure enthalpy changes, and computer programs such as LETAGROP,[28] MINIQUAD,[29] and SCOGS[30] for reducing data.

In biological fluids the concentration of metals is very much less than that of potential ligands. The ligands most successful in competing for the metals are multidentate ligands capable of displacing several other ligands, e.g.,

$$\text{Cu (acetate)}_3^- + \text{citrate}^{3-} = \text{Cu (citrate)} + 3\ \text{acetate}^-$$

The main driving force is the great gain in entropy. This effect, known as the chelate effect,[31] is of very great importance in all biological systems. The major low molecular weight ligands in biological systems are short chain peptides, amino acids, and polycarboxylic acids, all of them capable of forming chelates. In addition, proteins are major chelating agents. A metal bound to a protein not only has many of its coordination sites satisfied, but it is also removed from attack by water. A large number of metals normally exist in plasma almost wholly bound to protein. Exchange with the solution comes about not by dissociation of the hydrated metal from the protein, but rather the binding of another chelating ligand to the metal protein complex, forming a ternary state, and the dissociation of this, giving a low molecular weight complex and the free protein.[12]

Low molecular weight ternary complexes are also of great importance, although very much less data has been obtained for them than for binary complexes. Given a complex of one metal and one ligand, there is no special reason why a second ligand bound should be of the same type as the first. On statistical grounds alone ternary complexes are far more likely than binary complexes. These may also be additional stabilizing factors such as lower electrostatic change, or the second ligand may be a better geometrical fit if it is of a different type.[32]

One very important consequence of the stability of ternary complexes occurs in plasma. Copper (II) forms a complex with cysteine which is liable to disproportonation, the copper being reduced to copper (I) and the cysteine oxidized to cystine. In the presence of histidine the ternary complex copper (II)-histidine-cysteine is formed, in which the redox state of the copper is stable.[33]

For a system composed of several metals and several ligands, the problem of computing the concentration of all of the possible complexes is very great. A series of equations can be set up relating all of the possible species.[34,35] They are usually written in the form of Equation 10, using values of β either from tables of measured values or estimated from analogous compounds. Although it may seem better to disregard any complexes for which accurate data are not available, rather than attempt to estimate a stability constant, it is not necessarily so.[36] Any complex not included is equivalent to one with a stability constant of 0, which will usually be further from the true value than an estimate prepared using all the relevant data.

In addition to the equilibrium equations a second set exists, the mass balance equations, which relate the amount of each component in all of the species to the total amount of that component which is known to be present in the solution. The complete set of equations which must be satisfied are thus:

$$[H\,L_i] = \beta_{ioj}\ [H]\,[L] \tag{11}$$

$$[M_i\,L_j] = \beta_{oij}\ [M_i]\,[L_j] \tag{12}$$

$$[M_i\,L_j\,L_k] = \beta_{oijk}\ [M_i]\,[L_j]\,[L_k] \tag{13}$$

$$H^+ + \Sigma\,HL_i = H\ \text{total} \tag{14}$$

$$M_i + \Sigma_j\,M_i\,L_j + \Sigma_j\,\Sigma_k\,M_i\,L_j\,L_k = M_i\ \text{total} \tag{15}$$

$$L_i + \Sigma_j\,M_j\,L_i + \Sigma_j\,\Sigma_k\,M_j\,L_i\,L_k = L_i\ \text{total} \tag{16}$$

In these equations M_i represents the ith metal, L_i the jth ligand, and the subscripts in β represent the number of protons, metals, and ligands, respectively.

The equations are solved using the free metal and ligand concentrations as variables. The pH is generally fixed, as are all of the total concentrations. There have been several approaches to the problem.[37-39] Most of the major computer programs for this were written in the late 1960s, notably COMICS from Perrin[38] and HALTAFALL from Sillen.[39] While these are powerful tools, they are, in general, too limited for the present problem. HALTAFALL is extremely efficient in dealing with multiphase systems, but is too slow for very large numbers of complexes, while COMICS takes an inordinate amount of time for very large systems. The program ECCLES[40] uses many of the essential parts of COMICS, but it is able to switch the method of computation to suit the stage of the calculation. As in many situations, the optimum method alters as the calculation proceeds. ECCLES was originally written to accept 4000 species. It is about six times faster than COMICS using the same data, and requires somewhat less than half the core storage.

All of the computer calculations are effectively models of the real system. As such they are subject to the principal constraint of all models, that they are no better than the assumptions on which they are based. In applying any computer model to biological fluids such as plasma it must be critically assessed.

The major assumption is that the values obtained for the stability constants in vitro are applicable in vivo. Apart from the errors inevitably involved in estimating activity coefficients[41] and in extrapolating results obtained under one set of conditions to another, this assumption is probably safe. The errors associated with omitting complexes are described above, and the merits of estimating missing values can be readily assessed. The model assumes that all of the reactions are at equilibrium. For most metals this is a safe assumption. There are a few cases of kinetically stable forms, but these merely form a separate compartment, which must be taken into account in calculating the total metal present in the model. A more serious case would be if a species reacted with half times of the order of minutes, rather than either hours or milliseconds. At the present time no case of this has been shown.

The final assumption, which is in many ways the most serious, is that the distribution of metal ions among the possible ligands may be calculated without an accurate knowledge of the binding of the metal to the multitude of proteins which are present. Although the situation is very far from ideal, the calculations show that at the very low concentrations of free metal which are found in plasma, the percentage distribution of the metal among the different ligands is independent of the actual metal concentration. [40,42] This is due to the fact that metal complexes account for such an extremely small proportion of each ligand. The very low metal concentration which ensures this is due to the very strong binding to protein.

While present research in this field is aimed at filling the gap by the accurate measurement of the protein-metal binding constants,[43] results obtained without them can safely be applied to such questions as whether the metal complexes are changed or unchanged, and how the distribution will be affected by administering drugs which can chelate metals.[44]

The principal aim of most of the workers in this field is the simulation of all body fluids,[40,45] such as blood, cerebrospinal fluid, intercellular fluids, the gut contents at different positions, and the urine. Of these the most important and the best documented is blood. Accurate figures are known for the concentrations of amino acids, peptides, organic acids such as citrate, as well as all of the inorganic species. It is also a very constant medium in comparison with many of the other body fluids. There is a growing body of data on the distribution of metals between different proteins and the low molecular weight pool, and also knowledge of the extent to which the different species are in equilibrium.

Although each metal has its own very characteristic behavior, it is possible to infer some general principles from an examination of one metal alone. Copper has been a central feature of many of the modeling studies, as this is a metal for which a large number of stability constants are known, which has a well-defined binding pattern to blood proteins, and which is of medical importance in the treatment of hepatolenticular degeneration, Wilson's disease. In this condition there is a deficiency of ceruloplasmin, the main blood copper-protein, with a rise in the exchangeable copper and hence copper poisoning. The equilibrium of copper in blood has been studied both by May[40] and Perrin.[45]

The total copper in the blood is approximately 1.8×10^{-5} mol dm^{-3}. Almost 95% of this is held in ceruloplasmin, from which it cannot be readily removed and which plays no role in the equilibrium affecting the rest of the copper. Of the rest, most is bound to albumin,[46] binding at a specific site involving the terminal $-NH_2$ of the protein.[47] A dissociation constant has been measured for this complex,[48] although this value has been challenged (See Reference 22 or Reference 45). Using the value of Lau the free copper concentration in plasma would be about 10^{-19} mol dm^{-3}. Inserting this into a model of the blood equilibria, it is found that, depending on the pH, the low molecular weight copper is primarily found in a cysteine-histidine complex, a protonated form of this, or a bis(histidine) complex.[40] Of these only the first carries a charge, so copper in blood should be fully equilibrated with the tissues. This is due to the fact that small neutral species are generally capable of diffusing freely across membranes, while charged ones are not. The work of Lau indicates that copper is probably liberated from albumin via a ternary histidine complex, and all of the species found by the model with 1% or more of the copper contain histidine. This implies that all of the major copper equilibria in blood can be interpreted as being mediated by copper-histidine, rather than free copper.

It is possible to examine the effects of chelating agents on the metal distribution by adding them to the model, and to assess the ways in which they may work. In the treatment of Wilson's disease the main drugs used are penicillamine, $\beta\beta$-dimethyl-cysteine, and tirethylene tetramine (trien), although this has not undergone full clinical trials.

Penicillamine does not in fact act as a very effective chelating agent for copper (II).[49]

Instead, it almost certainly functions by reducing copper (II) to copper(I), itself being oxidized to oxidized penicillamine, tetramethyl cystine. Trien on the other hand is a very powerful chelating agent for copper. At drug concentrations of about 10^{-9} mol dm^{-2}, approximately 90% of the copper in low molecular weight complexes is present in a trien complex, while at higher levels it is capable of competing successfully with albumin for the metal.[44] Trien also binds zinc, but this does not become important until the drug concentration reaches about 5×10^{-5} *M*. Thus by careful control over dosage it should be possible to selectively remove copper.

Information of this sort is extremely difficult to obtain by animal studies. The drug concentrations in vivo can often be very difficult to measure, it is not usually possible to continuously monitor clearance from the blood, and animal studies are very slow. The use of modeling can show possible mechanisms and indicate the major factors to be observed in animal experiments.

IV. SUMMARY

For very many years biochemists, with a few exceptions, have been content to think of metals solely as adjuncts to certain enzymes, ignoring much of their general chemistry and the way in which this affects their position in the physiology of the whole organism. This approach has been challenged in recent years as more inorganic chemists apply the methods of their discipline to metals in biological systems.[50-52] Although the chief beneficiary of this has been enzymology, where the details of metal coordination are essential to an understanding of those enzymes containing them, there has been an extension into the whole area of metal transport, uptake, and excretion.

In blood the main effect has been to highlight the role of low molecular weight complexes as intermediates in exchange of metals between protein and protein and between blood proteins and the surrounding tissues. At the concentrations involved direct measurement is liable to disrupt the system to such an extent as to render the information gained worthless. So investigation of the equilibria involved depends on modeling the system using the most accurate data available from laboratory studies. The results obtained are no better than the data and the assumptions on which they are based and while these are subject to continued criticism and improvement, it is already possible to make definite predictions about the behavior of real biological systems from using mathematical models.

REFERENCES

1. **Park, J. B.,** *Biomaterials, an Introduction,* Plenum Press, New York, 1979, 203.
2. **Valentyne, J. R.,** *Current Aspects of Exibiology,* Pergamon Press, London, 1965, 1.
3. **Lehninger, A. L.,** *Biochemistry,* 2nd ed., Worth, New York, 1975, chap. 18.
4. **Holm, R. H.,** Identification of active sites in iron-sulphur proteins, in *Biological Aspects of Inorganic Chemistry,* Addison, A. W., Cullen, W. R., Dolphine, D., and James, B. R., Eds., John Wiley & Sons, New York, 1977, chap. 3.
5. **Williams, R. J. P., Moore, G. R., and Wright, P. E.,** Oxidation reduction properties of cytochromes and peroxidases, in *Biological Aspects of Inorganic Chemistry,* Addison A. W., Cullen, W. R., Dolphin, D., and James, B. R., Eds., John Wiley & Sons, New York, 1977, chap. 11.

6. **Loutie, R. and Vanquickenborne, L.**, The role of copper in haemocyanins, in *Metal Ions in Biological Systems*, Vol. 3, Sigel, H., Ed., Marcel Dekker, New York, 1974, chap. 6.
7. **Ochiai, E.**, *Bioinorganic Chemistry, An Introduction*, Allyn and Bacon, Boston, 1977, chap. 10.
8. **Wilson, P. W.**, The background, in *The chemistry and Biochemistry of Nitrogen Fixation*, Postgate, J. R., Ed., Plenum Press, New York, 1971, chap. 1.
9. **Poston, J. M. and Stadtman, T. C.**, Cobamides as cofactors: methyl cobamides and the synthesis of methionine, methane and acetate, in *Cobalamin*, Babior, B. M., Ed., John Wiley & Sons, New York, 1965.
10. **Ochiai, E.**, *Bioinorganic Chemistry, an Introduction* , Allyn and Bacon, Boston, 1977, 10.
11. **Underwood, E. J.**, *Trace Elements in Human and Animal Nutrition*, 4th ed., Academic Press, 1977, 25.
12. **Bates, G. W., Billups, C., and Saltman, P.**, The kinetics and mechanism of iron (III) exchange between chelates and transferrin. I, *J. Biol. Chem.*, 242, 2810, 1967.
13. **Davies, G. and Warnquist, B.**, Aspects of the chemistry of cobalt (III) in aqueous perchlorate solutions, *Coord. Chem. Rev.*, 5, 349, 1970.
14. **Gurney, R. W.**, *Ionic Processes in Solution*, Dover Publications, New York, 1953, 17.
15. **Cotton, F. A. and Wilkinson, G.**, *Advanced Inorganic Chemistry*, 3rd ed., Interscience, New York, 1972, 577.
16. **Cotton F. A. and Wilkinson, G.**, *Advanced Inorganic Chemistry*, 3rd ed. Interscience, New York, 1972, 596.
17. **Basolo, F. and Pearson, R. G.**, *Mechanisms of Inorganic Reactions*, John Wiley & Sons, New York, 1958, 395.
18. **Baes, C. F. and Mesmer, R. E.**, *The Hydrolysis of Cations*, Interscience, New York, 1976, 235.
19. **Baes, C. F. and Mesmer, R. E.**, *The Hydrolysis of Cations*, Interscience, New York, 1976.
20. **Cotton, F. A. and Wilkinson, G.**, *Advanced Inorganic Chemistry*, 3rd Ed., Interscience, New York, 1972, 861.
21. **May, P. M., Williams, D. R., and Linder, P. W.**, Biological significance of low molecular weight iron (III) complexes, in *Metal Ions in Biological Systems*, Vol. 7, Sigel, H., Ed., Marcel Dekker, New York, 1978, 43.
22. **Sinclair, P., Gibbs, A. H., Sinclair, J. F., and De Matteis, F.**, Formation of cobalt protoporphyrin in the liver of rats, *Biochem. J.*, 178, 529, 1979.
23. **Rossotti, F. J. C. and Rossotti, H.**, *The Determination of Stability Constants*, McGraw-Hill, New York, 1961, 9.
24. **Sillen, L. G. and Martell, A. E.** Stability Constants of Metal Ion Complexes, Suppl. No. 1., The Chemical Society, London, 1971.
25. **Martell, A. E. and Smith, R. M.**, *Critical Stability Constants*, Vol. 1, 3, and 4, Plenum Press, New York, 1974, 1976, 1977.
26. **Bjerrum, J.**, *Metal Ammine Formation in Aqueous Solution*, P. Haase and Son Forlag, Copenhagen, 1941.
27. **Baes, C. F. and Mesmer, R. E.**, *The Hydrolysis of Cations*, Interscience, New York, 1976, chap. 2.1.
28. **Ingri, N. and Sillen, L. G.**, High speed computers as a supplement to graphical methods. IV an ALGOL version of LETAGROP VRID, *Arkiv. Kemi*, 23, 97, 1964.
29. **Gans, P., Sabatini, A., and Vacca, A.**, An improved computer program for the computation of formation constants from potentiometric data, *Inorg. Chim. Acta*, 18, 237, 1976.
30. **Sayce, I. G.**, Computer calculation of equilibrium constants of species present in mixtures of metal ions and complexing agents, *Talanta*, 15, 1937, 1968.
31. **Beck, M. T.**, *Chemistry of Complex Equibilria*, Van Nostrand Reinhold, New York, 1970, 262.
32. **Sigel, H.**, Structural aspects of mixed-ligand complex formation in solution, in *Metal Ions in Biological Systems*, Vol. 2, Sigel, H., Ed., Marcel Dekker, New York, 1973, chap. 2.
33. **Greenstein, J. P. and Winitz, M.**, *Chemistry of the Amino Acids*, Vol 1, John Wiley & Sons, New York, 1961, 645.
34. **Beck, M. T.**, *Chemistry of Complex Equilibria*, Van Nostrand Reinhold, New York, 1970, 73.
35. **Rossotti, H.**, *Chemical Application of Potentiometry*, Van Nostrand, London, 1969, 127.
36. **May, P. M.**, Computer Simulation of Metal Ion Equilibria in Biochemical Systems, Models for Blood Plasma, Ph.D. thesis, University of Cape Town, South Africa, 1975, 88.
37. **Perrin, D. D.**, Multiple equilibria in assemblages of metal ions and complexing species: a model for biological systems, *Nature (London)*, 206, 170, 1965.
38. **Perrin, D. D. and Sayce, I. G.**, Computer calculations of equilibrium concentrations in mixtures of metal ions and complexing species, *Talanta*, 14, 833, 1967.
39. **Ingri, N., Kakolowicz, W., Sillen, L. G., and Warnquist, T. B.**, High speed computers as a supplement to graphical methods, V. Hattafall, a general program for calculating the composition of equilibrium mixtures, *Talanta*, 14, 1261, 1967.

40. **May, P. M., Linder, P. W., and Williams, D. R.,** Computer simulation of metal ion equilibria in biofluids: models for the low-molecular weight complex distribution of Calcium (II), Manganese (II), Iron (III), Copper (II), Zinc (II) and lead ions in human blood plasma, *J. Chem. Soc. Dalton Trans.*, 1, 588, 1977.
41. **Baes, C. F. and Mesmer, R. E.,** *The Hydrolysis of Cations,* John Wiley & Sons, New York, 1976, chap. 23.
42. **May, P. M., Linder, P. W., and Williams, D. R.,** Ambivalent effect of protein binding on computed distributions of metal ions complexed by ligands in blood plasma, *Experientia,* 32, 1492, 1976.
43. **Agarwal, R. P. and Perrin, D.,** Copper (II) and zinc (II) complexes of glycylglycyl-l-histidine and derivations, *J. Chem. Soc. Dalton Trans.,* 1, 53, 1977.
44. **May, P. M. and Williams, D. R.,** Computer simulation of chelaton therapy, *FEBS Lett.,* 78, 134, 1977.
45. **Perrin, D. D. and Agarwal, R. P.,** Multimetal-multiligand equilibria: a model for biological systems, in *Metal Ions in Biological Systems,* Vol. 2, Sigel, H., Ed., Marcel Dekker, New York, 1973, chap. 4.
46. **Peters, T. P.,** Serum albumin, *Adv. Clin. Chem.,* 13, 37, 1970.
47. **Bradshaw, R. A., Shearer W. T., and Gurd, F. R. N.,** Sites of binding of Copper (II) ion by peptide (1-24) of bovine serum albumin, *J. Biol. Chem.,* 243, 3817, 1968.
48. **Lau, S. J. and Sarker, B.,** Ternary coordination complex between human serum albumin, copper (II) and L-histidine, *J. Biol. Chem.,* 246, 5938, 1971.
49. **Perrin, D. D. and Agarwal, R. P.,** Multimetal - multiligand equilibria: a model for biological systems, in *Metal Ions and Biological Systems,* Vol. 2, Sigel, H., Ed., Marcel Dekker, New York, 1973, 186.
50. **Ochiai, E.,** *Bioinorganic Chemistry, an Introduction,* Allyn and Bacon, Boston, 1977.
51. **Williams, R. J. P. and Da Silva, J. R. R. F., Eds.,** *New Trends in Bioinorganic Chemistry,* Academic Press, London, 1978.
52. **Addison, A. W., Cullen, W. R., Dolphin, D., and James, B. R., Eds.,** *Biological Aspects of Inorganic Chemistry,* Interscience, New York, 1977.

Chapter 3

METAL ENZYME INTERACTIONS

Trevor Rae

TABLE OF CONTENTS

I. INTRODUCTION

It is somewhat paradoxical that while certain metals are essential trace elements for animals, in many cases the same elements are toxic when in excess. Metals play many diverse roles in the metabolism of animals; they may act as enzyme activators or be an integral part of an enzyme's structure or they may be involved in a physiological role such as osmoregulation.

The most abundant metals found in the body are sodium and potassium, the former being the major metallic ion of extracellular fluid while the latter is the most common intracellular metal. These two ions are important in maintaining osmotic pressure balances and, for instance, producing osmotic gradients in the kidney. It is the movement of these two ions which produces the difference in electrical potential across the cell membrane. In the case of nerve and muscle cells, the distribution of sodium and potassium can be altered transiently to form an action potential, the vital communications signal of the nervous system.

Calcium ions are also involved in the electrical events associated with contracting muscle fibers. Calcium is also a major structural element of the body, about 99% of total body calcium being contained in the skeleton.

In addition to a physiological role, many metals found in the body are, in minute quantities, activators or in some cases inhibitors of enzymes in biochemical reactions. Metals acting in such a role are the subject of the present review. They may be encountered by the body tissues as essential nutritional factors, they may be environmental pollutants, or they may be released by corroding prostheses.

Like other proteins, some enzymes contain within their structure a nonprotein prosthetic group. The prosthetic group may be a relatively simple organic molecule such as a derivative of a vitamin or a metal. The term prosthetic group is usually reserved for a tightly bound chemical group which is necessary for the integrity and hence activity of the molecule. In contrast to this, other enzymes are activated in the presence of nonprotein dialyzable cofactors. Such coenzymes, cofactors, or activators include both simple inorganic ions and complex organic molecules. According to Scrimgeour[1] approximately one quarter of the enzymes known require metallic cations to express their full catalytic activity.

Metal-enzyme complexes are usually divided into two groups. Metalloenzymes are those in which the metal is tightly bound and is an integral part of the molecule, and where removal of the metal usually results in the destruction of molecular structure. The other group comprises those in which the metal is loosely and reversibly bound to the protein. Under physiological conditions such binding usually leads to an increased activity and is referred to as a metal-activated enzyme. In some cases, however, the binding of a metal may bring about an inhibition of enzyme activity.

The mechanism by which metals confer activity on enzymes has been studied by various methods including enzyme kinetics, nuclear magnetic resonance, and electron spin studies. In some cases it has been possible to suggest the mechanism of enzyme action in relation to the bound metal.

Metals which are involved in enzyme activity fall into two main areas of the periodic table. First, there are the very reactive metallic elements sodium and potassium (group I) and magnesium and calcium (group II). Another group of elements which appear to be active in biological systems are vanadium, chromium, manganese, iron, cobalt, nickel, copper, and zinc of the first transition series of elements and molybdenum of the second.

Metal ions are Lewis acids; they are materials which can accept electron pairs, that is, they are electrophiles. They can form a σ bond (i.e., end-to-end overlap of atomic orbitals) by accepting a share in an electron pair. Some metals can also form π bonds (i.e., by lateral overlap of atomic orbitals) either by accepting an electron in a vacant lower energy orbital or by donating an electron from a higher energy filled orbital. Metals of the transition series are noted for their formation of coordination complexes; the central metal ion is surrounded by a group of ions or molecules known as ligands. For a general account of coordination chemistry, the reader is referred to Cotton and Wilkinson.[2]

Metals in enzyme catalysis have been extensively reviewed by Mildvan[3] who considered the reactions of metal complexes under four basic headings. These are
a) ligand substitution or addition at the metal atom, e.g.,

$$[M(A)_m] + B \rightarrow [M(A)_{m-1}B] + A \quad (1)$$

$$[M(A)_m] + nB \rightarrow [M(A)_m B_n] \quad (2)$$

b) reactions of the coordinated ligand, e.g.,

$$[M(A)_m] + C \rightarrow [M(A)_{m-1}D] + E \quad (3)$$

c) reactions of the complex as a whole, e.g.,

$$\ell\text{-}M(A)_m \underset{}{\overset{\text{isomerization}}{\rightleftharpoons}} d\text{-}M(A)_m \quad (4)$$

d) oxidation or reduction of the metal atom, e.g.,

$$[M(A)_m] \rightarrow [M(A)_m]^{n+} + ne^- \quad (5)$$

The majority of metal-activated enzymes probably have a simple stoichiometric relationship in their functional ternary complexes such that the ratio of enzyme active sites to bound metal to bound substrate is unity. There are three basic configurations of the ternary structure; they are

M—E—S	E—M—S	E—S—M
	E(—M)(—S), M—S	
e.g; carbonic anhydrase	muscle pyruvate kinase	muscle creatine kinase
I	II	III

where M represents the metal, E the enzyme, and S the substrate. Structure III is not possible for metalloenzymes.

Higher substrate complexes of the form E - M - S_1 - S_2 are possible such as

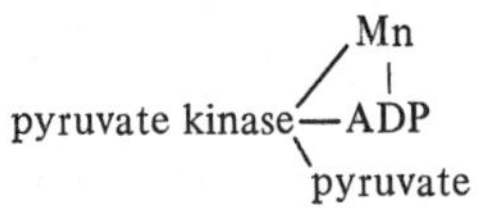

pyruvate kinase bonded to Mn, ADP and pyruvate, with Mn—ADP

and

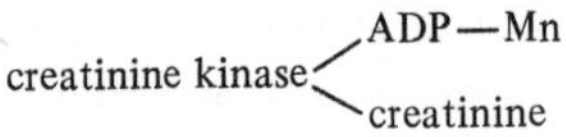

$$\text{creatinine kinase} \langle \begin{matrix} \text{ADP—Mn} \\ \text{creatinine} \end{matrix}$$

Higher metal complexes are probably also possible, such as

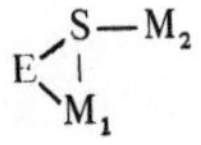

$$E \langle \begin{matrix} S—M_2 \\ | \\ M_1 \end{matrix}$$

where M_1 can be Ni^{++}, Fe^{++}, Ca^{++}, or Cu^{++} and M_2 is Mg^{++}, Mn^{++}, Fe^{++}, Co^{++}, or Ca^{++}.

Much of the initial evidence for a physiological role of metals in metabolism has been obtained from nutrition studies in which animals were fed diets containing a particular metal at a very low level. The occurrence of specific abnormalities may then have given an indication of the area of metabolism affected by the lack of metal. In some cases, however, the withdrawal of a metal from the diet has lead to substitution by another metal which maintains near normal enzyme activity. Sometimes it has proved impossible to correlate a metal-deficiency with a specific effect on an enzyme despite the development of gross abnormalities in the animal. In contrast, some metals have been fed or injected into animals in excess and the effects on enzyme activities studied. The role of specific metals as enzyme activators and inhibitors will now be considered.

II. MONOVALENT METALS

In general, living cells actively accumulate potassium ions, leaving sodium ions as the main metal ion of the extracellular fluid compartment. The observed activation of enzymes by these metals is governed to a large extent by the location of the former and hence to the major monovalent metal to which they are exposed. Suelter[4] has reported that over 60 enzymes are known to be dependent on K^+ for their full activity. Kachmar and Boyer[5] found that rabbit muscle pyruvate kinase was virtually inactive in the absence of K^+. Activity could be returned by the addition of Rb^+ or NH_4^+ but not by Li^+ or Na^+. This has subsequently led to the concept that K^+ is required to stabilize an active structure of the enzyme. Kachmar and Boyer[5] suggested that ionic radii were an important factor in activating enzymes. Thallous ions (Tl^+), for example, have been reported as being able to replace K^+. A comparison of the unhydrated ionic radii of a number of cations able to replace K^+ shows that they are all similar, i.e., Tl^+, 1.40 Å; K^+, 1.33 Å; NH_4^+, 1.48 Å; Rb^+ 1.48 Å. In general, neither cesium, sodium, nor lithium, whose respective ionic radii are 1.69, 0.95, and 0.60 Å, are able to activate enzymes usually dependent upon K^+ for their full activity.

Studies on the reassociation of the active tetramer of formyltetrahydrofolate synthetase have shown it to be dependent upon the presence of a number of ions,[6] the degree of effectiveness being in the order $NH_4^+ > Tl^+ > Rb^+ \simeq K^+ > Cs^+ > Na^+ > Li^+$. The mechanism of action of such cations is not known, but it has been suggested that they may either neutralize charges on the surface of the enzyme, thus preventing electrostatic repulsion of subunits, or are needed for the association of subunits. The main effects of monovalent ions are likely to be through tertiary or quaternary changes in protein structure, thus altering enzyme activity.

III. DIVALENT METALS

Many enzymes can be rendered inactive by dialyzing against a metal-free buffer but activity can often be restored by the addition of divalent metal ions. Since calcium and magnesium are the most common divalent metals under physiological conditions, it is often assumed that these are also the most active in vivo. However, enzymes can show wide variations in their degree and specificity of activation. In some cases other divalent cations, particularly of the first row of transition elements that also occur in vivo, can substitute for calcium and magnesium under in vitro experimental conditions. It is difficult, therefore, to assess the physiological importance of such a substitution. There are well-documented cases, however, in which specific transition metals form an essential part of a metalloenzyme, and these are considered in subsequent sections.

A large number of enzymes have been shown to be activated by magnesium and calcium, but it is outside the scope of this present review to consider them all. One well-characterized calcium metalloenzyme is bacterial alkaline phosphatase. This enzyme and the general field of phosphate transfer and activation by metal ions has been reviewed by Spiro.[7] Calcium-activated enzymes include the diverse group of hydrolases which hydrolyze macromolecules. Calcium may function in a role of adjusting the protein conformation or may be involved directly in catalysis as an enzyme-metal-substrate bridge complex. Reduced takaamylase, for example, requires a divalent cation (preferably calcium) to restore the appropriate three dimensional structure of the denatured enzyme.[8] Deoxyribonuclease from the pancreas, for instance, requires a weakly bound divalent cation for activity.[9]

A number of intracellular enzymes such as the kinases are activated by magnesium, probably through an ATP-Mg chelate. Creatine kinase, for example, catalyzes the phosphorylation of creatine using an ATP-Mg chelate as the source of phosphate.[10]

For a full account of the role of divalent cations in the activation of enzymes the reader is referred to Spiro[7] and Mildvan.[3]

IV. TRANSITION METALS

A. Chromium

Chromium has been shown to increase the growth rates of rats and mice fed a chromium-deficient diet[11,12] and a deficiency state has been recognized in man, usually associated with an impairment of glucose tolerance (see Mertz[13] and Schroeder[14]). Rats fed on a restricted chromium intake result in impaired growth, fasting hyperglycemia, glycosuria, and increased serum cholesterol levels, these events having been reviewed by Mertz et al.[15] Wacker and Vallee[16] and Kornicker and Vallee[17] have reported on the possible role of chromium in nucleic acid metabolism. There are few reports, however, on chromium as an enzyme activator. It has been reported as activating phosphoglucomutase in vitro, but probably its role in this case is as a substituting cation for the more important magnesium ion.[18] Chromium has been shown to stimulate the incorporation of labeled acetate into cholesterol[19] and the digestive enzyme trypsin has been reported as containing one atom of chromium per mole of enzyme.[20] A physiological role for chromium in enzyme reactions has yet to be fully described.

B. Cobalt

Probably the most important form in which cobalt is normally encountered in the body is in vitamin B_{12}. A cobalt (III) ion is situated at the center of an essentially planar corrin ring coordinated through the four nitrogen atoms. On one side of this ring, the cobalt ion is coordinated to the nitrogen of 5,6-dimethylbenzimidazole linked to a pentose sugar and phosphate group. On the other side of the plane of the corrin ring a number of different ligands may be present and it is these which give rise to the different chemical forms of vitamin B_{12}, the cobalamins. In the body one active form

Table 1
SUMMARY OF COPPER-CONTAINING METALLOENZYMES

Enzyme	Substrate	Source	Mole metal/ mole enzyme	Ref.
Superoxide dismutase	Superoxide anion $O_2^{\cdot-}$	Bovine erythrocytes	2 Cu 2 Zn	25
Cytochrome oxidase	Reduced cytochrome C	Bovine heart mitochondria	2 Cu 2 heme	29 30
Uricase	Purines, e.g., uric acid	Nonprimate mammalian liver and kidney	1 Cu	31
Tyrosinase	Tyrosine DOPA	Human and murine melanoma	?	32 33
Amine oxidase	Amines	Porcine plasma	3 Cu	35
Diamine oxidase	Amines	Porcine kidney	2 Cu	34
Dopamine β-hydroxylase	Dopamine	Bovine adrenal medulla	2 Cu^{++} variable Cu^{+}	36

of the vitamin is coenzyme B_{12} in which a 5-deoxyadenosine group is the ligand. It is involved in methionine biosynthesis[21] and in the conversion of methylmalonyl-coenzyme A to succinyl-coenzyme A by methylmalonyl-coenzyme A mutase.[22]

Experimental investigations into the toxicity of cobalt have led to the discovery that α-ketoglutarate dehydrogenase is susceptible to inhibition by a range of bivalent metals and the irreversible inhibition by cobalt has been well studied by Webb.[23] Cobalt also forms an oxygen-sensitive chelate with the reduced form of α-lipoic acid which is a coenzyme required in the formation of acetyl coenzyme A from pyruvate and in the formation of succinyl coenzyme A from α-ketoglutarate.[24] The above reactions are key steps in energy production via glycolysis and the tricarboxylic acid cycle from carbohydrate and thus inhibition of these reactions interferes with cellular energy production.

There do not appear to have been any cobalt-containing metalloenzymes described in vertebrate animals, neither does it seem to be important as an enzyme activator. In some cases, however, it can replace other divalent cations as activators of enzymes in vitro.[3]

C. Copper

Copper occurs in a number of oxidative metalloenzymes; a cyclical process of oxidation and reduction of the metal is often involved in the enzymatic activity. A summary of copper-containing metalloenzymes is presented in Table 1.

The reactions and composition of superoxide dismutase (SOD) have been extensively reviewed by Fridovich,[25] It catalyzes the destruction of superoxide radical ($O_2^{\cdot-}$) and involves a change in the oxidation state of copper as summarized below

$$E - Cu^{++} + O_2^{\cdot-} \longrightarrow E - Cu^{+} + O_2 \qquad (6)$$

$$E - Cu^{+} + O_2^{\cdot-} \xrightarrow{+2H^{+}} E - Cu^{++} + H_2O_2 \qquad (7)$$

The complete amino acid sequence of bovine erythrocyte SOD has been established and consists of two subunits of identical structure.[26] The two copper atoms are at opposite ends of the dimer and are coordinated with four histidine residues. The zinc atoms are close to the copper and are complexed with three histidine and one aspartate residues. SOD from mammalian sources usually contains copper and zinc while bacterial SOD reported so far contain manganese or iron[25] (see also relevant sections on these metals). The reaction mechanism and kinetic studies on SOD have been reported by Klug-Roth et al.[27] Rotilio et al.[28] have reported on a pulse radiolysis study of bovine SOD and gained further evidence for the involvement of copper at the active site of this enzyme.

Cytochrome oxidase contains copper and iron in the form of heme and is part of complex IV of the electron transport system of mitochondria. Beinert et al.[29] have studied the oxidation state of copper in cytochrome oxidase using a paramagnetic resonance technique. At least some of the copper is present in the cupric state and is reduced to the cuprous ion by substrate or other chemical reducing agents. In such reactions the oxidation state of the heme component parallels that of the copper. Morrison et al.[30] found that their preparation of cytochrome oxidase also appeared to contain loosely bound copper whose removal did not affect enzyme activity. The firmly bound copper, however, was in the same ratio as the heme content of the molecule.

Uricase, another copper-containing enzyme, is a key enzyme in the catabolism of nitrogenous compounds in general and purines in particular. A review of the activity of this enzyme has been written by Mahler[31] who at the time of writing could find no direct evidence for copper being at the active site, although circumstantial evidence did support this view.

Tyrosinase, isolated from human and murine melanomas, has been shown to be dependent upon copper for its full activity.[32,33] The role of copper and the content in the enzyme, however, have still to be determined.

Another group of enzymes which contain copper are the amine oxidases. Yamada et al.[34] have prepared a crystalline specimen of diamine oxidase from porcine kidney and found it to contain 2 mol of copper per mole of enzyme. In contrast to this, Buffoni et al.[35] found that amine oxidase from porcine plasma contained three copper atoms per mole of protein.

The mixed function oxidase dopamine β-hydroxylase utilizes molecular oxygen in the oxidative process. Friedman et al.[36] have provided evidence for copper undergoing cyclical changes in its oxidation state during the hydroxylation reaction. A mechanism for the action of copper in this enzyme was also advanced as was a role for the essential cofactor ascorbate. Variable quantities of Cu^{+} were also reported as being present.

D. Iron

Iron is a component of the vital blood pigment hemoglobin which transports oxygen in the red blood cells of vertebrates. Other iron-porphyrin metalloproteins participate in redox reactions such as the electron transport of the cytochromes. The enzymes of which iron is a component are summarized in Table 2.

Mahler[37] has reported on the properties of the NADH dehydrogenase of pig heart muscle. It contains four atoms of nonheme iron for each flavin molecule of which there is one for each enzyme molecule. The role of iron in catalysis by this enzyme is unclear as various metal-chelators were found to inhibit the enzyme to a greater or lesser degree. Removal of iron, however, produced a decrease in enzyme activity. The valence state of the iron varied according to the mode of preparation, but excess NADH was able to reduce all the iron to the ferrous state. It was also reported that excess cytochrome c never produced more than two atoms of ferric iron in a molecule of the enzyme. It was concluded that some of the iron may be important in binding the flavin prosthetic group to the enzyme. Reviews of the dehydrogenases have been published by Singer[38] and by Hatefi.[39]

Table 2
SUMMARY OF IRON-CONTAINING METALLOENZYMES

Enzyme	Substrate	Source	Mole metal/mole enzyme	Ref.
NADH dehydrogenase	NADH	Porcine heart	4 Fe	37
Succinate dehydrogenase	Succinic acid	Porcine heart	8 Fe	40
		Bovine heart	?	41
Aldehyde oxidase	Aldehydes	Rabbit liver	8 Fe 2 Mo	42
Xanthine oxidase	Purines, e.g., hypoxanthine xanthine	Bovine milk	8 Fe 2 Mo	43
		Mammalian liver	4 Fe 2 Mo	93
		Chicken liver	8 Fe 2 Mo	44
Sulfite oxidase	SO_3^{--}	Bovine liver	2 heme 2 Mo	56 57 58
Cytochrome oxidase	Reduced cytochrome C	Bovine heart mitochondria	2 heme 2 Cu	29 30

Succinate dehydrogenase, isolated from pig heart, is reported as containing eight atoms of iron and four acid-labile sulfur atoms for each molecule of enzyme.[40] It is believed that the sulfur is the ligand through which the iron is attached to the enzyme. Bovine heart succinate dehydrogenase has been described by Kearney et al.[41] as being a flavoprotein, the flavin component being firmly bound forming almost a prosthetic group.

Among the other enzymes which contain iron are those which also contain molybdenum; they are aldehyde oxidase and xanthine oxidase. Aldehyde oxidase has been described as containing eight atoms of iron, two of molybdenum, two molecules of flavin adenine dinucleotide, and one or two molecules of coenzyme Q_{10}.[42] Xanthine oxidase from milk has been found to contain about two atoms of molybdenum, eight of iron, and two molecules of flavine adenine dinucleotide for each molecule of enzyme.[43] Xanthine oxidase from chicken liver has a similar metal content and in addition was reported as having eight labile sulfur atoms; the internal electron transport system of this enzyme has also been described.[44]

E. Manganese

While there appear to be only two manganese-containing metalloenzymes, there are many other enzymes whose activity can be increased by manganese ions. Scrutton et al.[45] and Mildvan et al.[46] have described the presence of tightly bound manganese in pyruvate carboxylase (PC) isolated from chicken liver. This enzyme catalyzes the reaction

$$\text{pyruvate} + HCO_3^- + ATP \underset{Mg^{++}}{\overset{\text{acetyl CoA}}{\rightleftharpoons}} \text{oxaloacetate} + ADP + P_i \tag{8}$$

Table 3
SUMMARY OF MANGANESE-CONTAINING METALLOENZYMES

Enzyme	Substrate	Source	Mole metal/mole enzyme	Ref.
Pyruvate carboxylase	Pyruvate and oxaloacetate	Chicken liver mitochondria	2.5—4.3 Mn	45,46
		Turkey liver	0 Mg 3.5 Mn	
				47
		Calf liver	0—1.8 Mg 1—1.4 Mn	
Superoxide dismutase	Superoxide radical $O_2^{\cdot-}$	*Escherichia coli*B	1.6—1.8 Mn	48
		Streptococcus mutans	2 Mn	87
		Chicken liver mitochondria	2.3 Mn	49

Evidence has been provided[46] that the bound manganese plays a role in the transcarboxylation part of the reaction and the activating divalent metal ion participates in the formation of the enzyme-biotin$\sim CO_2$ intermediate. Calf liver pyruvate carboxylase contains a mixture of magnesium and manganese; the former can also functionally replace the manganese of chicken liver PC in birds fed on diets low in manganese.[47] The possible role of manganese in PC has been considered by Mildvan et al.[46]

SOD isolated from *Escherichia coli,* also a manganese metalloenzyme, has been described by Keele et al.[48] The catalytic mechanism is probably similar to that of the copper-zinc SODs in that the metal is alternately reduced and oxidized during successive reactions with superoxide.[25] The SOD of chicken liver mitochondria is also manganese metalloenzyme.[49] Some features of manganoenzymes are summarized in Table 3.

In contrast to the small number of manganoenzymes there are many enzymes which are activated by manganese, at least in vitro. Manganese-activated enzymes have been considered by Mildvan[3] and include muscle pyruvate kinase, adenylate kinase, creatine kinase, arginine kinase, xylose and arabinose isomerases, histidine deaminase, and enolase. The activation is not, however, necessarily specific since magnesium and other divalent cations can often substitute for manganese. It is difficult, therefore, to comment on the importance of this cation as an activator of enzymes in vivo.

Manganese has been found to prevent some skeletal abnormalities in experimental animals. Events associated with chondrogenesis appear to be most affected, while in manganese-deficient animals the incorporation of 35S and mucopolysaccharide synthesis was found to be decreased. The generalized observations on manganese-deficient animals may be explained by a requirement for Mn^{++} by glycosyl-transferase enzymes, in particular those required for chondroitin synthesis. Consequently two enzyme systems have been well studied in the chick, the polymerase responsible for the chain lengthening of the polysaccharide and the galactosyl transferase system which incorporates galactose into the trisaccharide which links the polysaccharide to the protein molecule. The role of manganese in mucopolysaccharide metabolism has been reviewed by Leach[50] and the kinetics of galactosyltransferase by Morrison and Ebner.[51-53]

F. Molybdenum

Molybdenum has been identified as playing a role in a number of enzymatic redox reactions, although the exact nature of its binding site and the function of the metal is

Table 4
SUMMARY OF MOLYBDENUM-CONTAINING METALLOENZYMES

Enzyme	Substrate	Source	Moles of metal/ mole enzyme	Ref.
Nitrate reductase	NO_3^-	*Neurospora crassa*	1—2 Mo	54
Nitrogenase	N_2	*Clostridium*	1,2 Mo	92
		Azotobacter		88
				89
Sulfite oxidase	SO_3^{--}	Bovine liver	2 Mo	56
			2 heme	57
				58
Aldehyde oxidase	Aldehydes	Rabbit liver	2 Mo 8 Fe	42
Xanthine oxidase	Purines, e.g., hypoxanthine	Chicken liver	2 Mo 8 Fe	44
		Bovine milk	2 Mo	94
				95
		Mammalian liver	2 Mo 4 Fe	93

not fully understood. Sulfite oxidase, aldehyde oxidase, and xanthine oxidase are major molybdenum-containing enzymes. The former catalyzes the last stage in oxidative metabolism of sulfur-containing amino acids and aldehyde oxidase catalyzes the oxidation of aldehydes, while xanthine oxidase is involved in the formation of uric acid from purine bases. Bacterial nitrogenases are molybdenum-containing enzymes and are responsible for the fixation of atmospheric nitrogen. The properties of nitrate reductase isolated from the mold *Neurospora crassa* have been reported by Garrett and Nason.[54] A summary of the molybdenum-containing enzymes is presented in Table 4.

In most cases, molybdenum in enzymes appears to take part in an electron transfer process in which it undergoes cyclic changes in its oxidation state. The most common oxidation state is Mo(VI), and during reaction with its substrate the metal in xanthine oxidase is reduced to Mo(V) and may possibly be reduced further to Mo(IV), although this is unstable in aqueous solution. The oxidation state in the fully reduced enzyme is unknown, but electron spin resonance studies suggest that there may be an equilibrium between Mo(V) and Mo(IV) (see Spence[55]). It seems likely that the function of molybdenum in xanthine oxidase and aldehyde oxidase is to transfer electrons between enzyme and coenzyme or between substrate and iron, the other metal present in many molybdoenzymes. A sequence of events for the passage of electrons in the active complex has been presented by Spence[55] and is as follows:

$$\text{Substrate} \rightarrow \text{Mo(VI)/Mo(V)} \rightarrow \text{FAD/FADH} \rightarrow \text{Fe}^{+++}/\text{Fe}^{++} \rightarrow \text{O}_2 \quad (9)$$

where FAD is a flavosemiquinone. The role of various components of sulfite oxidase in an electron transport system has been described by Cohen et al.[56,57,58] and Rajagopalan et al.[42]

The purification, properties, and functional role of sulfite oxidase have been studied in detail by Cohen et al.[56,57,58] A review of the oxidation state and binding sites of molybdenum-containing enzymes has been published by Spence.[55] Bray and Swann[59] have reviewed the role of molybdenum in biological systems, consideration being given to the role of the metal in enzyme function from a mechanistic point of view.

G. Nickel

Nickel has only relatively recently been considered as an essential trace element.[60] An induced nickel deficiency in chicks becomes manifest as impaired liver metabolism and morphology. This includes a reduced ability to oxidize α-glycerophosphate and swelling of the mitochondria. Nickel-deficient rats were also described as having a reduced hepatic capacity to oxidize α-glycerophosphate. In contrast to this, an increased dietary intake of nickel in mice led to a decrease in the activity of cytochrome oxidase and isocitrate dehydrogenase in the liver, isocitrate dehydrogenase in the kidney, and cytochrome oxidase and malic dehydrogenase in the heart.[61] Weanling rats fed a diet containing nickel up to 1000 ppm exhibited lower activities of cytochrome oxidase and alkaline phosphatase.

In vitro cxperiments with nickel, for example, have shown it to activate acetyl coenzyme A synthetase[62] and phosphoglucomutase.[63] These studies have not, however, shown nickel to be a specific activator. Acetyl coenzyme A synthetase from beef heart mitochondria, for example, has a double divalent cation requirement. In the first part of the reaction, magnesium, manganese, iron, cobalt, or calcium have been shown to be active, and in the second part divalent iron, nickel, cadmium, or copper.[62] Nickel may be effective in enzyme activation, therefore, only in a role of fulfilling a requirement for a divalent cation.

H. Tungsten

Callis et al.[64] have remarked on the observation that despite similarities in the formation constants of complexes with molybdenum (VI) and tungsten (VI) and in atomic radii and electronegativities, biological systems show a marked preference for molybdenum. There appears to be only one documented active tungsten metalloenzyme, that is formate dehydrogenase from *Clostridium thermoaceticum*.[65]

Others have found that tungsten can replace molybdenum in some enzymes with a consequent loss of enzyme activity. Johnson et al.[66] have shown that rats given tungsten in their drinking water exhibited lowered activities of xanthine oxidase and sulfite oxidase. Molybdenum added to such drinking water in concentrations as low as 1/100th that of tungsten protected against this enzyme defect. It was found that sulfite oxidase from the tungsten-treated rats showed a low native molybdenum electron paramagnetic resonance signal for this metal and that the livers from these rats were low in molybdenum. Further work[67] showed that purified sulfite oxidase from livers of tungsten-treated rats contained no molybdenum and that 35% of molybdenum-free molecules contained tungsten with no enzyme activity. Similar results were also found when tungsten substituted for molybdenum in xanthine oxidase.[68] As far as can be established, tungsten has no role in the activation of mammalian enzymes. As its deleterious effects are easily overcome by trace amounts of molybdenum, tungsten is unlikely to be responsible for the inhibition of the enzyme systems identified so far.

I. Vanadium

Although vanadium has been identified as an essential trace element for the chick[69] and the rat[70] and its role in nutrition has been reviewed by Hopkins[71](see also Chapter 12 of this volume), a specific function for this metal in enzyme activation has not been revealed. There is some evidence to suggest that vanadium may be able to substitute for molybdenum in those enzymes required for nitrogen fixation by bacteria (see Section IV.F).[72,73] Vanadium is active as a catalyst in the nonenzymic oxidation of catechol amines.[74] It also inhibits cholesterol synthesis by the inhibition of the microsomal enzyme system squalene synthetase.[19,75,76]

J. Zinc

Zinc is an essential cofactor for the enzymes carbonic anhydrase, carboxypeptidase, and many dehydrogenases, alcohol dehydrogenase probably being one of the best studied.

Carbonic anhydrase was the first zinc metalloenzyme to be discovered. The primary function is to catalyze the hydration of carbon dioxide to allow the transportation of the latter inside erythrocytes as bicarbonate ions. The failure of chelating agents to inhibit its activity has lead to opinion that zinc is not directly coordinated to carbon dioxide.[3]

Carbonic anhydrase has also been shown to catalyze a number of other hydrolysis and hydration reactions involving carbonyl groups. These have been reviewed by Coleman,[77] who has also extensively reviewed the action of this enzyme.

Carboxypeptidase is a single polypeptide chain of 307 amino acid units and contains one atom of zinc at the active site. It is an exopeptidase and is specific for the cleavage of carboxy-terminal amino acids. The structure is well characterized and has been described by Ludwig and Lipscomb[78] who also considered the relationships of structure and function of this enzyme. Mildvan[3] has reviewed the catalytic properties of this enzyme from a mechanistic point of view.

Many dehydrogenases also contain zinc; they are frequently NAD-linked and the metal does not usually undergo oxidation-reduction reactions. Suggestions have been made that zinc forms a bridge to coordinate the NADH to the substrate, and these have been considered by Mildvan.[3] A role for zinc in the mechanism of action of alcohol dehydrogenase[83] has been suggested by Mildvan and Weiner[79,80] and Weiner.[81] Although bovine liver glutamate dehydrogenase[82] and lactic dehydrogenase[83] have been reported as containing zinc, subsequent investigations have indicated the absence of significant concentrations of this ion.[84]

The role of zinc in the catalytic activity of alcohol dehydrogenase has been well studied. Drum et al.[85] have described the presence of four zinc atoms, two of which were "free" in that they could be exchanged with extramolecular zinc and are situated at the active site. The remaining two metal atoms were described as being "buried" and, while not playing a part at the active site of the enzyme, are essential for the quaternary structure of the enzyme molecule.

Zinc has also been identified in bovine blood SOD, an enzyme responsible for the destruction of the toxic free radical superoxide ($O_2^{\bar{\cdot}}$). Rotilio et al.[86] have provided evidence that zinc is not involved in the catalytic activity of this enzyme (this role appears to be effected through copper, see Section IV.C), as with alcohol dehydrogenase zinc is responsible for stabilizing the protein structure. A summary of zinc metalloenzymes is given in Table 5.

V. HEAVY METALS

Cadmium, lead, and mercury have a strong affinity for a number of ligands and can thus act at a number of biochemical sites. They all bind strongly to phosphates, sulfhydryl, cysteinyl, and histidyl residues of proteins, purines, pteridines, and porphyrins. Such binding often leads to the inhibition of enzymes. In some cases, however, heavy metals can substitute for other metals in metalloenzymes with a subsequent increase in enzyme activity. From the physiological point of view, the most important effect of excessive intake of these heavy metals is the inhibition of enzymes, particu-

Table 5
SUMMARY OF ZINC-CONTAINING METALLOENZYMES

Enzyme	Substrate	Source	Mole metal/mole enzyme	Ref.
Carbonic anhydrase	CO_2hydration	Bovine erythrocytes	1—1.5 Zn or 1 Zn	90 77
Carboxypeptidase A	Peptide bonds	Bovine pancreas	1 Zn	78
Alcohol dehydrogenase		Horse liver	2 Zn or 4 Zn	91 85
Superoxide dismutase	Superoxide anion (O_2^-)	Bovine blood	2 Zn 2 Cu	86

larly those which contain sulfhydryl groups or disulphide bridges. The biochemical effects of mercury, cadmium, and lead have been extensively reviewed by Vallee and Ulmer.[96]

A. Cadmium

As far as is known, cadmium does not have an essential biological function in mammalian biochemistry although the metalloenzyme alcohol dehydrogenase has been prepared in an active form in which zinc has been substituted by cadmium.[97,98] In contrast to this, the replacement of zinc by cadmium in bacterial alkaline phosphatase produces an inactive form of the enzyme.[99,100] Vallee and Ulmer[96] have reviewed a large number of enzyme systems either activated or inhibited by cadmium in vitro. Those activated include alkaline and acid phosphatase, carboxypeptidase, cholinesterase, glucose-6-phosphate dehydrogenase, malic dehydrogenase, and phosphorylase. Some of the above enzymes have also been described as being inhibited by cadmium and include carbonic anhydrase, acid and alkaline phosphatase, cholinesterase, cytochrome oxidase, glucose-6-phosphate dehydrogenase, and succinic dehydrogenase.

Oxidative phosphorylation in rat liver mitochondria in vitro is uncoupled by low concentrations of cadmium ions[101] and can be reversed by chelating agents and by some other metal ions. The cadmium ions, however, appear to be bound at sites other than those involved in oxidative phosphorylation.[102,103]

B. Lead

Lead has been shown to alter the activity of a number of enzymes. Vallee and Ulmer[96] have reviewed this topic and among those enzymes reported to have a higher activity in the presence of lead are alkaline phosphatase and lactic dehydrogenase. Enzymes listed as being inhibited by lead included acid phosphatase, carbonic anhydrase, some ATPases and succinic dehydrogenase. Some enzymes, such as glucose-6-phosphate dehydrogenase, have been reported by some workers to be inhibited by lead and by other workers to be enhanced by lead. The concentrations of lead required to experimentally inhibit enzymes in vitro is generally high in the range 1 to 10 mM, much higher than is likely to result from lead poisoning. Some enzymes, however, are inhibited by much smaller amounts of lead and are physiologically important.

An excessive uptake of lead from the gut leads to a state of anemia apparently due to an effect on heme synthesis caused by an inhibition of several enzymatic steps. The enzymes most sensitive to inhibition by lead in the above pathway are δ-aminolevulinic acid dehydratase, which converts δ-aminolevulinic acid to porphobilinogen, and ferrochelatase, which inserts iron into protoporphyrinogen.[96]

Other enzymes which may be inhibited by small amounts of lead include the erythrocyte membrane-bound (Na^+ + K^+)ATPase. The incubation of human erythrocytes in vitro results in the blood cells losing intracellular potassium and gaining extracellular sodium; an inhibition of (Na^+ + K^+)ATPase may cause this disturbance.[104,105] Similarly, sodium ion reabsorption in the kidney tubule is affected in lead-treated animals due to the inhibition of (Na^+ + K^+)ATPase.[96]

C. Mercury

Both ionic and organic derivatives of mercury interact strongly with the sulfhydryl groups and disulfide bridges of proteins. Such interactions have been extensively reviewed by Boyer.[106] Although mercury may lack specificity in inhibiting sulfur-containing enzymes, it may show selectively in the whole animal by being taken up by specific organs and may be concentrated in specific cell components. Fowler et al.[107] for instance, described the biochemical changes in the renal lysosome and microsome system of rats given methyl mercury hydroxide in their drinking water. Dense, granular, mercury-containing lysosomes were found in the renal tubule cells. Microsomal β-glucuronidase was inhibited while the levels of acid phosphatase were unchanged. Similarly, lysosomal β-glucuronidase was inhibited while lysosomal acid phosphatase showed an increased activity.

In addition to their action on sulfur-containing groups in enzymes, some organomercurial compounds bind to other sites; p-chloromercuribenzoate (PCMB), for example, binds to the active site of the apoenzyme carboxypeptidase although there is no sulfur present.[108] Some organomercurials also show specificity in action. Bowman and Landon, for instance, investigated the net movement of potassium in rat kidney slices.[109] The organic mericurial diuretic, meralluride, impaired the activity of kidney (Na^+ + K^+)ATPase activity while the nondiuretic PCMB had no effect.

ACKNOWLEDGMENTS

The author thanks Dr. M. F. Heath of the Strangeways Research Laboratory, Cambridge for his help and advice in the preparation of this manuscript. The author would also like to acknowledge the financial support of the Department of Health and Social Security, London.

REFERENCES

1. **Scrimgeour, K. G.,** *Chemistry and Control of Enzyme Reactions,* Academic Press, London, 1977, chap. 11.
2. **Cotton, F. A. and Wilkinson, G.,** *Advanced Inorganic Chemistry, A Comprehensive Text,* 3rd ed., Interscience, New York, 1972, chap. 19.
3. **Mildvan, A. S.,** Metals in enzyme catalysis in *The Enzymes,* Vol. 2, 3rd ed., Boyer, P. D., Ed., Academic Press, New York, 1970, chap. 9.
4. **Suelter, C. H.,** Enzymes activated by monovalent cations. Patterns and predictions for these enzyme-catalyzed reactions are explored, *Science,* 168, 789, 1970.
5. **Kachmar, J. F. and Boyer, P. D.,** Kinetic analysis of enzyme reactions. II. The potassium activation and calcium inhibition of pyruvic phosphorase, *J. Biol. Chem.,* 200, 669, 1953.
6. **Harmony, J. A. K., Shaffer, P. J., and Himes, R. H.,** Cation- and anion-dependent reassociation of formyltetrahydrofolate synthetase subunits, *J. Biol. Chem.,* 249, 394, 1974.
7. **Spiro, T. G.,** Phosphate transfer and its activation by metal ions; alkaline phsophatase, in *Inorganic Biochemistry,* Vol. 1, Eichhorn, G. L., Ed., Elsevier, Amsterdam, 1973, chap. 17.

8. **Friedman, T. and Epstein, C. J.,** The role of calcium in the reactivation of reduced taka-amylase, *J. Biol. Chem.,* 242, 5131, 1967.
9. **Price, P. A., Stein, W. H., and Moore, S.,** Effect of divalent cations on the reduction and re-formation of the disulfide bonds of deoxyribonuclease, *J. Biol. Chem.,* 244, 929, 1969.
10. **Cohn, M.,** Magnetic resonance studies of enzyme-substrate complexes with paramagnetic probes as illustrated by creatine kinase, *Q. Rev. Biophys.,* 3, 61, 1970.
11. **Schroeder, H. A., Vinton, W. H., and Balassa, J. J.,** Effect of chromium, cadmium and other trace metals on the growth and survival of mice, *J. Nutr.,* 80, 48, 1963.
12. **Schroeder, H. A.,** Chromium deficiency in rats: a syndrome simulating diabetes mellitus with retarded growth, *J. Nutr.,* 88, 439, 1966.
13. **Mertz, W.,** Chromium occurrence and function in biological systems *Physiol. Rev.,* 49, 163, 1969.
14. **Schroeder, H. A.,** The role of chromium in mammalian nutrition, *Am. J. Clin. Nutr.,* 21, 230, 1968.
15. **Mertz, W., Toepfer, E. W., Roginski, E. E., and Polansky, M. M.,** Present knowledge of the role of chromium, *Fed. Proc. Fed. Am. Soc. Exp. Biol.,* 33, 2275, 1974.
16. **Wacker, W. E. C. and Vallee, B. L.,** Nucleic acids and metals. I. Chromium, manganese, nickel, iron, and other metals in ribonucleic acid from diverse biological sources, *J. Biol. Chem.* 234, 3257, 1959.
17. **Kornicker, W. A. and Vallee, B. L.,** Metallocinium cations, nucleic acids and proteins, *Ann. N.Y. Acad Sci.,* 153, 689, 1969.
18. **Stickland, L. H.,** The activation of phosphoglucomutase by metal ions, *Biochem. J.,* 44, 190, 1949.
19. **Curran, G. L.,** Effect of certain transition group elements on hepatic synthesis of cholesterol in the rat, *J. Biol. Chem.,* 210, 765, 1954.
20. **Langenbeck, W., Augustin, M., and Schaefer, C.,** On the active metal ions of trypsins (in German), *Hoppe Seyler's Z. Physiol. Chem.,* 324, 54, 1961.
21. **Guest, J. R., Friedman, S., Woods, D. D., and Lester Smith, E.,** A methyl analogue of cobamide coenzyme in relation to methionine synthesis by bacteria, *Nature (London),* 195, 340, 1962.
22. **Stadtman, E. R., Overath, P., Eggerer, H., and Lynen, F.** The role of biotin and vitamin B_{12} coenzyme in propionate metabolism, *Biochem. Biophys. Res. Commun.,* 2, 1, 1960.
23. **Webb, M.,** The biological action of cobalt and other metals. IV. Inhibition of α-oxoglutarate dehydrogenase, *Biochim. Biophys. Acta,* 89, 431, 1964.
24. **Webb, M.,** The biological action of cobalt and other metals. III. Chelation of cations by dihydrolipoic acid, *Biochim. Biophys. Acta,* 65, 47, 1962.
25. **Fridovich, I.,** Superoxide dismutases, *Adv. Enzymol.,* 41, 35, 1974.
26. **Steinman, H. M., Vishweshvar, R. N., Abernethy, J. L., and Hill, R. L.,** Bovine erythrocyte superoxide dismutase. Complete amino acid sequence, *J. Biol. Chem.,* 249, 7326, 1974.
27. **Klug-Roth, D., Fridovich, I., and Rabani, J.,** Pulse radiolytic investigations of superoxide catalyzed disproportionation. Mechanism for bovine superoxide dismutase, *J. Am. Chem. Soc.,* 95, 2786, 1973.
28. **Rotilio, G., Bray, R. C., and Fielden, E. M.,** A pulse radiolysis study of superoxide dismutase, *Biochim. Biophys. Acta,* 268, 605, 1972.
29. **Beinert, H., Griffiths, D. E., Wharton, D. C., and Sands, R. H.,** Properties of the copper associated with cytochrome oxidase as studied by paramagnetic resonance spectroscopy, *J. Biol. Chem.,* 237, 2337, 1962.
30. **Morrison, M., Horie, S., and Mason, H. S.,** Cytochrome c oxidase components. II. A study of the copper in cytochrome c oxidase, *J. Biol. Chem.,* 238, 2220, 1963.
31. **Mahler, H. R.,** Uricase, in *The Enzymes,* Vol. 8, 2nd ed., Boyer, P. D., Lardy, H., and Myrbäck, K. Eds., Academic Press, New York 1963, chap. 9.
32. **Lerner, A. B., Fitzpatrick, T. B., Calkins, E., and Summerson, W. H.,** Mammalian tyrosinase: the relationship of copper to enzymic activity, *J. Biol. Chem.,* 187, 793, 1950.
33. **Kertész, D.,** The phenol-oxidizing enzyme system of human melanomas; substrate specificity and relationship to copper, *J. Natl. Cancer Inst.,* 14, 1081, 1954.
34. **Yamada, H., Kumagai, H., Kawasaki, H., Matsui, H., and Ogata, K.,** Crystallisation and properties of diamine oxidase from pig kidney, *Biochem. Biophys. Res. Commun.,* 29, 723, 1967.
35. **Buffoni, F. and Blaschko, H.,** Benzylamine oxidase and histaminase: studies of a crystalline preparation of the amine oxidase of pig plasma, *Biochem. J.,* 89, 111P, 1963.
36. **Friedman, S. and Kaufman, S.,** 3,4-Dihydroxyphenylethylamine β-hydroxylase. Physical properties, copper content, and role of copper in the catalytic activity, *J. Biol. Chem.,* 240, 4763, 1965.
37. **Mahler, H. R. and Elowe, D. G.,** Studies on metalloflavoproteins. II. The role of iron in diphosphopyridine nucleotide cytochrome c reductase, *J. Biol. Chem.,* 210, 165, 1954.
38. **Singer, T. P.,** Flavoprotein dehydrogenase of the electron-transport chain (survey), in, *The Enzymes,* Vol. 7, 2nd ed., Boyer, P. D., Lardy, H., and Myrbäck, K., Eds., Academic Press, New York, 1963, chap. 15.

39. **Hatefi, Y.**, The pyridine nucleotide-cytochrome c reductases in *The Enzymes*, Vol. 7, 2nd ed., Boyer, P. D., Lardy, H., and Myrbäck, K., Eds., Academic Press, New York, 1963, chap. 20.
40. **Zeylemaker, W. P., Der Vartanian, D. V., and Veeger, C.**, The amount of non-haem iron and acid-labile sulphur in purified pig-heart succinate dehydrogenase, *Biochim. Biophys. Acta*, 99, 183, 1965.
41. **Kearney, E. B.**, Studies on succinic dehydrogenase. XII. Flavin component of the mammalian enzyme, *J. Biol. Chem.*, 235, 865, 1960.
42. **Rajagopalan, K. V., Fridovich, I., and Handler, P.**, Hepatic aldehyde oxidase. I. Purification and properties, *J. Biol. Chem.*, 237, 922, 1962.
43. **Hart, L. I. and Bray, R. C.**, Improved xanthine oxidase purification, *Biochim. Biophys. Acta*, 146, 611, 1967.
44. **Rajagopalan, K. V. and Handler, P.**, Purification and properties of chicken liver xanthine oxidase, *J. Biol. Chem.*, 242, 4097, 1967.
45. **Scrutton, M. C., Utter, M. F., and Mildvan, A. S.**, Pyruvate carboxylase. VI. The presence of tightly bound manganese, *J. Biol. Chem.*, 241, 3480, 1966.
46. **Mildvan, A. S., Scrutton, M. C., and Utter, M. F.**, Pyruvate carboxylase. VII. A possible role for tightly bound manganese, *J. Biol. Chem.*, 241, 3488, 1966.
47. **Scrutton, M. C., Griminger, P., and Wallace, J. C.**, Pyruvate carboxylase. Bound metal content of the vertebrate liver enzyme as a function of diet and species, *J. Biol. Chem.*, 247, 3305, 1972.
48. **Keele, B. B., McCord, J. M., and Fridovich, I.**, Superoxide dismutase from Escherichia coli B. A new manganese-containing enzyme, *J. Biol. Chem.*, 245, 6176, 1970.
49. **Weisiger R. A. and Fridovich, I.**, Superoxide dismutase. Organelle specificity, *J. Biol. Chem.*, 248, 3582, 1973.
50. **Leach, R. M.**, Role of manganese in mucopolysaccharide metabolism, *Fed. Proc. Fed. Am. Soc. Exp. Biol.*, 30, 991, 1971.
51. **Morrison, J. F. and Ebner, K. E.**, Studies on galactosyltransferase. Kinetic investigations with N-acetylglucosamine as the galactosyl group acceptor, *J. Biol. Chem.*, 246, 3977, 1971.
52. **Morrison, J. F. and Ebner, K. E.**, Studies on galactosyltransferase. Kinetic investigations with glucose as the galactosyl group acceptor, *J. Biol. Chem.*, 246, 3985, 1971.
53. **Morrison, J. F. and Ebner K. E.**, Studies on galactosyltransferase. Kinetic effects of α-lactalbumin with N-acetylglucosamine and glucose as galactosyl group acceptors, *J. Biol. Chem.*, 246, 3992, 1971.
54. **Garrett, R. H. and Nason, A.**, Further purification and properties of Neurospora nitrate reductase, *J. Biol. Chem.*, 244, 2870, 1969.
55. **Spence, J. T.**, Reactions of molybdenum coordination compounds: models for biological systems, in *Metal Ions in Biological Systems*, Vol. 5, Sigel, H., Ed., Marcel Dekker, New York, 1976, chap. 6.
56. **Cohen, H. J. and Fridovich, I.**, Hepatic sulfite oxidase. Purification and properties, *J. Biol. Chem.*, 246, 359, 1971.
57. **Cohen, H. J. and Fridovich, I.**, Hepatic sulfite oxidase. The nature and function of the heme prosthetic groups, *J. Biol. Chem.*, 246, 367, 1971.
58. **Cohen, H. J., Fridovich, I., and Rajogopalan, K. V.**, Hepatic sulfite oxidase. A functional role for molybdenum, *J. Biol. Chem.*, 246, 374, 1971.
59. **Bray, R. C. and Swann, J. C.**, Molybdenum-containing enzymes, *Struct. Bonding (Berlin)*, 11, 107, 1972.
60. **Nielsen, F. H. and Ollerich, F. H.**, Nickel: a new essential trace element, *Fed. Proc. Fed. Am. Soc. Exp. Biol.*, 33, 1767, 1974.
61. **Whanger, P. D.**, Effects of dietry nickel on enzyme activities and mineral contents in rats, *Toxicol. Appl. Pharmacol.*, 25, 323, 1973.
62. **Webster, L. T.**, Studies of the acetyl coenzyme A synthetase reaction. III. Evidence of a double requirement for divalent cations, *J. Biol. Chem.*, 240, 4164, 1965.
63. **Ray, W. J.**, Role of bivalent cations in the phosphoglucomutase system. I. Characterization of enzyme-metal complexes, *J. Biol., Chem.*, 244, 3740, 1969.
64. **Callis, G. E. and Wentworth, R. A. D.**, Tungsten vs. molybdenum in models for biological systems, *Bioinorg. Chem.* , 7, 57, 1977.
65. **Ljungdahl, L. G. and Andreesen, J. R.**, Tungsten, a component of active formate dehydrogenase from *Clostridium thermoaceticum, FEBS Lett.*, 54, 279, 1975.
66. **Johnson, J. L., Rajagopalan, K. V., and Cohen, H. J.**, Molecular basis of the biological function of molybdenum. Effect of tungsten on xanthine oxidase and sulphite oxidase in the rat, *J. Biol. Chem.*, 249, 859, 1974.
67. **Johnson, J. L., Cohen, H. J., and Rajagopalan, K. V.**, Molecular basis of the biological function of molybdenum. Molybdenum-free sulphite oxidase from livers of tungsten-treated rats, *J. Biol. Chem.*, 249, 5046, 1974.

68. **Johnson, J. L., Waud, W. R., Cohen, H. J., and Rajagopalan, K. V.,** Molecular basis of the biological function of molybdenum. Molybdenum-free xanthine oxidase from livers of tungsten-treated rats, *J. Biol. Chem.*, 249, 5056, 1974.
69. **Hopkins, L. L. and Mohr, H. E.,** The biological essentiality of vanadium, in *Newer Trace Elements in Nutrition*, Mertz, W. and Cornatzer, W. E., Eds., Marcel Dekker, New York, 1971, 195.
70. **Schwarz, K. and Milne, D. B.,** Growth effects of vanadium in the rat, *Science*, 174, 426, 1971.
71. **Hopkins, L. L. and Mohr, H. E.,** Vanadium as an essential nutrient, *Fed. Proc. Fed. Am. Soc. Exp. Biol.*, 33, 1773, 1974.
72. **Takahashi, H. and Nason, A.,** Tungstate as a competitive inhibitor of molybdate in nitrate assimulation and in N_2 fixation by *Azotobacter, Biochim. Biophys. Acta*, 23, 433, 1957.
73. **Anderson, A. J.,** Role of molybdenum in plant nutrition, in *A Symposium on Inorganic Nitrogen Metabolism*, McElroy, W. D. and Glass, B., Eds., Johns Hopkins Press, Baltimore, 1956, 3.
74. **Martin, G. M., Benditt, E. P., and Eriksen, N.,** Vanadium catalysis of the dihydroxyphenylalanine and 5-hydroxyindoles, *Nature (London)*, 186, 884, 1960.
75. **Curran, G. L. and Costello, R. L.,** Reduction of excess cholesterol in the rabbit aorta by inhibition of endogenous cholesterol synthesis, *J. Exp. Med.*, 103, 49, 1956.
76. **Azarnoff, D. L. and Curran, G. L.,** Site of vanadium inhibition of cholesterol biosynthesis, *J. Am. Chem. Soc.*, 79, 2968, 1957.
77. **Coleman, J. E.,** Carbonic anhydrase, in *Inorganic Biochemistry*, Vol. 1, Eichhorn, G. L., Ed., Elsevier, Amsterdam, 1973, chap. 16.
78. **Ludwig, M. A. and Lipscomb, W. N.,** Carboxypeptidase A and other peptidases, in *Inorganic Biochemistry*, Vol. 1, Eichhorn, G. L., Ed., Elsevier, Amsterdam, 1973, chap. 15.
79. **Mildvan, A. S. and Weiner, H.,** Interaction of a spin-labelled analog of nicotinamide-adenine dinucleotide with alcohol dehydrogenase. II. Proton relaxation rate and electron paramagnetic resonance studies of binary and ternary complexes, *Biochemistry*, 8, 552, 1969.
80. **Mildvan, A. S. and Weiner, H.,** Interaction of a spin-labelled analogue of nicotinamide adenine dinucleotide with alcohol dehydrogenase, *J. Biol. Chem.*, 244, 2465, 1969.
81. **Weiner, H.,** Interaction of a spin-labelled dinucleotide with alcohol dehydrogenase. I. Synthesis, kinetics and electron paramagnetic resonance studies, *Biochemistry*, 8, 526, 1969.
82. **Adelstein, S. J. and Vallee B. L.,** Zinc in beef liver glutamic dehydrogenase, *J. Biol. Chem.*, 233, 589, 1958.
83. **Vallee, B. L. and Wacker, W. E. C., Zinc,** a component of rabbit muscle lactic dehydrogenase, *J. Am. Chem. Soc.*, 78, 1771, 1956.
84. **Scrutton, M. C.,** Metal enzymes, in *Inorganic Biochemistry*, Vol. 1, Eichhorn, G. L., Ed., Elsevier, Amsterdam, 1973, chap. 14.
85. **Drum, D. E., Li, T.-K., and Vallee, B. L.,** Zinc isotope exchange in horse liver alcohol dehydrogenase, *Biochemistry*, 8, 3792, 1969.
86. **Rotilio, G., Calabrese, L., Bossa, F., Barra, D., Agrò, A. F., and Mondovi, B.,** Properties of the apoprotein and role of copper and zinc in protein conformation and enzyme activity of bovine superoxide dismutase, *Biochemistry*, 11, 2182, 1972.
87. **Vance, P. G. and Keele, B. B.,** Superoxide dismutase from Streptococcus mutans. Isolation and characterization of two forms of the enzyme, *J. Biol. Chem.*, 247, 4782, 1972.
88. **Vandecasteele, J. and Burris, R. H.,** Purification, and properties of the constituents of the nitrogenase complex from *Clostridium pasteurianum, J. Bacteriol.*, 101, 794, 1970.
89. **Burns, R. C., Holsten, R. D., and Hardy, R. W. F.,** Isolation by crystallization of the Mo-Fe protein of *Azotobacter nitrogenase, Biochem. Biophys. Res. Commun.*, 39, 90, 1970.
90. **Lindskog, S.,** Purification and properties of bovine erythrocyte carbonic anhydrase, *Biochim. Biophys. Acta*, 39, 218, 1960.
91. **Vallee, B. L. and Hoch, F. L.,** Zinc in horse liver alcohol dehydrogenase *J. Biol. Chem.*, 225, 185, 1957.
92. **Mortenson, L. E., Morris, J. A., and Jeng, D. Y.,** Purification, metal composition and properties of molybdoferredoxin and azoferredoxin, two of the components of the nitrogen-fixing system of *Clostridium pasteurianum, Biochim. Biophys. Acta*, 141, 516, 1967.
93. **De Renzo, E. C.,** Chemistry and biochemistry of xanthine oxidase, *Adv. Enzymol.*, 17, 293, 1956.
94. **McGartoll, M. A., Pick, F. M., Swann, J., and Bray, R. C.,** Properties of xanthine oxidase preparations dependent on the proportions of active and inactivated enzymes, *Biochim. Biophys. Acta*, 212, 523, 1970.
95. **Hart, L. I., McGartoll, M. A., Chapman, H. R., and Bray, R. C.,** The composition of milk xanthine oxidase, *Biochem. J.*, 116, 851, 1970.
96. **Vallee, B. L. and Ulmer, D. D.,** Biochemical effects of mercury, cadmium and lead, *Annu. Rev. Biochem.*, 41, 91, 1972.

97. **Drum, D. E.**, Enzymatically and optically active cobalt and cadmium alcohol dehydrogenases, *Fed. Proc. Fed. Am. Soc. Exp. Biol.*, 29, 608, 1970.
98. **Druyan, R. and Vallee, B. L.**, Exchangeability of the zinc atoms in liver alcohol dehydrogenase, *Fed. Proc. Fed. Am. Soc. Exp. Biol.*, 21, 247, 1962.
99. **Lazdunski, C., Petitclerc, C., and Lazdunski, M.**, Structure-function relationships for some metalloalkaline phosphatases of *E. coli*, *Eur. J. Biochem.*, 8, 510, 1969.
100. **Applebury, M. L., Johnson, B. P., and Coleman, J. E.**, Phosphate binding to alkaline phosphatase, metal ion dependence, *J. Biol. Chem.*, 245, 4968, 1970.
101. **Jacobs, E. E., Jacob, M., Sanadi, D. R., and Bradley, L. B.**, Uncoupling of oxidative phosphorylation by cadmium ion, *J. Biol. Chem.*, 223, 147, 1956.
102. **Sanadi, D. R., Langley, M., and White, F.**, α-Ketoglutaric dehydrogenase. VII. The role of thioctic acid, *J. Biol. Chem.*, 234, 183, 1959.
103. **Fluharty, A. L. and Sanadi, D. R.**, On the mechanism of oxidative phosphorylation. VI. Localization of the dithiol in oxidative phosphorylation with respect to the oligomycin inhibition site, *Biochemistry*, 2, 519, 1963.
104. **Hasan, J., Vihko, V., and Hernberg, S.**, Deficient red cell membrane (Na^+ + K^+) — ATPase in lead poisoning, *Arch. Environ. Health*, 14, 313, 1967.
105. **Hernberg, S., Vihko, V., and Hasan, J.**, Red cell membrane ATPase in workers exposed to inorganic lead, *Arch. Environ. Health*, 14, 319, 1967.
106. **Boyer, P. D.**, Sulfhydryl and disulfide groups of enzymes, in *The Enzymes*, Vol. 1, 2nd ed., Boyer, P. D., Lardy, H., and Myrbäck, K., Eds., Academic Press, New York, 1959, 511.
107. **Fowler, B. A., Brown, H. W., Lucier, G. W., and Krigman, M. R.**, The effects of chronic oral methyl mercury exposure on the lysosome system of rat kidney, morphometric and biochemical studies, *Lab. Invest.*, 32, 313, 1975.
108. **Vallee, B. L., Coombs, T. L., and Hoch, F. L.**, The "active site" of bovine pancreatic carboxypetidase A, *J. Biol. Chem.*, 235, PC45, 1960.
109. **Bowman, F. J., and Landon, E. J.**, Organic mercurials and net movements of potassium in rat kidney slices, *Am. J. Physiol.*, 213, 1209, 1967.

Chapter 4

STRUCTURE AND FUNCTION OF METALLO-DNA IN THE LIVING CELL

Etienne Guille, Jeanine Grisvard, and Igor Sissoeff

TABLE OF CONTENTS

I. INTRODUCTION

Some metals are required in small concentrations for biological activities including enzymatic catalysis, oxidation-reduction and transport processes, membrane function, nerve conduction, muscle contraction, and functioning of subcellular organelles such as mitochondria. They can also play an important role in biological structures.

Metal deficiency in animals, plants, and bacteria produces characteristic diseases which reflects altered biochemical function. Thus, some metals are essential for living cells and their requirement is absolute for each essential element. In spite of differences between the various organisms the actually recognized essential elements include iron, iodine, copper, manganese, zinc, cobalt, molybdenum, selenium, chromium, tin, vanadium, fluorine, silicon, nickel, and arsenic.[1]

On the other hand, excesses of essential metals and all other metals are toxic and lead to several pathological perturbations including neoplastic diseases.[2,3] For instance, iron excess produces hemochromatosis while copper excess is implicated in Wilson's disease. Many of the toxic effects of metals in biological systems appear to occur at similar biochemical loci as do essential metals function.[4]

In addition to the deficiency and intoxication, the imbalance of metal ions may also lead to biological dysfunction. Indeed metal ion competition for a crucial biological site is probably involved in numerous cases of deficiency or intoxication.

In general, the effects of trace metals are related to their concentration. It is possible for a given element to find a continuous progression from a deficiency state, to an essential biological function, then to an imbalance when its excess interferes with the function of another metal, then to pharmacologically active concentrations and finally to toxic and even life threatening concentrations.[5] As a result, the list of essential metals is continuously growing to the detriment of those known only for their toxic effects.

Until now, metals were generally thought to play their most important role as cofactors for enzymes. The fact that metal ions found in DNA molecules are not bound at random and that their binding is correlated to the physiological state of the cell[6] has led us to reexamine the functions of the essential metal and to try to clear up the molecular mechanism underlying their role.

The purpose of this review is to describe the physiochemical properties of DNA-metal complexes and to try to understand the physiological function(s) of various ternary complexes present in the living cells. We shall also try to show if perturbations induced in these complexes by various physiological and pathological conditions are correlated to the function of essential metals and/or competition with toxic ones.

II. THE METAL ION-DNA COMPLEX

A. Metal Occurrence in Nuclei and DNA Samples

1. Natural Occurrence

For a long time, numerous authors have noticed the natural occurrence of metals in nuclei of various cells. For example, numerous metals have been observed in rat hepatic cells,[7] aluminum in the nuclei of various cellular types detected by ion microprobe analysis,[8] and lead in nuclei and chloroplasts of leaf cells.[9,10]

From these observations, although not exhaustively referenced, it must be pointed out that a nuclear localization does not necessarily imply binding to nucleic acids. Thus, a nuclear accumulation of iron in insects has been attributed to an iron-protein complex.[11] Nevertheless, except in special cases of intranuclear sequestration, the widespread occurrence of metals in nuclei allows us to suppose that they could reach the genetic material of the cell, i.e., DNA.[12]

In DNA preparations, the existence of metal ions on the DNA molecule has been pointed out, either after a systematic analysis or accidentally, in order to explain abnormalities in physical analysis generally due to paramagnetic impurities.[13-16] By the early 1950s work had been done on metal ion content of nucleic acid preparations[17,18] and the main conclusions were that, whatever the origin of the DNA, some metal ions (Cr^{3+}, Ni^{2+}, Fe^{2+}) are firmly bound to the DNA molecule and could be considered as a normal or intrinsic constituent of it. Since then, progress in analysis of trace metals in biological samples has allowed more accurate determinations.[19] By neutron activation analysis, Cr^{3+}, Sb^{2+}, Fe^{2+}, Zn^{2+}, and Co^{2+} have been found in DNA from lymphocytes,[20,21] and furthermore, the content of these trace metal ions indicates a specificity according to the tissue. The occurrence of metals in DNA preparations is not restricted to eukaryotic organisms since Fe^{2+}, Mn^{2+} and Cu^{2+} are present in DNA from various enterobacteria.[22] An exhaustive analysis of 32 elements, including metals, has been done on thymus and salmon sperm DNA.[23] Whatever their origin, the DNA samples contain metal ions in variable amounts and these metals have been considered as intrinsic elements rather as impurities coming from extraction procedures. Cd^{2+}, Pb^{2+}, and Cu^{2+} have been found in reiterative DNA sequences of DNA molecules isolated from various organisms (plants, animals, bacteria).[6,24,25]

The significance of the amount of metal ions present in the DNA molecule is now beginning to be understood in relation to various physiological or pathological states.[11,19,20]

Numerous metal ions have also been detected in RNA preparations and generally in greater amounts than in DNA samples.[23,26,27]

2. Radioactive Metals

Nuclear localization of metals has been studied with the use of radioactive metals. Taking into account that radioactive metals are used as tracers, they do not introduce any perturbation in the cells and their presence in the cells could be assimilated to a natural uptake. The highest $^{65}Zn^{2+}$ uptake has been ascribed to the nuclear fraction of rat liver,[28,29] and in a comparative study, the ^{65}Zn distribution in the nucleoli was found to be different in in vitro and in vivo studies.[30] Iron seems to play a fundamental role in the mitotic process, localized in the nucleolus during the interphase, it moves towards chromosomes during the metaphase step.[32] There is a preferential accumulation of ^{75}Sc in the nuclei of meristematic cells from autoradiography of root apexes.[32] Utilization of ^{25}Al has shown not only a specific incorporation of this metal in the nucleus of hepatic cells, but also that a DNA-Al complex exists, since radioactivity is found on DNA after its extraction.[33]

Thus, metals present in diet for animals or in soil for plants can reach, in some physiological conditions, the nucleus of the cells, and then bind to the chromosomal DNA and/or concentrate in some cases into intranuclear bodies.

3. Metal Occurrence after Experimental Supplementation

The effects of metal excess have been studied by numerous authors and often this metal excess has been determined to have an intranuclear localization. We shall examine now some relevant results, but since the toxicity of a metal and the tolerance of an organism towards it is sometimes determined during an experimental supplementation, the results must be analyzed with caution. Beryllium, which is carcinogenic for animals and induces chromosomal and mitotic abnormalities,[34] is found in nuclei and, particularly in nucleoli,[35] it binds preferentially to the nuclei of rat liver cells that have high mitotic activity.[36] Aluminum concentrates in the nuclei of neurons of cats and rats after intracranial application of aluminum salts and the neurotoxic effect may be due to an alteration of the nucleic acid metabolism.[37,38] In vivo, cupric ions are found in the chromatin of mouse liver cells in a variable amount in the heterochromatic part and in a fixed amount in the euchromatic one, whereas in vitro, and at high ionic strength, heterochromatin binds more copper than does euchromatin.[39,40] Bismuth gives intranuclear inclusion bodies in proximal renal tubular cells from rats poisoned with bismuth.[41] In human or experimental animals there are intranuclear inclusion bodies in kidney, liver, and brain composed of a lead-protein complex.[42,43] When cadmium, cobalt, and nickel are injected into rat muscles, most of these metals (70 to 90%) are bound to the cell nucleus. Nickel is associated in part with nucleolar DNA and mostly with nuclear RNA.[44] After application of cadmium to rat lateral prostate in vitro, the metal is found in nuclei of necrotic epithelial cells and basal cells with high mitotic activity.[45] Work has also been done with plants. On meristematic cells of *Allium*, about 40 metals have been tested and their biological effects described.[46] By using an electron microprobe X-ray analyzer, cobalt, zinc.[47] and aluminum[48] have been localized in the nuclei of *Zea mays* cells. Metallic nuclear inclusion bodies are also found in plants that are tolerant for lead[49] and in green algae in environment rich in copper.[50] Thus by metal addition at amounts which could sometimes be lethal, numerous authors show that metal could accumulate at least partially in the cell nucleus and for some cases could have a differential affinity towards hetero- and euchromatin.

The isolation of nuclear DNA also gives useful information about the binding of the metal added in vivo. A study of the effect of cadmium on nucleic acid and protein synthesis in rat liver cells gives results which could be interpreted as a direct binding of cadmium on DNA.[51] Similar findings have been obtained for chromium in hamster cell cultures.[52] Aluminum absorbed by pea roots is found in the cell wall and associated with nucleic acids[53] and more precisely, with DNA of a deoxyribonucleoprotein extract.[54]

From these facts, although scattered, it is possible to draw some conclusions. Metal ions seem to be intrinsic elements of the DNA molecule with a nonrandom distribution along the molecule. In vivo added metals can reach the nucleus of many cells and since DNA has a complexing ability,[55] there is no doubt that it could bind metals which may then influence the cell machinery.

B. In Vitro Studies of Metal-DNA Interactions

1. Elements of the Complex

a. DNA

It is necessary here to define the different DNA structural levels that metal ions can modify (Figure 1).

The primary structure is determined by the DNA nucleotidic sequence. The secondary structure is that of a double helix, each chain forming a right-hand helix. Hydrogen bond interaction between complementary bases (A-T and G-C base pairs) is essential for the DNA stability. Any perturbation of this interaction should modify the DNA stability. So, a strengthening of the interaction would stabilize the double helix, whereas a weakening would destabilize it and could lead to denaturation, where DNA is converted into two coiled polynucleotidic chains. This helix-coil transition is reversible (renaturation) under some conditions.

DNA denaturation can be carried out in several ways: by heat (thermal denaturation) which increases thermal movements of the two strands, by low pH (acid denaturation) by which amine functions are protonated, weakening H-bonds, by high pH (alkaline denaturation) where ionization of G and T bases also gave weakened H-bonds, and by low salt concentration that increases repulsion between the two strands as a consequence of a decrease in counter ion content. In double-helix stability all these parameters are interrelated.

The most widely used denaturation method is the thermal one usually studied by spectrophotometric measurements of the 260-nm hyperchromicity of a diluted DNA solution as a function of temperature. The curve obtained is characterized by two

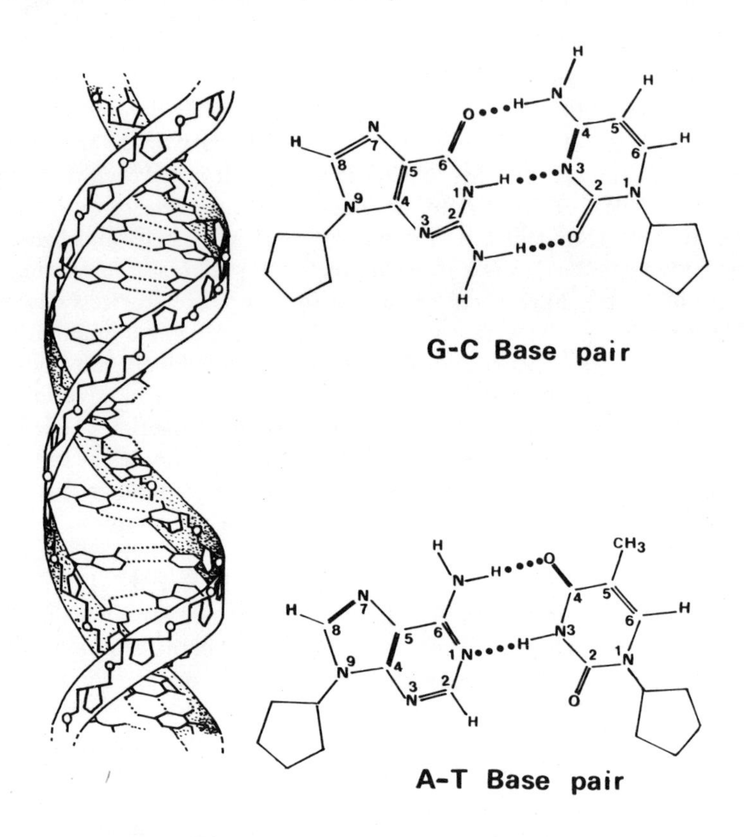

FIGURE 1. DNA double-helix scheme; the numbering of atoms used in the text is indicated in A-T and G-C base pairs.

parameters: the temperature at which 50% of the bonds are broken, Tm (melting temperature), and the slope at melting temperature. Stabilization would increase Tm and destabilization decrease Tm.

The tertiary structure of DNA is represented by the folding of the double-helix as a rigid rod or as a super-helix. DNA has a highly rigid structure in the double-helix form, but the molecule is very long and still keeps very small flexibility at every point and a worm-like structure has been generally accepted.

The quaternary structure results in the laterally or end-to-end binding of DNA molecules. During in vitro experiments, DNA molecules are generally assumed to be free from one another, and under some circumstances molecules could aggregate and sometimes precipitate.

In vitro studies have been done on natural or synthetic DNAs. Natural DNA is isolated from eukaryotic and prokaryotic cells using various extraction procedures, and experiments on whole DNA give no difference related to their origin as long as they have the same nucleotidic composition expressed as G + C%. Because of their known nucleotidic sequence and properties,[56] synthetic DNAs or polydesoxyribonucleotides such as poly dA-dT, poly dAT, and poly dG-dC give useful information. In aqueous solution, DNA is in the double-helix B form[57] at room temperature, at neutral pH, and at ionic streangth greater than 10^{-3} *M*. To get more information, either native double-helix DNA or denatured, coiled DNA has been used.

b. Metal Ions

For in vitro studies on metal ion-DNA interaction, numerous metal ions have been used including monovalent alkali metal ions (Na^+, K^+, . . .) divalent alkaline earths (Ca^{2+}, Mg^{2+}, . . .) metals of the transition series (Fe^{2+}, Co^{2+}, Ni^{2+}, Cu^{2+}, Zn^{2+}, Cr^{3+}, Mn^{2+} . . .), and even some lanthanides (Eu^{3+}, Tb^{3+}). From their known electron structure, it can be predicted that alkali metal ions can produce ionic or electrostatic binding, whereas metal ions of the transition series with their incomplete inner electron structure realize covalent coordination binding.

Metal ions are positively charged, and are, therefore, electrophiles. They react with nucleophile or negatively charged sites having a high electronic density. Since such sites are found in the DNA molecule[58] one could, without experimental work, put forward the following possible binding sites: oxygen atoms of phosphate residues located on the backbone of each strand, and oxygen and nitrogen atoms of the nucleic bases of DNA (adenine, thymine, guanine, cytosine) which maintain the two strands in register through hydrogen bonds. These sites are N_7, N_3, and NH_2 at C_2 in adenine, N_3, N_7,NH_2 at C_6 in guanine, N_3 and O_4 in thymine, and N_3 and NH_2 at C_4 in cytosine (Figure 1).

2. The Metal Ion-DNA Complexes

Background information was obtained from studies with nucleotides, nucleosides, and bases which are simpler models.[59] They indicate that effectively some metal ions bind preferentially to the phosphates while the others coordinate with the heterocyclic bases.[60] These properties of nucleic acid constituents have often been applied to the polymer itself. However, in double-stranded DNA only N_7(G), N_7(A), and O_4(T) are readily available, N_7 sites being the most reactive electron donor groups.

a. Binding Sites

Depending on the metal ion, two or several coordination bonds are available. A single metal ion could, therefore, bind simultaneously two phosphates, two bases, or a phosphate and a base (Figure 2).

i. Metal Ions which Bind Preferentially to the Phosphates

To this group belong almost all metal ions studied so far and principally alkali Na^+, K^+,[61] alkali earth Ca^{2+},[62-66] and Mg^{2+}[66,67] with no direct involvment of the nitrogen bases. Some other metal ions at low ionic strength fit into this group including Mn^{2+},[68] Cu^{2+},[69,70] and $Tb.^{3+}$.[71] The interaction is of an electrostatic nature, and metal ions form an ionic cloud around the backbone of negatively charged phosphate residues.[61] This cloud is highly mobile so that two types of binding sites are under discussion: diffuse binding and site binding.[63,64] In fact, only a few percent of alkali metal is involved in a real interaction with the DNA,[61] and since in a site type binding the metal ion must go to a specific binding place, the term "ion condensation" is preferred.[72] There is a strong association between these counterions and polynucleotides[73,77] which decreases along the alkali series[78] (for instance, the stronger binding constants for Ca^{2+} and Mg^{2+} are about 10^3 *M*). Kinetics studies show that this interaction is of a cooperative type.

ii. Metal Ions which Bind to the Bases

In spite of steric hindrance, metal ions can reach the double-helix inside and interact with the bases; they can form two types of chelate either by insertion or by intercalation.[79]

Insertion — In this case, chelation occurs between the two bases of a given base pair, disrupting the hydrogen bonds. To this group belong metals such as Ag^+,[80,81,84,87] Cu^{2+},[70] and Au^{3+}.[88]

Hg^{2+} binds preferentially to N_3 of thymine so that a specificity for A-T base pair is indicated with high constant of linear complex formation $\simeq 10^{18}$,[89,95] but other results

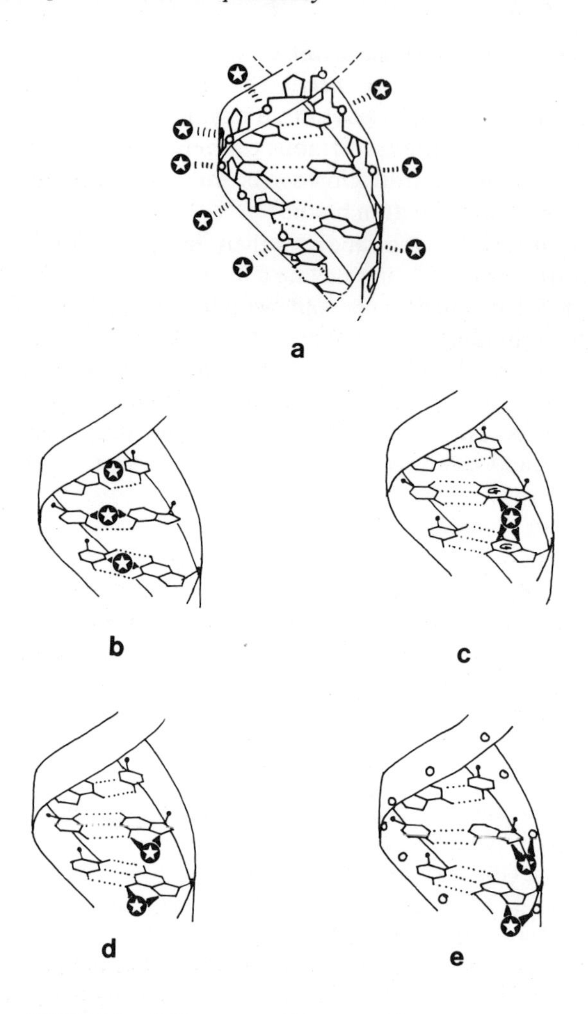

FIGURE 2. Different possible metal-binding sites in DNA molecule **a**, metal binding to phosphate groups; **b, c, d,** metal binding to the bases (b = insertion; c = intercalation); e = metal binding to both phosphate and base.

suggest[96] a chelate between two thymine belonging to opposite strands. A new type of double-helix thus appears[97,98] so that the two chains become cross-linked.

At high rf values, Ag^+ insertion occurs preferentially with thymine bases[99,100] and with an intrinsic binding constant of 4.10^4 *M*.[80] Profound changes in conformation are observed and a new unknown double-stranded helix appears.[83,101]

As a result of linear chelate formation, hydrogen bonds are broken and protons are released. This type of chelation is strongly dependent on pH and only slightly dependent upon ionic strength.

Intercalation — In the case of intercalation, the chelation occurs between two adjacent bases belonging to the same strand, either by charge transfer (π bonding) or chelation between two purines. This is the case of Cu^{2+}; at low rf values[102] the chelation is effected between two adjacent G-C pairs where the two guanines are in the same strand. Intercalation has also been proposed for Ag^+,[83] Fe^{2+},[103] Zn^{2+} and Cd.[2+79] These chelates are not dependent on ionic strength and pH since they maintain the stacking of nucleotides.

iii. Metal Ions which Bind to Both Phosphate and Base

Divalent cations could form macrochelate[79] or mixed chelate structures between one atom of a base and the oxygen of a phosphate residue in the same strand. Such macrochelates occur with divalent cations of transition metals (Mn^{2+}, Fe^{2+}, Co^{2+}, Cu^{2+}, Zn^{2+}).

For Cu^{2+} at high rf value, a chelate is formed between the phosphate group and N_7 of guanine, or N_3 of cytosine at room temperature and between the phosphate group and N_3 of adenine after thermal opening of an A-T base pair[102,111] with a binding constant of about 10^5 *M*.[107]

Mn^{2+} forms a macrochelate between a phosphate group and N_7 of guanine.[112,115] N_7 is the best ligand for such macrochelates, and metals could be ordered according to their affinity for one of the two ligands implicated: nitrogen of base and oxygen of phosphate. We could then find any intermediate state between strong chelation (Cu^{2+}) to pure electrostatic binding (Mg^{2+}). Owing to the electrostatic character of the binding to phosphate, macrochelates are very sensitive to ionic strength.

Finally, an order of the relative phosphate binding ability to base site binding ability in native DNA has been established,[59,115] as indicated in Table 1.

b. Different Complexes According to Metal Ion Concentration

As indicated above, it is possible for a given metal ion to bind at several sites on the DNA molecule and several complexes can even be formed simultaneously or consecutively.

Ca^{2+} complexes with phosphate groups with a stoichiometry of one-half[66,116] the phosphate groups implicated being adjacent on the same chain or belonging to the two chains. As calcium concentration increases a 1/1 complex appears.[64]

At low ionic strength, neutral pH, and low temperature, two successive DNA-Cu^{2+} complexes are formed as Cu^{2+} concentration increases. Cu^{2+} initially binds in a noncooperative manner to phosphates and DNA is in a C-like form. Then, Cu^{2+} forms a second bond with a nonadjacent base site on the same strand or another strand.[117] At high ionic strength, two complexes are simultaneously formed (the sandwich type chelate) by increasing Cu^{2+} concentration.[102] Another modification has been found by *in situ* reduction of Cu^{2+} to Cu^{+}, with the passage from sandwich-type to other chelates.[118]

Mn^{2+} binds to the phosphate groups and also to the electron donor groups of the bases. Two types of sites with different affinities for Mn^{2+} have been found in DNA, one with a $K \simeq 10^4$ *M* (phosphate site) and the other $K = 10^3$ *M* (base site).[119]

The results of studies on Ag^{+}-DNA complexes indicate that Ag^{+} binds to DNA in two successive complexes. As already shown, Ag^{+} is intercalated for $rf < 0.1$ with preference for guanine and inserted for $rf > 0.5$ between A-T or G-C base pairs.[80,82,83,100]

A similar case is found for Hg^{2+}-DNA complexes. A first complex of stoichiometry 1/2 begins, followed, when saturated, by a second complex at high rf.[84,120]

For Mg^{2+} binding, it appears that the primary mode of binding yields a maximum of 1 mg^{2+}/2 DNA-P and it has been proposed that at higher Mg^{2+} level, there is a 1/1 complex with one positive charge per nucleotide residue so that DNA-Mg might be considered as a polycation.[63,64]

All these results indicate that when metal ion concentration varies, different binding sites can be formed leading to quite different effects on DNA structure, as will be shown later.

3. DNA Structural Modifications Induced by Metal Ions

Metal ions exert various influences on the conformation of nucleic acids at primary, secondary, tertiary, and quaternary structure levels. As the structure of DNA is

Table 1
METALS

Presence, Biological Effect

Metals	Fe	Zn	Pb	Cu	Al	Cd	Sn	Mn	Mo	Ni	Au	Cr	Co	Cs	U	Be	V	Hg	Tl	Pd	Ti	Sb
Presence in man	+	+	+	+	+	+	+	+	+	+	+	+	+	+	+	+						
Essential	+	+		+			+	+	+	+		+	+				+					
Biomethylation			+				+				+							+	+	+	+	+
Carcinogenic	+	+	+			+				+		+	+			+					+	
Chromosomal aberrations			+	+	+	+				+		+						+				+
Presence in DNA	+	+	+	+		+		+		+		+	+									+

Binding Sites on DNA Molecule

⟵ Metal ions increasing affinity ⟶

Phosphate	Li^{+} (Fe^{3+})	Na^{+}	K^{+}	Rb^{+}	Cs^{+}	Mg^{2+}	Ca^{2+}	Sr^{2+}	Ba^{2+}
Phosphate and base	Co^{2+} (Fe^{2+})	Ni^{2+}	Mn^{2+}	Zn^{2+}	Cd^{2+}	Pb^{2+}	Cu^{2+}		
Base	Ag^{+}	Hg^{2+}							

thought to be an important factor for its biological activity, all modifications produced by metal ions must be considered.

a. Primary Structure Modification

The DNA primary structure could be modified in relation to metal ion interaction by breaking, sequence modification, or lengthening of the DNA molecule.

In vitro DNA cleavage by metal ions seems to be unlikely, contrarily to what happens with RNA.[121,122] This could be explained by the absence of 2'OH group in desoxyribose that would be involved in the degradation mechanism.[123] Nevertheless, it should be noted that among the metal ions tested, Pd^{2+} and Zn^{2+} have been shown to eventually induce some cleavage of the DNA molecule[124,125] and the production of new DNA fractions after DNA centrifugation in the presence of Ag^{+} ions could be explained as a rupture of the DNA molecule.[6]

DNA damage and degradation induced by metal compounds have been demonstrated in vivo in bacteria. Clearly, in vitro results must be extrapolated with caution to in vivo events.[126]

The DNA lengthening possibility will be invoked when discussing quaternary structure.

b. Secondary Structure

Metal interaction with the phosphate residue of the DNA molecule results in a stabilization of the double-helix through a reduction of the electrostatic repulsion between the charged phosphate groups located in the two strands. Conversely, as metal ion concentration decreases ($< 10^{-4}$ *M*), repulsion increases and denaturation occurs. This is the well-known denaturation by dilution of DNA.[127]

In the case of macrochelates (Fe^{2+}, Co^{2+}, Ni^{2+}, Cu^{2+}, Zn^{2+}) chelation between N_7 and phosphate promotes a locally denaturated state by distortion of the base, but conversely the electrostatic binding to phosphate would stabilize it. This explains the different effects observed with transition metal ions.

Strong perturbations of the conformation by insertion (Ag^{+}, Hg^{2+}) are generally detected as the metal ion substitutes coordination binding for the previous H-bonds. When intercalation occurs, as with Cu^{2+}, Ag^{+}, Cd^{2+}, and Fe^{2+}, H-bonds between base pairs are not modified. Nevertheless, a stabilization can be noted as two consecutive (adjacent) bases are held together. This fact is important in explaining some discrepancies observed in the effects of metal ions during renaturation processes.

The Tm variations as a function of rf at low ionic strength give useful information on the different stabilities produced by metal ions[105,128] (Figure 3). For alkali metal ions, stabilization is shown as rf increases, whereas at low rf, transition metal ions stabilize (regions A). As rf increases (regions B), Tm decreases to an extent depending on metal ion. Except for phosphate binding realized by all metal ions, in region A intercalation of transition metals occurs simultaneously with macrochelation, and in region B predominance of macrochelation leads to a decrease of Tm.[102]

Upon cooling, renaturation is observed: the complementary chains again form hydrogen bonds with one another. In the presence of alkali metal ions especially Mg^{2+}, renaturation could be due to the fact that the two chains remain in close proximity. In the case of other metal ions (Hg^{2+}, Zn^{2+}, Ni^{2+}, Co^{2+}, Mn^{2+}), cross-links between the two strands would promote rapid renaturation.[105,108,128] For the DNA-Cu^{2+} complex, renaturation could be achieved only by increasing the ionic strength.[129]

Spectrophotometric acid titration in the presence of Zn^{2+} leads to the conclusion that the metal ion suppresses protonation of a G-C base pair so that an acid denaturation occurs at less acid pH.[130]

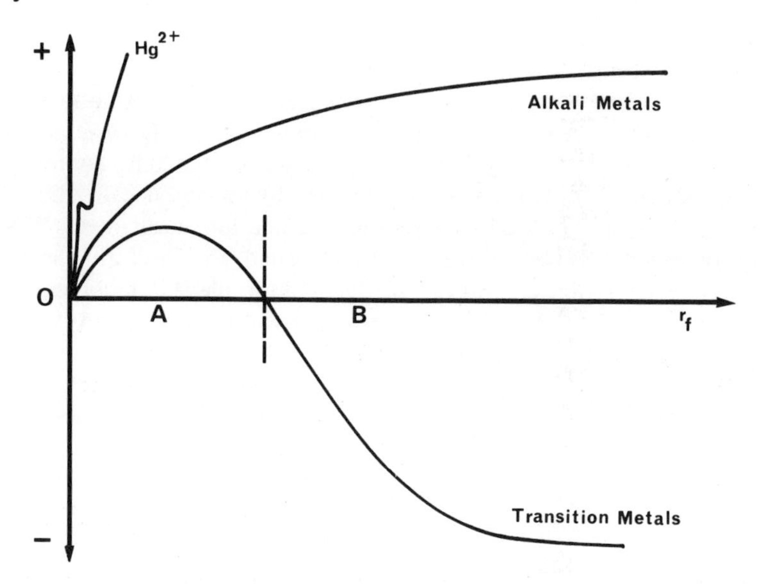

FIGURE 3. Double-helix stability according to the rf of different metal ions (rf = Metal/DNA-P).

In the experiments cited above, extreme changes in environmental conditions (i.e., temperature greater than 70°C, pH < 3 or > 12) are used, but it should be remembered that if, for instance, the ionic strength decreases, thermal denaturation can occur at room temperature, a more physiological condition.

To conclude, stabilization or destabilization of the double helix of DNA can be obtained according to the nature and the concentration of metal ions.

c. Tertiary Structure

As previously seen, almost all metal ions interact with phosphate residues, so the electrostatic repulsion between charged groups in one strand is reduced, giving rise to a decrease of DNA rigidity. With metal ions that are inserted (Hg^{2+}, Ag^{+}, Cu^{2+}) instead of an hydrogen bond between the two bases of a pair, locally denatured sites occur and the whole DNA molecule becomes less rigid.[131,132]

d. Quaternary Structure

Speculation on the role of metal ions in in vitro supramolecular structure has been presented.[120] Aggregation could be explained by charge neutralization along the DNA backbone chains, realized by all the metal ions studied. Whereas Na^{+} aggregates only denatured DNA[133,134] in this way, other alkali metal ions, and especially divalent ones, promote intermolecular cross-links (Mg^{2+},[135,136] Hg^{2+},[137] Be^{2+}, Al^{3+}, Cu^{2+},[138]) because they could easily link together two negatively charged groups.

The bound water modification at the site of metal binding or between cross-linked molecules leads to a progressive precipitation following an excessive exclusion of water molecules. Some precipitation has been observed with numerous multivalent cations: Mg^{2+},[135,139] Be^{2+}, Al^{3+}, Cu^{2+},[138] Cr^{3+}, In^{3+}, and Pb^{2+} are more effective in precipitating denatured DNA.[141,142]

4. Specificity of Metal Ion-DNA Interaction

As discussed before, temperature, pH, and ionic strength are factors that influence metal ion binding by modifying the type and site of binding. Other parameters which may be involved in the metal binding are the specificities towards DNA base composition, base nature, and nucleotidic sequence.

There is no apparent specificity of alkali metal ions towards base nature or base sequence since they interact only with phosphates. In fact, although Sr^{2+}, Ba^{2+}, Mg^{2+}, and Ca^{2+} preferentially stabilize A-T rich DNA, Ca^{2+} shows identical binding for A-T and G-C rich DNA regions and this might be explained only by solvent effects.[65] With regard to transition metal ions, Cu^{2+} preferentially binds to G-C rich DNA.[102] It is also the case of Ag^{+} at low rf values,[143] whereas at high rf Ag^{+} shows similar affinity for A-T and G-C rich sites.[99] Hg^{2+} ions have been shown to prefer A-T rich regions of the DNA molecule.[90]

The specificity of transition metals is obviously related to the fact that they bind to the bases. Thus, N_7 of guanine is a preferential binding site for Ag^{+} and Cu^{2+}.[83,107] N_3 of thymine has the highest affinity for Hg^{2+}.[144] For other metal ions, the precise binding sites are not so well known, and they tend to bind the bases with the affinity in the order $G > A > C > T$.[145]

For some metal ions (Hg^{2+}, Cu^{2+}) nucleotide sequence influence has been postulated but still only at the dinucleotide level. At high ionic strength and low rf, an intercalation chelate can be formed with Cu^{2+} ions only if there is a superposition for four purine bases,[132] among which it is necessary that two guanine belong to the same strand so that binding should be specific to a GpG sequence. With Hg^{2+} a chelate is formed between two thymine and would be specific of a TpT sequence.[146] To our knowledge, no specificity towards longer sequence has been demonstrated. Many of the in vitro experiments have been performed on rather simple sequence of synthetic polynucleotides. Nevertheless, it should be noted that reiterative DNA could be separated from bulk DNA by Cs_2SO_4 density gradient centrifugation if Ag^{+} or Hg^{2+} ions are added,[6,147] so that one must consider a wider nucleotidic specificity. This could be explained by the fact that the specificity does not reside in a metal ion-nucleotidic interaction by itself, but also in the nature of the binding site conformation which depends on the nature of the neighboring bases or DNA sequences and particularly on ionic strength and pH values.

III. PHYSIOLOGICAL PROPERTIES OF THE DNA-METAL COMPLEXES

A. The Dynamics of Metals in the Living Cell

When metals go into a cell, they are loaded by a carrier molecule which makes their transport across the cytoplasmic membrane easier. For instance, iron acquisition by many microbial cells is achieved by means of iron chelating substances named siderochromes. One major category of siderochrome is constituted by the secondary hydroxamic acids.[148]

If the metal is useful or essential for the cell, the optimal metal concentration for the regulative functions is homeostatically controlled and a metal excess induces the synthesis of sequestration sites. This synthesis is carried out directly if the metal is toxic. The various compartments of the metal repartition in the cell are illustrated by the schemes in Figure 4.

1. The Sequestration Sites

The metal excess is bound by specific chromatic areas and induces the synthesis of mRNA and then the translation in polypeptidic chains constituting the metallothioneins (Figure 5). These cytoplasmic proteins isolated from a variety of animal cells

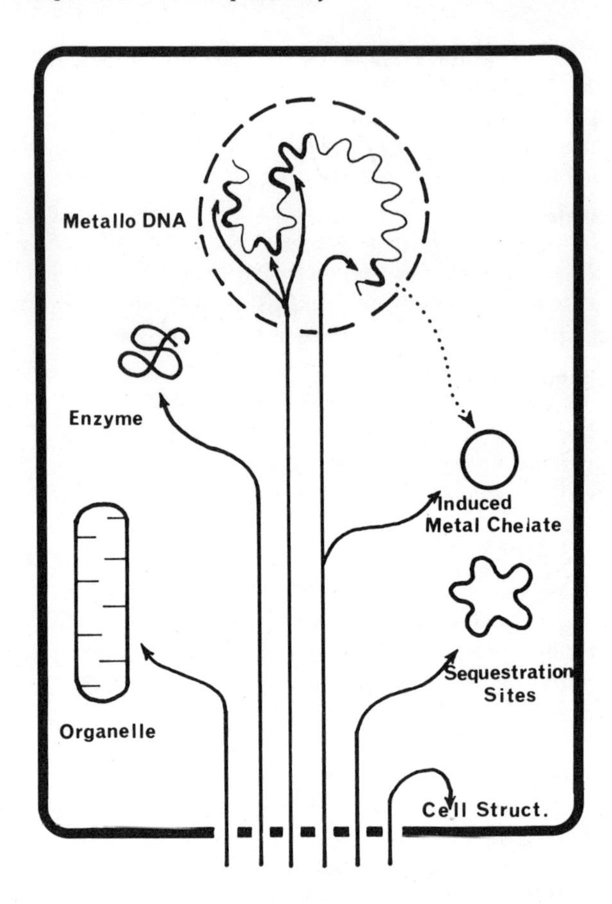

FIGURE 4. Different possible cellular localization of metal ions.

reveal an unusually high cysteine content ranging from 26 to 33 mol % and an absence of aromatic amino acids. Metallothioneins from various origins have been shown to contain cadmium, zinc, and traces of mercury and copper.[149] The metal-free protein has a molecular weight of 10,500 and apparently has 26 sulfhydryl residues available for metal binding. Different animal species appear to synthesize very similar or identical proteins for this purpose.

The metallothioneins may have a metal storage role in the homeostatic control mechanism or alternatively they may function as a mechanism to prevent free metal ions from exerting cytotoxic effects. In rabbits the concentration of these proteins appears to be increased by repeated exposure to cadmium. In rat liver nuclei, Bryan and Hidalgo[150] have shown that the disappearance of ^{115}Cd labeling in nuclei is correlated with the appearance of a cytoplasmic Cd-binding protein. Nonspecifically bound Cd is free to enter the nucleus while specifically bound Cd remains in the cytoplasm.

2. The "Active" Sites for Metal Binding

a. The Metalloenzymes

In general, the role of transition metals concerns catalysis, either redox or superacid, and this function is performed in the active centers of enzymes. Metal ion may be permanently bound to the active site of the metalloenzyme (Fe^{2+} in hemoglobin for instance) or may be part of a coenzyme (Co^{2+} in vit B_{12} coenzyme). Numerous review articles in this field have been recently published.[151,152]

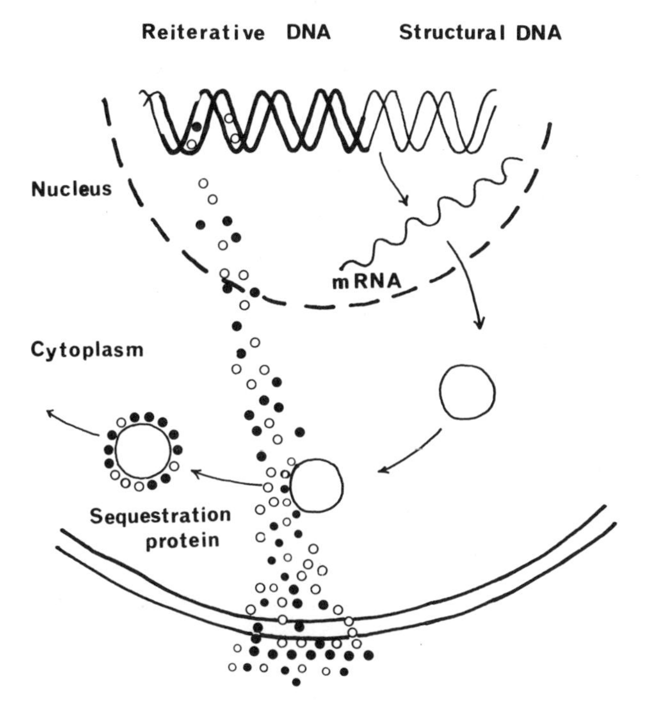

FIGURE 5. Scheme of the induction by metal ions of protein-chelates specific of them, the metallothioneins (• and O: metal ions).

In the case of chromatin area, DNA polymerases from animal, bacterial, and viral sources contain stoichiometric amounts of tightly bound zinc and require added Mg^{2+} or Mn^{2+} for catalysis.[153,154] The enzyme-bound Zn^{2+} appears to be involved in the binding to the DNA template-primer. In the same way, copper ions are required for the biological function of some enzymes.

b. The Metallo-DNA

Metals present in the chromatin areas have not only access to the active sites of the enzymes implied in nucleic acid metabolism, but also to histones and to the nucleic acid matrix itself. We have particularly paid attention to this last compartment that we named metallo-DNA and we shall describe its physiological role(s) later.

c. The Other Sites

Sequestration sites, metalloenzymes, and metallo-DNA are the major terminal compartments where metals are bound. However, other cell components may also be terminal compartments and this may be the case of some membrane components such as polysaccharides.[155] Furthermore, it exists with numerous low molecular weight compounds that may be implicated in the transport of metals from one site to another.

A metal bound to one compartment can be exchanged with another metal or move to another compartment. These variations of repartition are not only dependent on the metal nature, but also on the competition between several metals for the same binding site or even on the conditions of the microenvironment. For instance, Ca^{2+} influx can displace metal ions from their sequestration sites and free metals may be able to bind to another metal compartment of the cell. Finally, a subtle equilibrium may exist depending on each ion concentration, pH, and ionic strength values in each compartment of the cell.

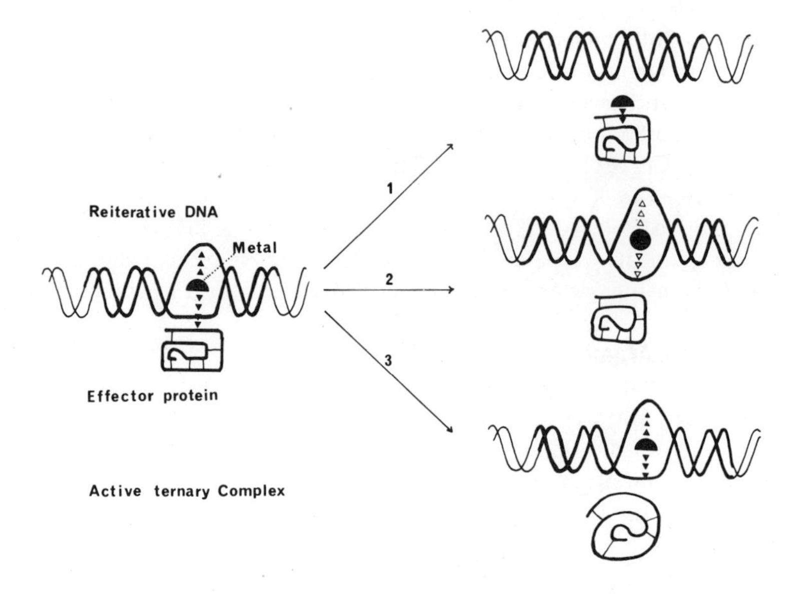

FIGURE 6. Possible modifications of the ternary DNA-metal-ligand complex. (1) Different or modified reiterative DNA sequence; (2) different or modified metal; (3) different or modified effector protein.

3. Ternary Complex: Reiterative DNA Sequence-Metal-Ligand

It is well known that beside its template function, the genome contains sequences which play a regulatory role in the processes of DNA synthesis, DNA amplification, and mRNA synthesis. These DNA regions are involved in the strand identification, enzyme-binding sites, and include operator and promotor sequences.[156] In the case of bacteria and phages, numerous regulative DNA sequences have been elucidated[157] for eukaryotic cells and numerous facts favor the attribution of all these functions to the reiterative DNA sequences intermingled among the unique DNA sequences which code for mRNA.[6]

We propose that the conformational variations induced in reiterative DNA sequences by metals, according to the modification of the microenvironment, play a decisive role in the binding of effector molecules such as DNA and RNA polymerases and reverse transcriptases. In the case of polymerases implicated in the nucleic acid metabolism, metal ion requirements are well known and are generally devoted to the enzyme molecule itself and not to the nucleic matrix.[158] To show the obviousness of the existence and the role(s) of such ternary complexes, we shall describe some conditions where they are disturbed and modified.

B. Modifications Induced in Ternary Complex

A priori, three kinds of modifications can take place (Figure 6), metal exchange, reiterative DNA sequence modification, and ligand modification or exchange.

1. Metal Modifications

a. The Mutagenic Metals

A number of metal compounds have been shown to be mutagenic in the *Salmonella typhimurium* reversion assay.[150] They include compounds of chromium, iron, manganese, platinum, and selenium.[160,161]

Nishioka[162] has tested the potential mutagenicity of 56 metal compounds by using their capacities to inhibit DNA repair. Among the metals tested, As, Cd, Cr, Hg, Mn, and Mo were positive while Be, Co, Cu, Fe, Pb, and Ni were not.

Sirover and Loeb[163] have developed an in vitro assay to screen for potential mutagenicity of carcinogenic metal compounds and can measure the perturbations in the fidelity of DNA synthesis. Among 31 metals tested, 12 give a positive test (decreased fidelity by at least 30%), of which 8 are metals known to be mutagenic or carcinogenic (Ag, Be, Cd, Co, Cr, Mn, Ni, and Pb). The others (Cu, Fe, and Zn) do not give clear-cut results.

Compounds of Al, Sb, As, Cd, Cu, Pb, Hg, Ni, and Te induce chromosomal damage in higher organisms. As, Cd, Cr, Fe, Pb, Mn, and Ni are reported as being carcinogens in animals or man while Se and Pt have been described as carcinostatic agents.[2]

b. Type of Induced Mutations

Manganese salts have been shown to induce both point and frameshift mutations in microorganisms.[164] Chromium also appears to induce both frameshift and point mutations depending on the strains and assay systems used. Thus salts of chromate and dichromate are frameshift mutagens in *Salmonella*, reverting strains containing either a reiterative G-C sequence near the site of the histidine mutation or a run of C's at the site of mutation.[165] Thus chromate and bichromate apparently show a preference for G-C rich regions of the bacterial chromosome. In contrast, selenium and platinum both appear to be point mutagens.[160] Similar experiments have been done with other in vitro systems such as reverse transcriptase.

Mutation induction has often been used as a screening test for carcinogens. Toxic metal carcinogenicity may be interpreted by a progressive accumulation of errors in the replication of the DNA sequence, followed by modification or absence of recognition between the reiterative DNA sequence and the essential metal or between the reiterative DNA sequence-essential metal complex and the ligand.

2. Modifications of the Reiterative DNA Sequences: Possible Intervention of Mutagens and Carcinogens

No simple common denominator exists to predict the mutagenicity or the carcinogenicity of a compound. However, it now appears that most if not all chemical carcinogens are metabolized to ultimate carcinogens which are strong electrophiles. These electron deficient atoms are highly reactive with nucleophilic (electron rich) molecules of the tissue, such as the bases of nucleic acids. The precarcinogen 2-acetylaminofluorene (2-AAF) is transformed into a proximate carcinogenic metabolite: N-hydroxy, 2-AAF is activated in vivo in target tissues to electrophilic forms which react with nucleic acids and proteins to yield residues of 2-AAF and 2-aminofluorene covalently bound to nucleophilic components (guanine bases and methionine residues).

Carcinogens are covalently bound to the C_8 position of guanine in various nucleic acids. This binding is associated with a specific conformational change in nucleic acid termed "base displacement."[166] The susceptibility of guanine residues is a function of ionic environment and of the nucleic acid secondary structure. It might be asked whether any guanine residue along the DNA helix is susceptible to the attack by 2-AAF or whether a supplementary specificity is imposed by the nucleotide position, the nature of its neighboring nucleotides, or the conformation of the DNA sequence where it is located. In a more general way, some regions of the genetic material take up configurations which protect or preferentially expose particular sites to attack by specific mutagens and carcinogens.

According to the type of DNA replication, carcinogen binding induces single base-pair errors, frameshift mutations, or deletions. Alternatively, it is possible that the replication mechanism simply stops when it encounters an AAF-modified region and cannot proceed until DNA repair of the lesion occurs. Which of the above processes will prevail may depend on the local base sequence at the site of the AAF modification, on the nature of polymerase which may have different response when encountering the site of AAF modification, and on the efficiency and fidelity of DNA repair mechanisms.

We are going to describe successively some frameshift mutagens, chelating mutagens and carcinogens, and a general case.

a. The Frameshift Mutagens

Numerous chemical carcinogens, including aflatoxin B, polycyclic hydrocarbons, acetylaminofluorene, and dimethylnitrosamine, are mutagens giving rise to frameshift mutations in repeats of identical nucleotides present in the DNA of *Salmonella.*[159] The reiterative DNA sequences are assumed to be subject to strand slippage during recombination or repair replication. In the T_4 e gene there is a spontaneous frameshift mutant that reverted to wild type at a very high frequency. It was found that the mutation occurred in a run of 6 A-T base pairs and that it reverted by a frameshift.[167] Both spontaneous and ICR 196-induced frameshift mutations in the histidinol deshydrogenase gene of *Salmonella typhimurium* occurred preferentially in runs of G-C base pairs.[168] The hydroxylamine-induced hot spots in the lysozyme gene of bacteriophage T_4B, studied by Ravin and Artemiev[169] can be interpreted by the heightened reactivity of G-C base pair adjacent to runs of A-T base pairs. In the same way, in the case of base pair substitution, there is evidence that the two neighboring bases affect the mutability of a given site.[170]

In all these cases, mutations are induced in reiterative DNA sequences present along a unique sequence. If the same phenomenon can take place in reiterative DNA sequences belonging to regulative areas in eukaryotic chromomers, it can be imagined that the conformation of these sequences may be progressively modified by addition or deletion of nucleotides (slippage mechanism). It will result in a modification of recognizing processes between reiterative DNA sequence and metal and/or in accordance with the state of the microenvironment, a modification in the recognizing processes between the complex reiterative DNA sequence of bacteria and phages give information of this effect.[157]

These results show that it is highly probable that the mutation (or reversion) sites are not only single base pairs but in fact are dependent on the nucleotidic arrangement of a short nucleotidic sequence.

b. Mutagens and Carcinogens with Chelating Abilities

If the carcinogen or the mutagen is a chelating agent, it would preferentially recognize DNA sites where metal ions are bound, or it would reach the susceptible site in DNA in the chelated form. Whatever the case, a new ternary complex reiterative DNA sequence-metal-carcinogen or mutagen will be found which is able to modify the functioning of the regulative unit. Numerous examples of such compounds are known. The cupric chelate of N-hydroxy,2-acetylaminofluorene, [Cu (N-HO,AAF)$_2$], induces the formation of numerous neoplasms including osteosarcomas in rat. Moutschen-Dahmen et al.[171] found that the chromosome breaking activity of ethylmethane sulfonate (EMS) was better when EMS was dissolved in water contaminated by copper and/or zinc ions and that this reaction was dependent on the pH and the ionic strength. 4-Nitroquinoline-1-oxide, known to cause covalent strand breakage in DNA, increases integration of SV 40 DNA into cellular DNA and increases transformation of CHO cells by SV 40.[172] This compound is very similar to 8-hydroxyquinoline, a copper chelating agent, suggesting that a metal may be implicated in its mode of action. Some examples of mutagens and carcinogens with chelating abilities are given in Table 2.

Table 2
MUTAGENS AND CARCINOGENS

		In vitro					In vivo				
	Metals	Binding sites	Ternary complex	Tm	Strand scission	Synthesis inhib	Strand scission	Repair	Synthesis inhib	Biological effects	Ref.
Nickel	Ni	Ph/Base				RNA proteins				Antimitotic	2, 173—176
Chromium	Cr	Ph				DNA, RNA proteins				Mutagen	2, 177—180
Cadmium	Cd	Ph/Base			+						2
Iron	Fe	Ph/Base									2
Lead	Pb	Ph/Base									2
Beryllium	Be									Carcinogen in rabbit bones	2, 181
Cobalt	Co	Ph/Base									2
Zinc	Zn	Ph/Base									2
Titanium	Ti										2
Manganese	Mn	Ph/G									2, 164
Ethyl methane sulfonate	Cu, Zn				+					Mutagen	171
cis-5,6-Dihydro-6-hydroperoxy-5-hydroxythymine	Cu, Co Mn, Fe Zn									Mutagen	182
Ascorbic acid	Cu, Fe	Bases			+		+	+		Mutagen	183
N-hydroxy-2-acetylamine fluorene	Cu									Carcinogen (rodents)	184

Note: Abbreviations: Ph = phosphate, inhib = inhibition.

c. Other Cases

The specificity of the nonchelating carcinogens may arise because of the nature of the nucleotide or the period of the mitotic cycle and it is possible that the carcinogen may have no specificity at all. In this case, the consequences of the compound binding will be different according to the nature of the DNA sequence implicated. In the majority of cases, it is probable that the major part of the breaks in the DNA helix are repaired if the repair processes function normally. On the other hand, if the breaks occur on a great number of identical neighboring sites such as those present in constitutive heterochromatin areas or more generally in clustered reiterative DNA sequences, functional repair processes might have great difficulties to correct all the mistakes. This hypothesis may explain at least partly the greatest sensitivity of these chromatin areas to numerous mutagens and/or carcinogens. Nevertheless, it is possible that much of the observed DNA binding occurs at low affinity, nonspecific sites which may in fact obscure a smaller number of high affinity sites.

In conclusion, it seems that mutagens and/or carcinogens are able to select a nucleotide or in fact a short nucleotidic sequence according to their chelating ability. The changes induced by their binding in one of the components of the ternary complex involve profound modifications in the recognizing processes.

3. Ligand Modification or Exchange

In the case of chromatin areas, several ligand modifications or exchanges are well known. Biochemical and genetic studies showed that some temperature-sensitive bacteriophage T_4 mutants (mutators and antimutators) having single base substitutions in gene 43, the gene coding for DNA polymerase,[185] can exhibit increased or decreased mutation frequencies. It is conceivable that the mutated polymerase has acquired a change in its ionic environment that affects the ionization of the nucleotidic bases and thus their base-pairing properties. In fact recent results assign these modifications and the role of proofreading enzyme to the $3' \rightarrow 5'$ deoxyribonuclease activity associated with these DNA polymerases.[186]

In the case of nonhistone chromosomal proteins, different metabolic modifications such as methylation, acetylation, and phosphorylation can profoundly affect their conformation and can thus play a fundamental role in the recognition by the DNA molecule.[6]

Other well-known examples that may also be mentioned here include modification of RNA polymerase specificity by σ-like factors or ϱ-like factors and modifications of repressor conformation by effectors.

It is clear that numerous modifications of binding can take place according to the nature and/or the state of the ligand.

C. Interactions Between Various Ligands and the DNA Molecules

Since the ternary complex reiterative DNA sequence-metal-ligand may be altered in various ways that would lead to mutations and/or carcinogenicity, we are going to look at all the possibilities of linkage between DNA-metal complexes and the various metal ligands either added or naturally present in the living cells.

The modifications described after treatment with mutagens and/or carcinogens have led us to study the properties of antitumorous substances (Table 3).

1. The Antineoplastic Compounds

a. Thiosemicarbazones

The thiosemicarbazones possess a wide spectrum of antifungal, antibacterial, and antiviral activities. Thus, the N-methyl-isatin derivative (M-IBT) inactivates the transforming ability of Rous sarcoma virus (RSV).

Table 3
ANTITUMORAL SUBSTANCES

		In vitro					In vivo			
	Metals	Binding sites	Ternary complex	Tm	Strand scission	Synthesis inhib	Strand scission	Synthesis inhib	Biological effects	Ref.
Thiosemicarbazones	Cu, Zn Fe, Se Ni, Co	Bases	+		+	DNA Reverse transc		DNA RNA Proteins	Antiviral antibact antifung	187—193
8-Hydroxyquinolines	Cu	Bases	+		+	Reverse transc			Antiviral	194
Isoniazide	Cu	Bases			+	Reverse transc			Carcinogen for ro- dents	193—195
α-Picolinic acid	Fe, Zn Co							DNA	Antimitotic	196—200
cis-Dichlorodiamine plati- num (II)	Pt	GpG	+					DNA, RNA Proteins		201—210
Rhodium (II) Carboxylate	Rh									211
					ANTIBIOTICS					
Polypeptides										
Bleomycin	Cu, Zn Co	A-T			+	DNA Ligase	+		Antimitotic Antibact	211—219
Phleomycin	Cu	A-T		+				RNA Proteins	Antiviral	220—221
Neocarcinostatin		A-T		+	+					222—225
Anthracyclins Family										
Daunomycin	Cu	G				DNA polym RNA polym		DNA, RNA		226—229
Adriamycin		Bases				DNA polym		DNA, RNA	Antiviral	
Chromomycin Family										
Chromomycin A_3	Mg	Ph/G		+		RNA polym		RNA		230—231
Mithramycin	Mg	Ph/G				RNA polym		RNA		227—230

Table 3 (continued)
ANTITUMORAL SUBSTANCES

	Metals	In vitro					In vivo			Ref.
		Binding sites	Ternary complex	Tm	Strand scission	Synthesis inhib	Strand scission	Synthesis inhib	Biological effects	
Olivomycin	Mg	Ph/G				RNA polym		RNA		227—230
Nogalomycin										227—230
Aminoquinone										
Streptonigrin	Cu, Mn Zn				+		+	DNA		232—235
Actinomycin D		G								227—236
Halogen Pyrimidines										
5-Fluoro-uracil	Cu									237—238
Other substances										
Ethambutol	Cu	G-C			+			RNA		239
Tilorone, HCl	Mg	A-T				DNA polym			Antimitotic	240
L. Mimosine	Cu, Fe									241

Note: Abbreviations: inhib = inhibition, transc = transcriptase, Ph = phosphate, antibact = antibacterial, antifung = antifungal, polym = polymerase.

These semicarbazones are strong chelators of copper and of other cations of the first transition series.[188] Some viruses refractory to M-IBT alone are susceptible to the M-IBT-Cu combination.[187] Levinson et al. found that the contact inactivation of RSV by semicarbazones was correlated with the ability of these compounds to bind copper ions. Recently, the finding that M-IBT-copper complexes, but not M-IBT alone, bind to nucleic acids[189] allowed a hypothesis of a mode of action for the drug to be made: it would form a ternary complex with metal and DNA thus preventing the binding of the enzymes. The nature of the metal involved in the ternary complex DNA-metal-drug seems to depend on the drug. Therefore, copper is implicated for M-IBT; in the case of 2-pyridine thiosemicarbazone, cobalt is the metal involved.[193] In the case of M-IBT where Cu^{2+} must be present for the RSV inactivation, it is interesting to note that Cu^{2+} ions are present in chick embryo cell DNA, preferentially bound to G-C rich sequences and that their content varies according to the mitotic cycle.[242]

Some of thiosemicarbazones such as 2-pyridine and 1-formyl-isoquinoline derivatives are potent growth inhibitors of a variety of rodent tumors.[243]

b. 8-Hydroxyquinoline (8-HQ)

8-HQ is the only one of seven isomeric monohydroxyquinolines that has antibacterial activity and is also the only one that forms chelate complexes with divalent metal ions including copper.[193,194] It was shown that 8-HQ inhibits the RNA-dependent DNA polymerase of avian myeloblastosis virus and, as with several of its derivatives, it inactivates the RSV transforming ability. In the presence of copper these drugs bind to DNA which may thus be the site of their action.

c. α-Picolinic Acid

α-Picolinic acid, which is a chelating agent of Fe and Zn, profoundly affects iron metabolism in cultured cells.[199] Normal cells treated with picolinic acid showed no toxic effects, whereas cytotoxicity was observed in all transformed cells that were blocked in G_1, in G_2, or at random. The block in transformed cells was dependent upon the transforming virus and independent of the species or origin of the cell line.[197] The authors ascribe the selectivity of picolinic acid towards transformed cells to different trace metal requirements for normal and transformed cells.

Recently, Fernandez-Pol[200] has isolated a highly specific iron-binding ligand termed siderophore-like growth factor (SGF) from mutant Balb/3T3 cells transformed by SV 40 and adapted to grow in the presence of picolinic acid. The striking specificity of SGF for iron and the excellent correlation of stimulation of iron uptake and DNA synthesis by SGF indicate that the same factor regulates both iron uptake and DNA synthesis.

d. The Platinum Compounds

These antitumoral agents are interesting because of their low toxicity. Among numerous potential binding sites, the PDD (*cis*-platinum (II)-diamine-dichloride) seems to act in vivo by binding to DNA.[202,205] In vitro, PDD binds preferentially to guanine then to adenine and then to cytosine: it seems that there is little or no binding to thymine. It has also been shown that the rate of PDD binding increases proportionally with the G + C content of DNA.[206,209]

PDD induces chromosomal aberrations in Chinese hamster cells after either in vivo or in vitro exposure. The induced break points are nonrandom, being preferentially localized either in heterochromatin areas which are late replicating or in DNA sequences replicating at the beginning of the S phase.

e. Bleomycin

Bleomycin A_1 to A_6 and B_1 to B_5, discovered by Umezawa et al.,[215] are a group of antibiotics produced by *Streptomyces verticillus;* they are peptidic in nature and chelate copper. Bleomycin treatments of diverse cell types result in DNA strand breakage and inhibition of DNA synthesis and cell division. These and other properties of bleomycin have recently been reviewed.[213] DNA breaks induced in drug-treated cells are believed to be the primary cause of cell death. Discovery of the selective antitumor effect of bleomycin on squamous cell carcinomas led to its present widespread use in the clinical treatments of cancers.

Bleomycin A_2 decreased the Tm of all DNAs tested when incubated in media containing 2-mercaptoethanol at 37°C before the Tm determination. The effect of bleomycin A_2 on Tm is reduced by Cu^{2+}, Zn^{2+}, and Co^{2+} and is completely reversed when the antibiotic is saturated by copper. EDTA inhibits the bleomycin A_2 effect on the Tm and also inhibits the DNA breakage by bleomycin and organic reducing agents.[212]

The parameters of bleomycin binding to DNA in the presence of metal ions may differ from those obtained previously in their absence. Bleomycin cleaves DNA at GC and GT sequences and to a lesser extent at TA sequences with a degradative activity enhanced by mercaptoethanol. In the presence of ferrous ions, bleomycin cleaves DNA at TT, AT, and TA, as well as at GC and GT sequences.[217] It should be interesting to test if, in vivo, metal ions play a role in the sequential discrimination shown by bleomycin in vitro[219] and particularly if bleomycin cuts between the nucleotide-metal complexes and if the amplitude of the Tm lowering is dependent on the quantity of metals bound to the tested DNA.

f. Antitumoral Antibiotics

Bleomycin is the only known antibiotic which decreases the Tm value of DNA. Phleomycin, which resembles bleomycin in chemical and biological properties, increases the DNA Tm regardless of the presence or absence of 2-mercaptoethanol. Metal ions are implicated in the binding of streptonigrin and chromomycin to DNA[231,233] and may also be implicated in the antibiotic action which is enhanced by transition metal divalent cations. Moreover, most antibiotics are chelating substances, thus penicillin, tetracycline, streptomycin, cycloserine, and almost every common antibiotic are copper binding ligands.

Thus, it seems that, at least in those cases where we have sufficient information, the antitumoral compounds may act via the formation of a ternary complex DNA-metal-drug which could play different kinds of roles illustrated in the schemes of Figure 7.

They bind metals, preventing them reaching their ultimate binding sites where they should be susceptible to produce the good conformation either at the level of the active site of an enzyme or on DNA sites. This kind of chelating substance may thus take part in eliminating excess of essential and toxic metals by the detoxification organs such as liver and kidney. They bind the metals which may help them to recognize the good action site where they should prevent the binding of the effector ligand, for example, by steric hindrance or in inducing simple- and/or double-stranded breaks (illustrated by cases 2 and 3 on Figure 7). This action may also be realized by recognizing the metal already bound on the active site.

Antitumoral substances may be classified into two groups depending upon the values of stability constants of the complexes that they formed with metals:

1. Chelates of high effective stability which do not release their metals to the acceptor sites of the cells, since their stability constants are greater than those of the natural acceptors. These chelates may be eliminated by detoxification organs, as they are often toxic for the healthy cells.

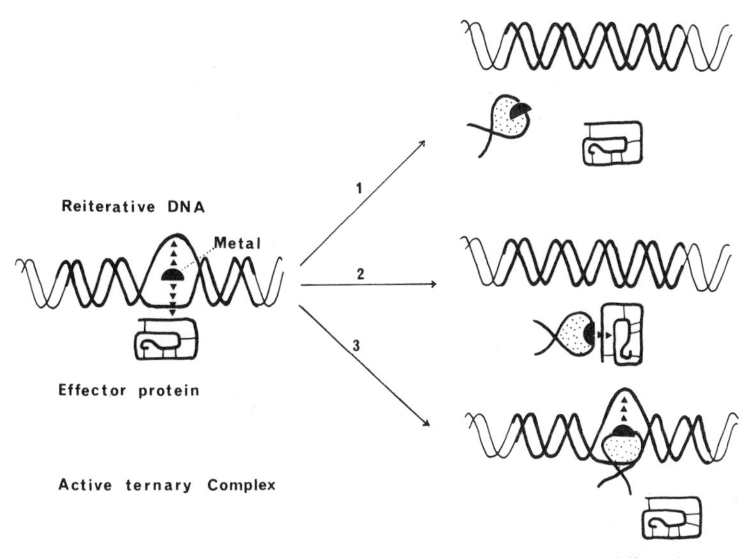

FIGURE 7. Modifications of the functioning of the DNA-metal-ligand complex induced by chelating agents. (1) Metal sequestration by chelating agent; (2) New ternary chelating agent metal-effector protein complex; (3) New ternary DNA-metal chelating agent complex

2. Chelates of low effective stability can release their metals to the acceptor site of the natural transfer systems for metals in the living cells.

The transfer of the chelated metals in the cell or their elimination by the cell depends on the chemical properties of the intact chelates. According to the nature of the chelate, there is either an enhancement or an inhibition of heavy metal uptake. For instance, the polyaminocarboxylic acids such as EDTA, CDTA, DTPA, and EGTA have broad chelating abilities but they are largely unspecific and tend to deplete body stores of essential metals. On the contrary, results with various thiosemicarbazones clearly show that they may serve as a vehicle to transport first transition series metals to the appropriate sites.

It seems probable that there is a very large number of chelating agents developed by living organisms in the course of their evolution that allow them to control the concentration of essential metals which are toxic when in excess and to tie up toxic elements which have no necessary function.

2. Metal Chelating Substances Currently Present in Living Cells

Studies on carcinogen and antitumoral substances have led us to look more precisely at the properties of metal ions ligands in living cells.

a. Nature of Metal Chelating Substances

Biological systems contain a multitude of metal ligands. In general when external metals or ligand-metal complexes enter an organism, a variety of substitution reactions may occur depending on respective affinity constants, concentrations, and solubilities in the various compartments of the cell. The oxidation-reduction potential of a metal is always altered by chelation. The change resulting from chelation could alter the subsequent affinity of the metal for another ligand or shift the cellular location of the chelate. It could also increase liposolubility and consequently increase membrane permeability. On the other hand, if chelation leads to polymerization, the mobility of a metabolite could be restricted or its biological function-severely altered.

A list of naturally occurring chelating molecules has been recently published[244] and two classes of compounds can be easily identified: substances such as proteins or functional moieties of protein molecules (vitamins, porphyrins, metalloenzymes) and substances which have access to the DNA molecules. In this second category, we found several amino acids (especially histidine, cysteine, serine), polyamines (spermine, spermidine, catecholamines), hormones (steroids), and nucleoproteins. In fact, the cell contains several compounds with chelating functions which are not evidently devoted to the two classes summarized above. Thus, in cultured mammalian cells, numerous serum factors have been found to be necessary for the cell growth. The length and variety of the list of these factors suggest that a common property may exist between some of them and we think that it could be their chelating abilities which would be the case, particularly for peptides and low-molecular-weight compounds. Some natural metal-chelating substances are well known: transferrin is the main iron carrier in mammalian systems. The metal transfer into the cells involves a specific transferrin receptor. Cells also possess a pool of low-molecular-weight chelates which may be the primary regulators of cellular iron metabolism.[245] A number of such cell metabolites may represent the agents by which iron is released from the transferrin during the process of cellular uptake. Zinc binding ligands of low molecular weight (8,700 to 15,000 daltons) have been also isolated from milk and intestinal cells.[246] Perrin and Agrawal[247] have shown that more than 80% of the copper in plasma is bound at pH 7.4 and 37°C to histidine and as histidine-cystine mixed complexes.

In a general manner, peptides have been clearly shown to bind metal ions with differing affinities depending on the amino acid sequence and on the metal.[248] Growth factors and hormones also have similar properties, so the formation of metal complexes involving steroid sex hormones is well established.[249] In plants, auxines, kinetines, and gibberellines have metal-chelating properties.[250] This is consistent with the fact that numerous chelating substances such as o-phenenthroline, α,α'-dipyridyl, diethyl-dithiocarbonate, and 8-hydroxyquinoline have the same effect as growth factors and particularly play a fundamental role in developmental steps.[251]

b. Existence of Metal-Transfer Chains (Figure 8)

In a hypothetical transfer chain for metal ions, polyamines, growth factors, and hormones should appear as specific metal ions carriers, binding them with sufficient affinity to prevent their binding by other numerous cellular ligands, such as polysaccharides, proteins, amino acids, nucleotides, and other sequestering sites. They should be able to lead the metal to specific sites of chromatin areas having higher affinities and thus may be responsible for conformational variations in DNA sequences. Thus steroid hormones bound to cytoplasmic proteins move into the nucleus and activate specific transcription. This result is in agreement with the chelating ability of steroids which would recognize the metal along the DNA helix or lead the metal to the proper site. It should be noted that in some known cases when a given metal is required in differentiation or tumorization processes, the hormone implicated in these same processes is a chelate of this metal.

The model involving the formation of a reiterative DNA-metal-growth factor (or growth factor-protein receptor) complex depending strictly on the microenvironment of the cell, has the advantage of explaining the specificity of hormone action.

We propose that metal ion-transfer chains exist in living cells that are constituted progressively during the evolution and adapted to lead metals from the periphery of the cell (membranes) to the ultimate action sites (characterized by the highest affinity

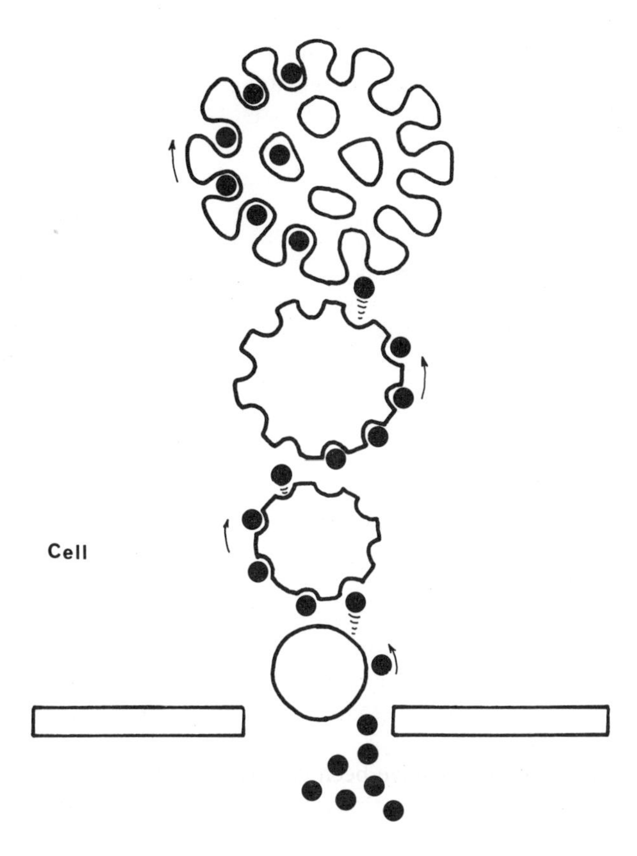

FIGURE 8. Scheme of the functioning of the metal-transfer chain (•, metals).

for the metal). These action sites would mainly be composed of active enzyme sites and regulative DNA sites.

D. The Concept of Metallo-DNA

The hypothesis of metallo-DNA arises from the following statements:

1. The differential structure of DNA: The existence of reiterative DNA sequences intermingled within unique ones can be revealed by the specific interactions of metal ions with regulative DNA sequences equated with some reiterative areas.
2. The binding of metal ions which takes part in the control, either positive or negative, of the regulative DNA sequence functioning is strictly dependent on the microenvironmental conditions of the implicated chromatic areas (ionic strength, pH values, transmembrane potential).
3. The variations of conformation induced in the reiterative DNA sequences are involved in the binding of effector molecules including proteins. It results in the formation of ternary complexes that would play fundamental roles in genetic regulation processes.
4. From the plasmic membrane, metals are loaded on acceptor ligand molecules such as amino acids, peptides, proteins, growth factors, and hormones and may thus reach the acceptor active site(s) on DNA molecules.
5. Mutagens, carcinogens, antibiotics, and antitumorous substances may interfere with the ternary complex function.

E. Metals and Constitutive Heterochromatin Areas

Since metal ions are preferentially bound to reiterative DNA sequences and since the clustered reiterative DNA sequences are essentially present in constitutive heterochromatin areas, it is interesting to consider more precisely the properties of these sequences and their function(s), taking into account the presence of metal ions at some steps of the mitotic cycle.

1. Properties of Heterochromatic Areas

Constitutive heterochromatin occupies identical positions in homologous chromosomes and may be characterized in most cases by the following properties:[252]

1. It stays in a condensed state during interphase, appearing as chromocenters.
2. Its DNA is late replicating during the S phase and is located near the nuclear periphery.[253,254]
3. It displays a faster replication of its DNA than do euchromatic areas, leading to a good synchrony of its replication.
4. It manifests a transcriptional inactivity and few mappable genes are localized in it.
5. It contains highly reiterative DNA sequences that often appear as satellite DNA in neutral CsCl; reiterative DNA sequences of constitutive heterochromatin may be either A-T rich, G-C rich, or may have nearly the same buoyant density as main band DNA.[255] Generally a clustering of regulatively A-T rich heterochromatin and relatively G-C rich euchromatin is observed.[256]
6. It has a greater content in metal ions than euchromatic areas.[6]
7. It is preferentially involved in spontaneous and induced chromosomal structural changes.[257]
8. It manifests differential staining with banding techniques.[258]

In fact, it turns out that no class of constitutive heterochromatin possesses all of these features.

Some of these properties are interdependent: sensitivity to numerous mutagens and/or carcinogens should be related to specific structural features of these regions. Therefore, if the agent is specific of a nucleotide present in the reiterative DNA sequences, the repair would be more difficult and aberrations may result. The synchrony of their replication in late S phase favors action of carcinogens at the end of the S phase. Lastly, we have shown that if drugs are chelating agents, they may recognize preferentially this kind of DNA sequence.

Formation of chromocenters, especially those resulting from nonhomologous association of telocentric chromosomes, can be explained by direct pairing of reiterative DNA sequences, and the same interpretation can be applied to the ectopic pairing. Translocation in heterochromatic regions could be favored because mitotic crossover may take place in reiterative DNA sequences that have a greater probability of recognizing each other than do the cistrons connected to them.

However, in these observations, we have only static or structural effects. It is difficult to clearly understand how constitutive heterochromatin can exert an influence on the metabolism of nucleic acids, regulate mitosis, change substances from nuclear membrane, and generally intervene in cell metabolism. In order to try to understand these facts, it is appropriate to describe a property of reiterative DNA sequences concentrated in heterochromatin areas i.e., the simultaneous presence of alternating A-T rich and G-C rich DNA sequences and of metal ions.

2. Existence of Alternating A-T Rich and G-C Rich Blocks Along the Chromosomal DNA

Regulatory regions in DNA have been suggested to be enriched in dA-T (cf lac operon)[157] or dG, dC (histone binding sites in calf thymus DNA).[259] Some of these regions would be at least partially transcribed since uridylate rich sequences have been observed within heterogeneous nuclear RNA molecules.[260] Electron microscopic studies of partially denatured Chinese hamster cell and chick fibroblast DNAs have revealed that small easily denatured regions (thus enriched in A-T base pairs) are distributed throughout chromosomal DNA in the vicinity of G-C rich ones.[261]

The nucleotide sequences of a number of regulatory sites (operators, promoters) of prokaryotic and eukaryotic genomes have been published: contiguous blocks of A-T followed by G-C base pairs, occur.[157,262]

3. Properties of Alternating Blocks Polymer: the Concept of Telestability

Thermal denaturation of synthetic double-stranded polymers with 10 to 15 A-T base pairs attached to 15 to 20 G-C base pairs was studied: a regional cooperative action exists in the melting of these polymers. For instance, the polymer dC_{20}, G_{20} melts in a single monophasic transition, the G-C portion of the polymer stabilizing the A-T portion and being itself destabilized by the A-T portion of the helix.[263]

These studies show that the ability of a portion of a natural DNA helix to breathe can be influenced by the nucleotide sequence of adjacent regions. In the same way, when a base pair-specific drug is used, the binding of the ligand to one region of a DNA molecule can influence the dynamic properties (e.g., breathing frequency) of the adjacent regions. The model of telestability has been applied to explain the functioning of catabolite gene activator protein (CAP) where it is shown that a protein binding to one region of DNA alters the binding of a second protein to adjacent regions. In the case of the lac operon, binding of the CAP protein to its interaction site could destabilize the neighboring G-C rich region. This would lower the melting temperature of the promoter site and allow formation of the open complex. In this model, the G-C rich region is a transducer, transmitting the effect of CAP binding to the entry site, some 14 base pairs away.[157] Numerous compounds including ethylene glycol, dimethylsulfoxide, sucrose, and 1,3-propane-diol are known to lower the DNA Tm and to specifically activate promoter regions.[264,266]

4. Application of Telestability to Constitutive Heterochromatic Areas

Some reiterative DNA sequences belonging to constitutive heterochromatin are enriched in metals, especially those of the first transition series. Techniques of metal determination (atomic absorption, anodic stripping voltammetry) are not sufficiently sensitive to prove that this is also true for reiterative DNA sequences scattered along the DNA thread where at least a part of them play a regulative function. However, in vitro conformational variations are induced by metals or metal ligands on specific DNA sequences and these conformational variations are frequently invoked to explain the mechanism of recognition between an effector protein and a nucleotide sequence. Alterations in the nucleotide sequence (via deletions or base-pair changes) of regions adjacent to an actual DNA-protein interaction site could influence the recognition of that site, particularly if the frequency of breathing is important for protein binding.

Numerous physiological facts from in vivo observations may be interpreted by the blocking or deblocking of reiterative DNA-metal complexes by specific chelating ligands. In the case of highly reiterative DNA sequences, the same interactions are not demonstrated. However, the presence of metals in these sequences can be contrasted with the existence of alternating blocks respectively A-T or G-C rich in these same sequences, taking into account the intervention of telestability processes in these blocks. The destabilization or stabilization effect induced by metal binding along a short DNA sequence could be transmitted at a distance via alternating blocks of different nucleotidic composition and modify the state and thus the activity of some neighboring unique DNA sequences. The effect produced would be dependent on the respective nature of the metal, of the DNA sequence, of the ligand which carries the metal and on the microenvironment of the chromatic areas. This application of the telestability model to constitutive heterochromatin may give a coherent molecular interpretation of numerous biological facts, already described for heterochromatic areas, such as:

1. The activity of a chromomere is dependent on its position along the chromosome and particularly on its distance from the centromere[267] (position effects, paramutation). The control exerted by the alternating of different reiterative DNA sequences could be dependent on the nature and the number of these sequences present in each controlling area.[255]
2. Heterochromatic areas are localized at the periphery of the nucleus at some moments of the mitotic cycle and they are thus the first DNA sequences to receive information from peripheric signals which have passed through the cytoplasm. Their structure allows them to transmit a possible induced perturbation to distant unique DNA sequences. Such signals can be made of variations in the ionic strength, in the ionic nature, and in the chelating ligands of the transfer chains for metal ions; they are, therefore, specific signals taking into account the real state of the microenvironment through the differential structure of the DNA sequences. Thus, there is an integration of elementary signals at the level of each chromomer (as for gal operon) by an elevated hierarchy signal that coordinates the first one. This may explain another role of heterochromatin, i.e., because of their late replication in the S phase, they control, directly or indirectly, the cellular division.

IV. PERTURBATIONS OF THE DYNAMIC OF METALS IN THE LIVING CELL: PATHOLOGICAL ASPECTS

We have seen that the ternary complex, reiterative DNA sequence-metal-ligand could be altered in several ways: by change of the metal, change of the DNA sequence, or change of the ligand molecule. If the perturbation induced by a variation in the microenvironment is maintained for a sufficient period and if the microenvironment does not return to its former state, numerous consequences may arise, some of which are described below.

A. Modifications of the Microenvironment Composition

We shall describe variations induced in the cell by modifications in the composition of the microenvironment, the ionic strength, the transmembrane potential, and the pH values.

1. Variations of Metal Balance: Zn/Cd Couple

When *Euglena gracilis* is grown on medium deprived of zinc, many metabolic disturbances are produced. It is clear that events critical to cellular division and nucleic acid metabolism are concerned. The intracellular content of Ca, Mg, Mn, Cr, Ni, and Fe is markedly increased; several biochemical processes, including tubulin polymerization[268] and RNA and DNA polymerase functioning,[269] are affected in different ways. Zinc-deficient *E. gracilis* cells contain twice as much DNA but the same amount of RNA as zinc-sufficient cells. RNA synthesis is depressed, although the mRNA fraction of Zn-deficient cells is increased and its base composition is significantly altered.

The biological effects are more spectacular when Zn-deficient cells are placed in culture medium containing suboptimal doses of cadmium: the cells become multinucleated and do not divide.[270,273] Similar alterations have been described in cells infected with some viruses (syncitium formation), in various neoplastic tissues (giant cell tumors), and in leukemic cells following treatment with some antineoplastic agents.[274,275]

Therefore, a decrease of zinc content in the nutritive medium, or a decrease of zinc with a concomitant increase of cadmium content, leads to the release of various metal ions from their compartments including sequestration sites. Knowing that toxic metals and excess of essential ones may accumulate in these places, it can be imagined that a perturbation, even a subtle one, in the composition of the nutritive medium may have dramatic consequences at various cellular levels if it is repeated or eventually maintained during a long time. The metals released from their sequestration sites become able to bind to numerous natural chelating substances present in the cells, such as the electron donor groups. It results, at least transitorily, in the formation of new reiterative DNA-metal and protein-metal complexes giving rise to new ternary complexes that would contribute to the blocking or deblocking of new chromomeres. The scheme described with some details for the Zn/Cd couple can be applied in the same way to other essential metal/toxic metal couples such as Mn/Ni.[276] Metal-metal interrelations have been recognized in numerous organisms and tissues and described as unbalance, antagonism, synergism, and conditioned deficiencies. They are discussed further by Magos in the next chapter.

2. Variations of Ionic Strength

Isolated polytene chromosomes and polytene chromosomes in isolated nuclei respond to changes in the electrolyte medium by changes in their puffing or swelling pattern. In salivary glands from *Chironomus* larvae, it is possible to induce activation of different puffs along three different chromosomes by incubation in media containing various combinations of NaCl, KCl, MgCl.[277] The same puffs are produced under ecdysone or juvenile hormone control during the normal development of the larvae. Through controlled shifts in the concentration of the intranuclear ions, this differential sensitivity could play an important role in the activation or inactivation of genetic loci.[278] This possibility is supported by information on the existence of and the extent of the nuclear electrolytic changes during development, including oogenesis and differentiation processes. Thus, intranuclear electrolytes may not only influence the pattern of chromosomal RNA synthesis, but may also control the DNA replication: a high nuclear Na^+ titre would specifically induce DNA synthesis.[279]

The differential response of chromosomal segments according to the ionic strength of the medium can be connected to the existence of specific DNA sequences, giving specific ternary complexes, DNA-metal-ligand, dependent on the ionic strength of the microenvironment. In vivo, the hormones implicated in this process may regulate ion influx and efflux and may bring the specific metal to the action site. The same events could take place during some developmental processes such as embryogenesis and differentiation.

3. Variations of the Transmembrane Potential

Setting intramolecular ionic concentration to values corresponding to a high electrical transmembrane potential (Em) level fully, but reversibly, blocks phase cultured cells in the G_1 stage.[280] Nonproliferative cell types such as neurons and muscle cells have a high negative membrane potential. Cone and Cone[281] have shown that even mature neurons can be induced to undergo mitosis by sustained depolarization. Of particular importance is the general correlation between membrane potential and mitotic activity that accompanies the malignant transformation of somatic cells.[282]

Since a functional correlation exists between the degree of mitotic activity of a cell and the ionic concentrations associated with the level of the cell Em potential, the situation is the same as that described in the preceding part.

4. The pH Variations

pH variations have been detected in various tissues; in cancer cells the pH of 6.5 is lower than in healthy ones (pH 7.2). We have seen in Chapter 2 that the complexes DNA-metal are dependent on the pH. Thus, the complexes of Ag^+ ions with N_3 of thymine and N_1 of anenine or with N_3 of cytosine and N_1 of guanine are favored by a pH increase.[82,83]

B. The Wound Healing

After a lesion or during the wound healing process, membranes are disrupted and numerous components of the cells may meet each other. New complexes including DNA-metal complexes may be formed. We are going to describe the role of metals in the wounding process in the case of plant and animal tissues.

1. The Lesion Process in Plant Tissues

During the lesion process in tissues of higher plants, a G-C rich satellite DNA is transitorily synthesized before the overall DNA synthesis occurs.[283] During this process, there would be a greater probability of meetings between DNA sequences and metal ions. We found that the G-C rich DNA satellite is enriched in metal ions such as cadmium and copper in comparison with main band DNA.[25] The presence of the G-C satellite DNA is correlated with the ulterior differentiation and/or to the neoplastic induction by the virulent bacteria *Agrobacterium tumefaciens.*[284]

2. The Wound Healing in Animal Tissues

In animal tissues, similar phenomena have been described. Purified DNA from hyperplastic nodules induced in liver cells by the hepatic carcinogen 2-fluorenylacetamide contains a minor DNA fraction (1 to 2%) having a buoyant density lower than that of the major DNA and which is absent in the DNAs extracted from livers of control animals and from liver regions surrounding the nodules.[285] This DNA is characterized by excessive branching and frequent presence of strand separation. While the presence of metal ions has not been demonstrated in this case, the formations of branched structures may be explained, at least in part, by the presence of metal ions.

Moreover, the requirement for metal ions during the wound healing process is well established for some animal tissues. Thus, zinc accelerates wound healing.[286,287] The studies of wound healing in zinc-deficient bovines and the effect of zinc treatment on man show an apparent beneficial effect of the metal.[288] Interference with growth factors and hormones (corticosteroids lower the zinc content) has also been observed.[289]

C. Implications of Metals and Metal-DNA Complexes in the Induction and Development of Neoplastic Cells

Interactions between genetic factors and certain trace metals (Mn, Cu, Zn) have been described during development in animals.[290] Thus, in the mice, a mutant gene, "pallid," leads to congenital ataxia and reduction of pigmentation. Addition of manganese salts to the diet produces the normal phenotype. Similarly, during the development of neoplastic cells, an intervention of metals can be described.

1. The Metal Content in Neoplastic Cell DNA

Numerous authors have shown that the metal content is different in DNA preparations isolated from healthy and neoplastic cells, whatever the tissues tested and the DNA extraction procedure used.[291,292] Generally, in tumorous tissue DNA, the zinc content is lower, and copper, cadmium, and lead contents are higher than in healthy tissue DNAs.

In Walker 256 carcinosarcoma and sarcoma M_1, the content in Zn, Co, Fe, Sb, and Sc is higher than that of liver from normal rats.[293] DNA from leukemic cells contains less Cr^{2+}, Fe^{2+} and Zn^{2+} than does that of lymphocytes from a normal donor, whereas Co^{2+} content is amplified 20 times in DNA from leukemic cells.[291] Differences were also found in the metal content of RNA and histones isolated from these various tissues. Metal content during the transplantation of sarcoma M was also investigated in DNA preparations.[293] Zinc content stays constant while other metal content (Fe, Sb, Cr, Co, and Sc) increases after the transplantation.

2. Metal Requirements of Neoplastic Cells

The general increase of metal content observed in tumorous cells, in comparison with healthy ones, is clearly illustrated by studies of ^{64}Cu incorporation in various organs of mouse injected with Elrhich or Krebs ascitic cells.[294] The major part of the radioactive metal is found in the ascitic cells which deviate, to their own advantage, the metabolic pathway of copper since the liver from injected mouse contains less ^{64}Cu than that from healthy animal[294] (liver is the regular major detoxification organ for metals in the healthy animal).

Numerous metals are required for the tumor growth as shown by experiments using culture media deficient in metal salts or using chelating agents. For instance, experimentally produce zinc deficiency has been used to inhibit the growth of various animal tumors including Walker 250 carcinosarcoma, Lewis lung carcinoma, L 1210 leukemia, P 388 leukemia, and 3-methyl-4-dimethyl-aminoazobenzene-induced hepatoma.[295,296] Chelating substances such as α-picolinic acid specifically destroy cancer cells while healthy ones are arrested in G_1 phase.[197]

The metal requirement of cancer cells seems to be dependent on the cell studied and on the metal nature. In some cases where a given metal appears to be present in lower amounts than in the control cells, it cannot be excluded that the metal may be actively used and then eliminated by the detoxification organs.

3. The Metal Transport in Neoplastic Cells

Numerous chelating substances accumulate in tumorous tissues. They include amino acids, peptides, proteins, growth factors, and hormones.[297,298] These compounds may be present or absent in corresponding healthy tissues. Thus, mutants of SV 40-transformed Balb/3T3 cells adapted to grow in presence of picolinic acid contain a highly specific iron-binding ligand. High levels of ferritin synthesis have been described in some cancer cells. The presence in serum from patients of isoferritins characteristic of some cancer and fetal tissues suggests that much of the ferritin may come from cancer cells.[299,300]

Cultured mammalian cells seem to have lower requirements for serum factors when they are transformed by virus or become neoplastic.[301,303] In fact, neoplastic cell requirements, especially for metal-binding ligands, might be more important and more specific than those of normal cells. In cell culture, these requirements are masked since transformed cells are able to synthesize all or part of their growth-requiring substances.

The genetic information for the synthesis of all these cancer-specific substances may be already present in the healthy tissue which would be able to synthesize them at some moments of the developmental program, or it may be introduced during the induction processes either by an exogenous molecule (virus) or by a subtle modification of the host genetic information.

4. The Specificity of Host Susceptibility

The reiterative DNA-metal complex may be the primary target of ultimate carcinogens whatever the nature of the carcinogen (chemicals, virus, radiations) and we have seen that the mode of action of numerous antineoplastic agents can be related to interference with the metals. In neoplastic cells, ionic strength,[304,305] transmembrane potential, and pH are modified. All these conditions favor the formation of new ternary complexes, reiterative DNA-metal-ligand, and the suppression of some complexes preexisting in healthy cells. These new complexes are formed transitorily during the lesion process and under the influence of several carcinogenic substances. Studies by Hitotsumachi et al.[306] indicate that the transformation of Syrian hamster cells by polyoma virus is associated with the gain of group 5 chromosomes and the transformed phenotype is suppressed when these chromosomes are lost and certain others gained. The authors have interpreted these results in terms of a balance of factors carried on specific chromosomes for the expression or suppression of transformation. Fibrosarcomas induced in Chinese hamsters and rats by viruses and by a polycyclic aromatic hydrocarbon are associated with reproducible nonrandom chromosome variation.[307] Fahmy and Fahmy[308] showed that deletions within the rRNA cistrons might play a significant role in cancer initiation. This last situation is very comparable to the existence of hot spots. Changes were detected in the deoxyadenylate regions of DNA extracted from 7,12-dimethylbenza-anthracene-induced and RSV-induced rat tumors in comparison with DNA extracted from normal rats.[309] Similar results were obtained in breast tumors and in chronic lymphocytic leukemic DNAs.[310,311] These findings satisfy the statement that regulative sequences in chromatin are the major targets of carcinogenic substances.

All these results are clear indications for a nonrandom nature of the interaction of several carcinogens with DNA in chromatin.

At the level of transcription, Ts'o[312] shows that no more than 4% of 12,500 mRNAs are different between normal and transformed embryonic fibroblasts from Syrian hamster. The maintenance of malignant transformation is accompanied by relatively small changes in the number of polysome-associated poly (A) mRNA sequences, and massive gene activation need not be invoked to account for the transformed state.

Nature of the Specificity — Clearly, differential mutability is a function of the biochemical and biological organization of the genome. The susceptibility to mutagenic and/or carcinogenic compounds seems to be dependent on the nature, place, and number of reiterative DNA sequences clustered in the constitutive heterochromatic areas or scattered along the chromosomes. Generally, there is a correlation between the aptitude to differentiate and the possibility to become neoplastic and in some cases it has been possible to link this correlation to a common step at the molecular level, the faculty of amplifying particular DNA sequences.[39,313] The chromatin areas generally implicated in these modifications frequently include the ribosomal cistrons, the transfer RNA cistrons, and the constitutive heterochromatin areas. The genome component which would be responsible for the synthesis of metal chelates and carriers and which may be linked to the preceding genomic constituents could also be added. It is probable that the chromosomal areas susceptible to various carcinogenic agents are implicated in the control of many fundamental metabolic processes in direct connection with proliferative capacities, such as ribosome synthesis, mitotic activity, and synthesis of chelating substances involved in the cell homeostasis. These areas have a common feature: they are reiterative DNA sequences.

5. The Induction of Cancer

Cellular occurring ternary complex, reiterative DNA sequences-metal-ligand, have properties in many ways similar to the complexes, nucleic acids-carcinogen studied in vitro by Kubinski et al.[319] Nucleic acids exposed to several carcinogens including alkylating agents, hydrocarbons, aromatic amines, and carcinogenic metals (Be, Cr, Ni), become sticky; they are able to bind to other macromolecules such as purified proteins,[315,318] DNA molecules,[314] to cellular membranes, to glass surfaces, nitrocellulose tubes, and filters.[317,318] These last results can explain the difficulties met in the study of DNA modifications induced by wounding and/or carcinogens such as the nonreproductibility of the experimental results. In these complexes the DNA sensitivity towards nucleases decreased markedly in all cases studied. It suggests that these complexes are able to enter recipient cells.

It is tempting to think that such complexes may be realized during lesion processes and more generally during penetration and transport of exogenous DNAs such as viral and plasmid DNAs.

V. CONCLUSION

Metallochelates found in biological systems fall into two main classes: those in which the metal plays an active part, such as metalloenzyme and electron or oxygen-transport systems, and those which function purely as metal storers and carriers to supply the former class.

Among all the cell compartments where metals have access, the enzymes have been particularly well studied. Numerous metals (iron, copper, zinc, manganese, molybdenum, and cobalt) are essential components of enzyme molecules. Metals influence various steps of metabolic pathways even if for some of them it is actually difficult to discern their real site and mode of action.

In this review, we have described physicochemical and physiological properties of metallochelates, i.e., DNA molecules where metals are bound at precise moments of the cell cycle.

In vitro studies show that metals induce conformational variations in the DNA molecule. These conformational variations are dependent on the metal, on the binding site nature, and on the characteristics of the microenvironment. The effects produced are specific and quantifiable with the metals of the first transition series.

In vivo studies confirm the existence of metallo-DNA. The metal-recognition site is a larger unit than an electron-donor group on a nucleotide residue or even a base pair. It may be short nucleotidic sequences scattered in the genome continuity where they take part in regulative functions or clustered in constitutive heterochromatic areas. The presence of metal ions in reiterative DNA sequences may give a molecular basis to the selection for genetic areas; this selection may be related to the chelating ability of the DNA-metal ligand such as growth factor hormones, polyamines.

Various kinds of perturbations may induce transient variations of membrane permeability towards different metal ions or the release of metal ions from some cellular compartments including sequestration sites. New ternary DNA-metal-ligand complexes may result. Chelating agents can therefore either increase or decrease the uptake of a given metal. According to their affinity constant for this metal, they may either sequester it, which may be a good thing if the metal is toxic but detrimental for the cell if it is essential, or carry the metal to a binding site of higher affinity leading to a new functional complex. If the new environmental conditions are maintained, a new steady-state will establish.

Some facts described during steps of oogenesis and differentiation show that it is not any ternary complex which can be formed in response to exogenous signals and according to the cell history. In the case of the induction of neoplastic state, even if preexistent ternary complexes are activated,[319] new ternary complexes may be formed under the action of carcinogenic substances.

ACKNOWLEDGMENTS

The authors thank Y. Coudray and A. Feau for helpful technical assistance.

REFERENCES

1. **Schwarz, K.**, Essentiality versus toxicity of metals, in *Clinical Chemistry and Chemical Toxicology of Metals,* Brown, S. S., Ed., Elsevier North-Holland Biomedical Press, Amsterdam, 1977, 3.
2. **Sunderman, F. W., Jr.**, Carcinogenic effects of metals, *Fed. Proc. Fed. Am. Soc. Exp. Biol.*, 37, 40, 1978.
3. **Heath, J. C. and Webb, M.**, Content and intracellular distribution of the inducing metal in the primary rhabdomyosarcomata induced in the rat by cobalt, nickel and cadmium, *Br. J. Cancer*, 21, 768, 1967.
4. **Vallee, B. L. and Ulmer, D. D.**, Biochemical effects of mercury, cadmium and lead, *Annu. Rev. Biochem.*, 41, 91, 1972.
5. **Ulmer, D. D.**, Metals — from privation to pollution, *Fed. Proc. Fed. Am. Soc. Exp. Biol.*, 32, 1758, 1973.
6. **Sissoëff, I., Grisvard, J., and Guillé, E.**, Studies of metal ions-DNA interactions: specific behaviour of reiterative DNA sequences, *Prog. Biophys. Mol. Biol.*, 31, 165, 1976.
7. **Edwards C., Olson, K. B., Heggen, G., and Glen, J.**, Intracellular distribution of trace elements in liver tissue, *Proc. Soc. Exp. Biol. Med.*, 107, 94, 1961.
8. **Truchet, M.**, Mise en evidence par microsonde electronique et par microanalyse ionique de localisations naturelles d'aluminium dans les noyaux de divers types cellulaires, *C. R. Acad. Sci. Ser. D*, 282, 1785, 1976.
9. **Ophus, E. M. and Gullvag, B. M.**, Localization of lead within leaf cells of *Rhytidiadelphus squarrosus* by mean of transmission electron microscopy and X ray microanalysis, *Cytobios,* 10, 45, 1974.
10. **Vecher, A. S., Bardyskev, M. A., Klinger, Y. E., and Nikol'shii, Yu. K.**, Distribution pattern of some trace elements in nuclei and chloroplast of polyploid forms, *Dokl. Akad. Nauk. S.S.S.R.*, 20, 1042, 1976.
11. **Gouranton, J.**, Accumulation de ferritine dans les noyaux et le cytoplasme de certaines cellules du mesenteron chez des Homoptères cercopides âgés, *C. R. Acad. Sci. Ser. D,* 264, 2657, 1967.
12. **Webb, M.**, Metabolic targets of metal toxicity, in *Clinical Chemistry and Toxicology of Metals,* Brown, S. S., Ed., Elsevier North-Holland Biomedical Press, Amsterdam. 1977, 51.
13. **Ptak, M., Ropars, C., and Douzou, P.**, Electron paramagnetic resonance of the nucleic acids, *J. Chim. Phys.*, 59, 659, 1962.
14. **Blois, M. S., Jr., Maling, J. E., and Taskovich, L. T.**, Electron spin resonances in deoxyribonucleic acid, *Biophys. J.*, 3, 275, 1963.
15. **Bljumenfeld, L. A.**, Ferromagnetisme des structures organiques, *Dokl. Akad. Nauk S.S.S.R.*, 148, 361, 1963.
16. **Walsh, W. M., Jr., Shulman, R. G., and Heindewreich, R. D.**, Ferromagnetic inclusions in nucleic acid samples, *Nature (London),* 192, 1041, 1961.
17. **Wacker, W. E. C. and Vallee B. L.**, Nucleic acids and metals. I. Chromium, manganese, nickel, iron and other metals in ribonucleic acid from diverse biological sources, *J. Biol. Chem.*, 234, 3257, 1959.
18. **Deranleau, D. A.**, Occurrence of Trace Metals in Deoxyribonucleic Acid, M.Sc. thesis, Stanford University, Palo Alto, CA, 1958.

19. **Andronikashvili, E. L., Mosulishvili, L. M., and Belokobyl'ski, A. E.**, Analyse instrumentale par activation d'albumines et d'acides nucléiques, *J. Radioanal. Chem.*, 19, 299, 1974.
20. **Andronikasvili, E. L., Belokobyl'skii, A. J., Mosulishvili, L. M., Kharabadze, N. E., and Shonya, N. I.**, Tissue specificity of the binding of trace elements with DNA *in vivo, Dokl. Akad. Nauk S.S.S.R.*, 227, 1244, 1976.
21. **Bala, Yu. M. and Livshits, V. V.**, Change in the structure of nucleic acids and their bonds with metals in leukemia, *Probl. Gematol. Pereliv. Krovi*, 21, 21, 1976.
22. **Demikhorskaya, A. A. and Zakharenko, N. I.**, Content of iron, manganese and copper in DNA of enteric group bacteria, *Mikrobiol. Zh. (Kiev)*, 38, 233, 1976.
23. **Hamilton, E. I. and Minski, M. J.**, Inorganic constituents present in DNA and RNA, *Sci. Total Environ.*, 1, 104, 1972.
24. **Sissoëff, I., Grisvard, J., and Guillé, E.**, Localisation préférentielle du cadmium sur des sequences itératives de l'ADN isolé de cultures de tissus de crown-gall de tabac, *C. R. Acad. Sci. Ser. D*, 280, 2389, 1975.
25. **Guille, E., Sissoëff, I., and Grisvard, J.**, Presence of metals in reiterative DNA sequences from eukaryotic cells, in *Clinical Chemistry and Toxicology of Metals*, Brown, S. S., Ed., Elsevier North-Holland Biomedical Press, Amsterdam, 1977, 83.
26. **Wacker, W. E. C., Gordon, M. P., and Huff, J. W.**, Metal content of TMV and TMV-RNA, *Biochemistry*, 2, 716, 1963.
27. **Wester, P. O.**, Trace elements in RNA from beef heart tissue, *Sci. Total Environ.*, 1, 97, 1972.
28. **Weser, U. and Bischoff, E.**, Incorporation of zinc-65 in rat liver nuclei, *Eur. J. Biochem.*, 12, 571, 1970.
29. **Weser, U. and Brauer, H.**, $^{65}Zn^{2+}$ uptake by rat liver nuclei in presence of some chelating agents, *Biochim. Biophys. Acta*, 204, 542, 1970.
30. **Bohnenkamp, W., Scholte, W., and Weser, U.**, Distribution of $^{65}Zn^{2+}$ in rat liver nuclei, *Int. J. Biochem.*, 3, 207, 1972.
31. **Robbins, E. and Pederson, T.**, Iron: its intracellular localization and possible role in cell division, *Proc. Natl. Acad. Sci. U.S.A.*, 66, 1244, 1970.
32. **Clarkson, D. T. and Sanderson, J.**, The uptake of a polyvalent cation and its distribution in the root apices of *Allium cepa* tracer and autoradiographic studies, *Planta*, 89, 136, 1969.
33. **Kushelevksy, A., Yagil, R., Alfasi, Z., and Berlyne, G. M.**, Uptake of aluminium ion by the liver, *Biomedicine*, 25, 59, 1976.
34. **Kharlamova, S. E. and Potapova, I. N.**, Distribution of berylllium in the liver and its cellular fractions, *Farmakol. Toksikol. (Moscow)*, 31, 357, 1968.
35. **Firket, H.**, Mise en évidence histochimique du béryllium dans les cellules cultivées, *in vitro, C. R. Soc. Biol.*, 147, 167, 1953.
36. **Truhaut, R., Boudene, C., and Le Talaer, J. Y.**, Preferential fixation of Be in the nuclei of livers with high mitotic activity, *Ann. Biol. Clin. (Paris)*, 23, 45, 1965.
37. **De Boni, U., Otvos, A., Scott, J. W., and Crapper, D. R.**, Neurofibrillary degeneration induced by systemic aluminum, *Acta Neuropathol.*, 35, 295, 1976.
38. **De Boni, U., Scott, J. W., and Crapper, D. R.**, Intracellular aluminium binding-histochemical study, *Histochemistry*, 40, 31, 1974.
39. **Hardy K. J. and Bryan, S. E.**, Localization and uptake of copper into chromatin, *Toxicol. Appl. Pharmacol.*, 32, 62, 1973.
40. **Bryan, S. E., Simons, S. J., Vizard, D. L., and Hardy, K. J.**, Interaction of mercury and copper with constitutive heterochromatin and euchromatin *in vivo* and *in vitro, Biochemistry*, 15, 1667, 1976.
41. **Fowler, B. A. and Goyer, R. A.**, Bismuth localization with in nuclear inclusions by X-ray microanalysis. Effects of accelerating voltage, *J. Histochem. Cytochem.*, 23, 722, 1975.
42. **Goyer, R. A. and Cherian, M. A.**, Tissue and cellular toxicology of metals, in *Clinical Chemistry and Chemical Toxicology of Metals*, Brown, S. S., Ed., Elsevier North-Holland Biomedical Press, Amsterdam, 1977, 89.
43. **Choie, D. D. and Richter, G. W.**, Lead poisoning: rapid formation of intranuclear inclusion, *Science*, 177, 1194, 1972.
44. **Webb, M., Heath, J. C., and Hopkins, T.**, Intranuclear distribution of the inducing metal in primary rhabdomyosarcomata induced in the rat by nickel, cobalt and cadmium, *Br. J. Cancer*, 26, 274, 1972.
45. **Chandler, J. A. and Timms, B. G.**, Effect of testosterone and Cd^{2+} on the rat lateral prostate *in vitro, J. Endocrinol.*, 69, 22, 1976.
46. **Levan, A.**, Cytological reactions induced by inorganic salt solutions, *Nature (London)*, 156, 751, 1945.
47. **Bajaj, J. P. S., Rasmussen, H. P., and Adams, M. W.**, Electron microprobe analysis of isolated plant cells, *J. Exp. Bot.*, 22, 749, 1971.

48. Rasmussen, H. P. The mode of entry and distribution of aluminum in *Zea mays*: Electron microprobe X-ray analysis, *Planta (Berlin)*, 81, 28, 1968.
49. Skaar, H., Ophus, E., and Gullwag, B. M., Lead accumulation within nuclei of moss leaf cells, *Nature (London)*, 214, 215, 1973.
50. Silverberg, B. A., Stokes, P. H., and Ferstenberg, L. B., Intranuclear complexes in a copper-tolerant green alga, *J. Cell. Biol.*, 69, 210, 1976.
51. Stoll, R., Bousquet, W. F., White, J. F., and Miya, T. S., Effect of cadmium on nucleic acids and protein synthesis in rat liver, *Toxicol. Appl. Pharmacol.*, 37, 61, 1976.
52. Tamino G., Interaction of chromium with nucleic acids of mammalian cells, *Atti. Assoc. Grenet. Ital.*, 22, 69, 1977.
53. Matsumoto, H., Hirasawa, E., Torikai, H., and Takahashi, E., Localization of absorbed aluminium in pea root and its binding to nucleic acids, *Plant Cell Physiol.*, 17, 127, 1976.
54. Matsumoto, H., Morimura, S., and Takahashi, E., Binding of aluminium to DNA in pea root nuclei, *Plant Cell Physiol.*, 18, 987, 1977.
55. Bauer, E., Berg, H., and Schütz H., Ionenstärkeeinflub anf das komplexbildungsvermögen der desoxyribonucleinsäure, *Faserforsch. Textiltech.*, 24, 58, 1973.
56. Wells, R. D., Larson, J. E., Grant, R. C., Shortle, B. E., and Cantor, C. R., Physicochemical studies on polydeoxyribonucleotides containing defined repeating nucleotide sequences, *J. Mol. Biol.*, 54, 465, 1970.
57. Watson, J. D. and Crick, F. H. C., A structure for deoxyribose nucleic acid, *Nature (London)*, 171, 737, 1953.
58. Pullman, B. and Pullman, A., *Quantum Biochemistry*, Interscience, New York, 27, 867, 1963.
59. Izatt, R. M., Christensen, J. J., and Rytting, J. H., Sites and thermodynamic quantities associated with proton and metal ion interaction with ribonucleic acid, deoxyribonucleic acid, and their constituent bases, nucleosides and nucleotides, *Chem. Rev.*, 71, 439, 1971.
60. Eichorn, G. L., *Inorganic Biochemistry*, Vol. 2, Elsevier, New York, 1973.
61. Reuben, J., Schporer, H., and Gabbay, E. J., The alkali ion-DNA interaction as reflected in the nuclear relaxation rates of ^{23}Na and ^{87}Rb, *Proc. Natl. Acad. Sci. U.S.A.*, 72, 245, 1975.
62. Felsenfeld, G. and Huang, S., The interaction of polynucleotide with cations, *Biochim. Biophys. Acta*, 34, 234, 1959.
63. Lyons, J. W. and Kotin, L., On the interaction of magnesium with deoxyribonucleic acid, *J. Am. Chem. Soc.*, 86, 3634, 1964.
64. Mathieson, A. R. and Olayemi, J. Y., The interaction of calcium and magnesium ions with deoxyribonucleic acid, *Arch. Biochem. Biophys.*, 169, 237, 1975.
65. Dix, D. E. and Straus, D. B., DNA helix stability. I. Differential stabilization by counter cations, *Arch. Biochem. Biophys.*, 152, 299, 1972.
66. Dove, W. F. and Davidson, N., Cation effects on the denaturation of DNA, *J. Mol. Biol.*, 5, 467, 1962.
67. Lyons, J. W. and Kotin, L., On the interaction of magnesium with deoxyribonucleic acid, *J. Am. Chem. Soc.*, 86, 3634, 1964.
68. Shulman, R. G. and Sternlicht, H., Nuclear magnetic resonance determination of divalent metal ion binding to nucleic acids and adenosine triphosphate, *J. Mol. Biol.*, 13, 952, 1965.
69. Frieden, E. and Alles, J., Subtle interactions of cupric ion with nucleic acid, *J. Biol. Chem.*, 230, 797, 1958.
70. Schreiber, J. P. and Daune, M., Fixation de l'ion cuivrique sur l'acide desoxyribonucleique, *C. R. Acad. Sci. Ser. D*, 264, 1822, 1967.
71. Jonusahot, G. R. and Mushrush, G. W., Terbium as a fluorescent-probe for DNA and chromatin, *Biochemistry*, 14, 1677, 1975.
72. Pörschke, D., Thermodynamic and kinetic parameters of ion condensation to polynucleotides. Outer sphere complex formed by Mn^{2+} ions, *Biophys. Chem.*, 4, 383, 1976.
73. Eisinger, J., Fawaz-Estrup, F., and Shulman, R. G., Binding of Mn^{2+} to nucleic acids, *J. Chem. Phys.*, 42, 43, 1965.
74. Cohn, M., Danchin, A., and Grunberg-Manago, M., Proton magnetic relaxation studies of manganous complexes of transfer RNA and related compounds, *J. Mol. Biol.*, 39, 199, 1969.
75. Willemsen, A. M. and Van Os, G. A. J., Interaction of magnesium ions with poly A and poly U, *Biopolymers*, 10, 945, 1971.
76. Krakauer, H., The binding of Mg^{2+} ions to polyadenylate, polyuridilate and their complexes, *Biopolymers*, 10, 2459, 1971.
77. Sander, C. and Ts'o, P. O. P., Interaction of nucleic acids. VIII. Binding of magnesium ions by nucleic acids, *J. Mol. Biol.*, 55, 1, 1971.
78. Ross, P. D. and Scruggs, R. L., Electrophoresis of deoxyribonucleic acid (III): effect of several univalent electrolytes on the mobility of deoxyribonucleic acid, *Biopolymers*, 2, 79, 1964.

79. **Daune, M.**, Interaction of metal ions with nucleic acids, *Metal Ion Biol.*, 3, 1, 1974.
80. **Yamane, T. and Davidson, N.**, On the complexing of deoxyribonucleic acid by silver I, *Biochim. Biophys. Acta*, 55, 609, 1962.
81. **Giachino, G. G. and Kearns, D. R.**, Nature of the external heavy atom effect on radioactive and non radioactive singlet-triplet transition, *J. Chem. Phys.*, 52, 2964, 1970.
82. **Jensen, R. H. and Davidson, N.**, Spectrophotometric, potentiometric and density gradient ultracentrifugation studies of the binding of silver ion to DNA, *Biopolymers*, 4, 17, 1966.
83. **Daune, M., Dekker, C., and Schachman, H. K.**, Complexes of silver ion with natural and synthetic polynucleotides, *Biopolymers*, 4, 51, 1966.
84. **Yamane, T. and Davidson, N.**, On the complexing of deoxyribonucleic acid by mercuric ion, *J. Am. Chem. Soc.*, 83, 2599, 1961.
85. **Eichhorn, G. L., Butzow, J. J., Clark, P., and Tarien, E.**, Interaction of metal ions with polynucleotides and related compounds. X. Studies on the reaction of Silver (I) with the nucleosides and polynucleotides and the effect of Silver (I) on the zinc (II) degradation of polynucleotides, *Biopolymers*, 5, 283, 1967.
86. **Gruenwedel, D. W. and Davidson, N.**, Complexing and denaturation of DNA by methylmercuric hydroxide. II. Ultracentrifugation studies, *Biopolymers*, 5, 847, 1967.
87. **Gruenwedel, D. W. and Davidson, N.**, Complexing and denaturation of DNA by methyl mercuric hydroxide. I. Spectrophotometric studies, *J. Mol. Biol.*, 21, 129, 1968.
88. **Pillai, C. K. S. and Nandi, U. S.**, Binding of gold (III) with DNA, *Biopolymers*, 12, 1431, 1973.
89. **Simpson, R. B.**, Association of methyl mercuric and mercuric ions with nucleosides, *J. Am. Chem. Soc.*, 86, 2059, 1964.
90. **Naudi, U. S., Wang, J. C., and Davidson, N.**, Separation of deoxyribonucleic acids by Hg (II) binding and Cs_2SO_4 density gradient centrifugation, *Biochemistry*, 4, 1687, 1965.
91. **Wang, J. C., Nandi, U. S., Hogness, D. S., and Davidson, N.**, Isolation of λ dg deoxyribonucleic acid halves by Hg (II) binding and Cs_2SO_4 density grandient centrifugation, *Biochemistry*, 4, 1967, 1965.
92. **Thomas, C. A.**, The interaction of mercuric chloride with sodium thymonucleate, *J. Am. Chem. Soc.*, 76, 6032, 1954.
93. **Dove, W. F. and Yamane, T.**, The complete retention of transforming activity after reversal of the interaction of DNA with mercuric ion, *Biochem. Biophys. Res. Commun.*, 3, 608, 1960.
94. **Katz, S.**, Low-angle light scattering and the molecular weight of deoxyribonucleic acid, *Nature (London)*, 195, 897, 1962.
95. **Eichhorn, G. L.**, Metal ions as stabilizers or destabilizers of the deoxyribonucleic acid structure, *Nature (London)*, 194, 474, 1962.
96. **Katz, S.**, The reversible reaction of Hg (II) and double-stranded polynucleotides: a step function theory and its significance, *Biochim. Biophys. Acta*, 68, 240, 1963.
97. **Walter, A. and Luck, G.**, Interactions of mercury (II) ions with DNA as revealed by CD measurements, *Nucleic Acid Res.*, 4, 539, 1977.
98. **Walter, A. and Smettan, G.**, Kinetics of the DNA-Hg(II) association and dissociation, *Stud. Biophys.*, 57, 157, 1976.
99. **Wilhelm, F. X. and Daune, M.**, Ag-DNA complexes studied by flow dichroism, *C. R. Acad. Sci.* Ser. D, 266, 932, 1968.
100. **Luck, K. F. S., Maki, A. H., and Hoover, R. J.**, Studies of heavy metal binding with polynucleotides using optical detection of magnetic resonance silver (I) binding, *J. Am. Chem. Soc.*, 97, 1241, 1975.
101. **Poletaev, A. I., Ivanov, V. I., Minchenkova, L. E., and Sachekiva, A. K.**, Thermodynamic parameters of DNA copper and DNA-silver complexes, *Mol. Biol. (Moscow)*, 3, 303, 1969.
102. **Richard, H., Schreiber, J. P., and Daune, M.**, Interaction of metallic ions with DNA. V. DNA renaturation mechanism in the presence of Cu^{2+}, *Biopolymers*, 12, 1, 1973.
103. **Bach, D. and Miller, I. R.**, Polarographic investigation of binding of Cu^{2+} and Cd^{2+} by DNA, *Biopolymers*, 5, 161, 1967.
104. **Zakharenko, E. T. and Moshkovskii, Yu. Sh.**, Binding of copper and cadmium ions by deoxyribonucleic acid and its degradation products, *Biophysics (USSR)*, 11, 1083, 1966.
105. **Eichhorn, G. L. and Shin, Y. A.**, Interaction of metal ions with polynucleotides and related compounds. XII. The relative effect of various metal ions on DNA helicity, *J. Am. Chem. Soc.*, 90, 7323, 1968.
106. **Eichhorn, G. L. and Clark, P.**, Interaction of metal ions with polynucleotides and related compounds. V. The unwinding and rewinding of DNA strands under the influence of copper (II) ions, *Proc. Natl. Acad. Sci. U.S.A.*, 53, 586, 1965.
107. **Ropars, C. and Viovy, R.**, Fixation de l'ion cuivrique sur l'acide desoxyribonucleique de thymus de veau, sur les nucleosides et les nucleotides correspondants: étude en résonance paramagnétique électronique. *J. Chim. Phys.*, 62, 408, 1965.

108. **Hiai, H.**, Effects of cupric ions on thermal denaturation of nucleic acids, *J. Mol. Biol.*, 11, 672, 1965.
109. **Tu, A. T. and Friederich, C. G.**, Interaction of copper ion with guanosine and related compounds, *Biochemistry*, 7, 4367, 1968.
110. **Zimmer, C., Luck, G., and Triebel, H.**, Conformation and reactivity of DNA. IV. Base binding ability of transition metal ions to native DNA and effect on helix conformation with special reference to DNA-Zn (II) complex, *Biopolymers*, 425, 1974.
111. **Förster, W.**, Modified scatchard plots and their application to the binding of copper (II) ions to native DNA, *Stud. Biophys.*, 45, 75, 1974.
112. **Schneider, P. W., Brintzinger, H., and Erlenmeyer, H.**, Structure of adenosine triphosphate-bivalent cation complexes. IV. Coordination derivatives of the adenine ring, *Helv. Chim. Acta*, 47, 992, 1964.
113. **Clement, R. M., Sturm, J., and Daune, M. P.**, Interaction of metallic cations with DNA. VI. Specific binding of magnesium ion and manganese ion, *Biopolymers*, 12, 405, 1973.
114. **Shulman, R. G. and Sternlicht, H.**, Study of metal-ion binding to nucleic acids by ^{31}P nuclear magnetic resonance, *J. Mol. Biol.*, 13, 252, 1965.
115. **Daune, M.**, Binding of divalent cations to DNA, *Stud. Biophys.*, 24, 287, 1970.
116. **Thomas, R.**, Denaturation of deoxyribonucleic acid, *Biochim. Biophys. Acta*, 14, 219, 1954.
117. **Forster, W., Bauer, E., and Berg, H.**, Association and dissociation kinetics of DNA-copper (II)-complexes, *Stud. Biophys.*, 57, 165, 1976.
118. **Minchenkova, L. E. and Ivanov, V. I.**, Influence of reductants on optical characteristics of the DNA-Cu^{2+} complex, *Biopolymers*, 5, 615, 1967.
119. **Jouve, H., Jouve H., Melgar, E., and Lizarraga, B.**, A study of the binding of Mn^{2+} to bovine pancreatic deoxyribonuclease I and to deoxyribonucleic acid by electron paramagnetic resonance, *J. Biol. Chem.*, 250, 6631, 1975.
120. **Thomas, C. A.**, The interaction of mercuric chloride with sodium thymonucleate, *J. Am. Chem. Soc.*, 76, 6032, 1954.
121. **Huff, J. W., Sivarama, S., Gordon, M. P., and Wacker, W. E. C.**, Action of metal ions on tobacco mosaic virus ribonucleic acid, *Biochemistry*, 3, 501, 1964.
122. **Werner, C., Krebs, B., Keith, G., and Dirheimer, G.**, Specific cleavages of pure tRNAs by plumbous ions, *Biochim. Biophys. Acta*, 432, 161, 1976.
123. **Eichhorn, G. L. and Butzow, J. J.**, Interaction of metal ions with polynucleotides and related compounds III. Degradation of polyribonucleotides by lanthanum ions, *Biopolymers*, 3, 79, 1965.
124. **Butzow, J. J. and Eichhorn, G. L.**, Interactions of metal ions with polynucleotides and related compounds. IV. Degradation of polyribonucleotides by zinc and other divalent metal ions, *Biopolymers*, 3, 95, 1965.
125. **Butzow, J. J. and Eichhorn, G. L.**, Different susceptibility of DNA and RNA to cleavage by metal ions, *Nature (London)*, 254, 358, 1975.
126. **Yagi, T. and Nishioka, H.**, DNA damage and its degradation by metal compounds, *Sci. Euger. Rev. Doshisha Univ.*, 18, 1, 1977.
127. **Thomas, R.** Recherches sur la structure et les fonctions des acides désoxyribonucleiques: études génétiques et chimiques, in *Actualités Biochimiques*, Vol. 21, Florkin, M. and Roche, J., Eds., Masson, Paris, 1962.
128. **Venner, H. and Zimmer, C.**, Studies on nucleic acids. VIII. Changes in the stability of DNA secondary structure by interaction with divalent metal ions, *Biopolymers*, 4, 321, 1966.
129. **Richard, H.**, *Physico-Chemical Properties of Nucleic Acids*, Vol. 3, Duchesne, J., Ed., Academic Press, New York, 1973, chap. 17.
130. **Zimmer, C.**, Interaction of Zn^{2+} ions with nature DNA, *Stud. Biophys.*, 35, 115, 1973.
131. **Wilhelm, F. X. and Daune, M.**, Etude des complexes Ag-DNA par dichroisme d'ecoulement, *C. R. Acad. Sci. Ser. D*, 266, 932, 1968.
132. **Schreiber, J. P. and Daune, M.**, Interaction of metallic ions with DNA. IV binding of cupric ion to DNA, *Biopolymers*, 8, 139, 1969.
133. **Doty, P.**, Polynucleotides and nucleic acids, *J. Polym. Sci.*, 55, 1, 1961.
134. **Doty, P., Marmur, J., Eigner, J., and Schildkraut, C. L.**, Strand separation and specific recombination in deoxyribonucleic acid, *Proc. Natl. Acad. Sci. U.S.A.*, 46, 461, 1960.
135. **Lyons, J. W. and Kotin, L.**, The effect of magnesium ion on the secondary structure of deoxyribonucleic acid, *J. Am. Chem. Soc.*, 87, 1781, 1965.
136. **Ott, G. S., Ziegler, R., and Bauer, W. R.**, The DNA melting transition in aqueous magnesium salt solutions, *Biochemistry*, 14, 3431, 1975.
137. **Katz, S.**, The reversible reaction of sodium thymonucleate and mercuric chloride, *J. Am. Chem. Soc.*, 74, 2238, 1952.
138. **Needham, A. E.**, The effect of beryllium on the ultraviolet absorbances spectrum of nucleic acids, *Int. J. Biochem.*, 5, 291, 1974.

139. **Bauer, W. R.**, Premelting unwinding of the deoxyribonucleic acid duplex by aqueous magnesium perchlorate, *Biochemistry,* 11, 2915, 1972.
140. **Danchin, A.**, Labelling of biological macromolecules with covalent analogs of magnesium. II. Features of the chromic Cr (III) ion, *Biochimie,* 57, 875, 1975.
141. **Aldridge, W. G.**, Use of indium trichloride for the separation of native, denatured and single-stranded deoxyribonucleic acid, *Nature (London),* 195, 284, 1962.
142. **Stevens, V.L. and Duggon, E. L.**, Deformation of deoxyribonucleate. II. Precipitation of heat-deformed DNA with millimolar lead ion, *J. Am. Chem. Soc.,* 79, 5703, 1957.
143. **Dove, W. and Yamane, T.**, The complete retention of transforming activity after reversal of the interaction of DNA with mercuric ion, *Biochem. Biophys. Res. Commun.,* 3, 608, 1960.
144. **Mansy, S., Wood, T. E., Sprowls, J. C., and Tobias, R. S.**, Heavy metal nucleotide interactions. Binding of methylmercury (II) to pyrimidine nucleosides and nucleotides. Studies by Raman difference spectroscopy, *J. Am. Chem. Soc.,* 96, 1762, 1974.
145. **Fiskin, M. and Beer, M.**, Determination of base sequence in nucleic acids with the electron microscope. IV. Nucleoside complexes with certain metal ions, *Biochemistry,* 4, 1289, 1965.
146. **Luck, G. and Zimmer, C.**, Thermisches schmelzen und optische rotation dispension von DNA-Hg^{2+} komplexen, *Eur. J. Biochcm.,* 18, 140, 1971.
147. **Corneo, G., Ginelli, E., and Polli, E.**, Different satellite deoxyribonucleic acids of guinea pig and ox, *Biochemistry,* 9, 1565, 1970.
148. **Arceveaux, J. E. L. and Byers, B. R.**, Inhibition of iron uptake and deoxyribonucleic acid synthesis of desferol in a mutant strain of *Bacillus subtilis, J. Bacteriol.,* 129, 1639, 1977.
149. **Winge, D. R., Premakvwar, R., and Rajagopalan, K. V.**, Metal-induced formation of metalothionein in rat liver, *Arch. Biochem. Biophys,* 170, 242, 1975.
150. **Bryan, S. E and Hidalgo, H. A.**, Nuclear ^{115}Cadmium: uptake and disappearance correlated with cadmium-binding protein synthesis, *Biochem. Biophys. Res. Commun.,* 68, 858, 1976.
151. **Wacker, W. E. C. and Vallee, B. L.**, The metallo proteins in *The Proteins,* Vol. 5, Neurath, H. Ed., Academic Press, New York, 1970, 129.
152. **Loeb, L. A.**, Eukaryotic DNA polymerase in*Enzymes,* Vol. 10, 3rd ed., Boyer, P. D., Ed., Academic Press, New York, 1974, 173.
153. **Slater, J. P.**, Zinc in DNA polymerase, *Biochem. Biophys. Res. Common.,* 44, 37, 1971.
154. **Auld, D. S., Kawaguchi, H., Livingston, P. M., and Vallee, B. L.**, RNA-dependent DNA polymerase (reverse transcriptase) from avian myeloblastosis virus: a zinc metallo-enzyme, *Proc. Natl. Acad. Sci. U.S.A.,* 71, 2091, 1974.
155. **Robbins, E., Fant, J., and Norton, W.**, Intracellular iron-binding macromolecules in HeLa cells, *Proc. Natl. Acad. Sci. U.S.A.,* 69, 3708, 1972.
156. **Yarus, M.**, Recognition of nucleotide sequences, in *Annual Review of Biochemistry,* Vol. 38, Snell, E. E., Ed., Annual Reviews, Palo Alto, 1969, 841.
157. **Dickson, R. C., Abelson, J., Barnes, W. M., and Reznikoff, W. S.**, Genetic regulation: the lac control region: the nucleotide sequence of the lac control region containing the promoter and operator is presented, *Science,* 187, 27, 1975.
158. **Valenzuela, P., Morris, R. W., Faras, A., Levinson, W., and Rutter, W. J.**, Are all nucleotidyl transferases metalloenzymes, *Biochem. Biophys. Res. Commun.,* 53, 1036, 1973.
159. **McCann, J., Choi, E., Yamasaki, E., and Ames, B. N.**, Detection of carcinogens as mutagens in the *Salmonella/*microsomes test: assay of 300 chemicals, *Proc. Natl. Acad. Sci. U.S.A.,* 72, 5135, 1975.
160. **Flessel, C. P.**, Metals as mutagens, in *Inorganic and Nutritional Aspects of Cancer,* Schrauzer, G. N., Ed., Plenum Press, New York, 1978, 117.
161. **Murray, M. J. and Flessel, C. P.**, Metal-polynucleotide interactions. A comparison of carcinogenic and non-carcinogenic metals *in vitro, Biochim. Biophys. Acta,* 425, 256, 1976.
162. **Nishioka, H.**, Mutagenic activities of metal compounds in bacteria, *Mutat. Res.,* 31, 185, 1975.
163. **Sirover, M. A. and Loeb, L. A.**, Infidelity of DNA synthesis *in vitro:* screening for potential metal mutagens or carcinogens, *Science,* 194, 1434, 1976.
164. **Orgel, A. and Orgel, L.**, Induction of mutations in bacteriophage T_4 with divalent manganese, *J. Mol. Biol.,* 14, 453, 1965.
165. **Venitt, S. and Levy, L.**, Mutagenicity of chromates in bacteria and its relevance to chromato-carcinogenesis, *Nature (London),* 250, 493, 1974.
166. **Levine, A. F., Fink, L. M., Weinstein, B. I., and Grunberger, D.**, Effect of N-2-Acetylaminofluorene modification on the conformation of nucleic acids, *Cancer Res.,* 34, 319, 1974.
167. **Okada, Y., Streisinger, G., Owen, J., Newton, J., Tsugita, A., and Inouye, M.**, Molecular basis of a mutational hot spot in the lysozyme gene of bacteriophage T_4, *Nature (London),* 236, 338, 1972.
168. **Isono, S. and Yourno, J.**, Non-suppressible addition frameshift in *Salmonella, J. Mol. Biol.,* 82, 355, 1974.

169. **Ravin, V. K. and Artemier, M. I.**, Fine structure of the lysozyme gene of bacteriophage T_4 B, *Mol. Gen. Genet.* 128, 359, 1974.
170. **Koch, R. E.**, The influence of neighboring base pairs upon base-pair substitution mutation rate, *Proc. Natl. Acad. Sci. U.S.A.*, 68, 773, 1971.
171. **Moutschen-Dahmen, J. and Moutschen-Dahmen, M.**, Influence of Cu^{2+} and Zn^{2+} ions on the effects of ethylmethanesulfonate (EMS) on chromosomes, *Experientia*, 19, 144, 1963.
172. **Hirai, K., Defendi, V., and Diamond, L.**, Enhancement of SV40 transformation and integration by 4-nitroquinoline, 1-oxide, *Cancer Res.*, 34, 3497, 1974.
173. **Sunderman, F. W., Jr. and Esfahani, M.**, Nickel carbonyl inhibition of RNA polymerase activity in hepatic nuclei, *Cancer Res.*, 28, 2565, 1968.
174. **Sunderman, F. W., Jr. and Leibman, K. C.**, Nickel carbonyl inhibition of induction of aminopyrine demethylase activity in liver and lung, *Cancer Res.*, 30, 1645, 1970.
175. **Sunderman, F. W., Jr.**, Metal carcinogenesis in experimental animals, *Food Cosmet. Toxicol.*, 9, 105, 1971.
176. **Basrur, P. K. and Gilman, J. P. W.**, Morphologic and synthetic response of normal and tumor muscle culture to nickel sulfide, *Cancer Res.*, 27, 1168, 1967.
177. **Fradkin, A., Janoff, A., Lane, B. P., and Kushner, M.**, *In vitro* transformation of BHK 21 cells grown in the presence of calcium bichromate, *Cancer Res.*, 35, 1058, 1975.
178. **Levis, A. G., Bianchi, V., Tamino, G., and Pegorano, B.**, Cytotoxic effects of hexavalent and trivalent chromium on mammalian cells *in vitro*, *Br. J. Cancer*, 37, 386, 1978.
179. **Levis, A. G., Buttignol, M., Bianchi, V., and Sponza, G.**, Effects of potassium dichromate on nucleic acid and protein synthesis and on precursor uptake in BHK fibroblasts, *Cancer Res.*, 38, 110, 1978.
180. **Mertz, W.**, Chromium occurrence and function in biological systems, *Physiol. Rev.*, 49, 163, 1969.
181. **Luke, M. Z., Hamilton, L., and Hollocher, T. C.**, Beryllium-induced misincorporation by a DNA polymerase. A possible factor in beryllium toxicity, *Biochem. Biophys. Res. Commun.*, 62, 497, 1975.
182. **Thomas, H. F., Herriott, R. M., Hahn, B. S., and Yang, S. Y.**, Thymine hydroperoxide as a mediator in ionising radiation mutagenesis, *Nature (London)*, 259, 341, 1976.
183. **Stich, H. F., Karim, J., Koropatnick, J., and Lo, L.**, Mutagenic action of ascorbic acid, *Nature (London)*, 260, 722, 1976.
184. **Stanton, M. F.**, Primary tumors of bone and lung in rats following local deposition of cupric-chelated N-hydroxy-2-acetylaminofluorene, *Cancer Res.*, 27, 1000, 1967.
185. **Speyer, J. F., Karam, J. D., and Lenny, A. B.**, On the role of DNA polymerase in base selection, *Cold Spring Harbor Symp. Quant. Biol.*, 31, 693, 1966.
186. **De Waard, A., Paul, A. V., and Lehman, I. R.**, The structural gene for deoxyribonucleic acid polymerase in bacteriophage T_4 and T_5, *Proc. Natl. Acad. Sci. U.S.A.*, 54, 1241, 1965.
187. **Levinson, W., Faras, A., Woodson, B., Jackson, J., and Bishop, J. M.**, Inhibition of RNA-dependent DNA polymerase of Rous sarcoma virus by thiosemicarbazones and several cations, *Proc. Natl. Acad. Sci. U.S.A.*, 70, 164, 1973.
188. **Kessel, D. and Elhinney, R. S.**, The role of metals in the antitumor action of 1,5-bisthiosemicarbazones, *Mol. Pharmacol.*, 11, 298, 1975.
189. **Mikelens, P., Woodson, B., and Levinson, W.**, Association of nucleic acids with complexes of N-methylisatin beta-thiosemicarbazone and copper, *Biochem. Pharmacol.*, 25, 281, 1976.
190. **Rohde, W., Mikelens, P., Jackson, J., Blackman, J., Whitcher, J., and Levinson, W.**, Hydroxyquinolines inhibit RNA-dependent DNA polymerase and inactive Rous sarcoma virus and herpes simplex virus, *Antimicrob. Agents. Chemother.*, 10, 234, 1976.
191. **Minkel, D. T., Saryan, L. A., and Petering, D. H.**, Structure-function correlations in the reaction of Bis(thiosemicarbazonate)copper complexes with Ehrlich ascites tumor cells, *Cancer Res.*, 38, 124, 1978.
192. **Fox, M. P., Bopp, L. H., and Pfau, C. J.**, Contact inactivation of RNA and DNA viruses by N-methyl isatin beta-thiosemicarbazone and $CuSO_4$, *Ann. N.Y. Acad. Sci*, 284, 533, 1977.
193. **Levinson, W., Rhode, W., Mikelens, P., Jackson, J., Antony, A., and Ramakrishnan, T.**, Inactivation and inhibition of Rous sarcoma virus by copper-binding ligands: thiosemicarbazones, 8-hydroxyquinolines and isonicotinic acid hydrazide, *Ann. N.Y. Acad. Sci.*, 284, 525, 1977.
194. **Albert, A. and Gledhill, W.**, The choice of a chelating agent for inactivating trace metals, *Biochem. J.*, 41, 529, 1947.
195. **Fallab, S. and Erlenmeyer, H.**, Zur bestimmung der bildungskonstanten des komplexes and isonicotinsâurehydrazid and Cu, *Helv. Chim. Acta*, 36, 6, 1953.
196. **Fernandez-Pol, J. A. and Johnson, G. S.**, Selective toxicity induced by picolinic acid in simian virus 40-transformed cells in tissue culture, *Cancer Res.*, 37, 4276, 1977.
197. **Fernandez-Pol, J. A., Bono, V. H., Jr., and Johnson, G. S.**, Control of growth by picolinic acid: differential response of normal and transformed cells, *Proc. Natl. Acad. Sci. U.S.A.*, 74, 2889, 1977.

198. **Fernandez-Pol, J. A.**, Transition metal ions induce cell growth in NRK cells synchronized in G_1 by picolinic acid, *Biochem. Biophys. Res. Commun.*, 76, 413, 1977.
199. **Fernandez-Pol, J. A.**, Iron: possible cause of the G_1 arrest induced in NRK cells by picolinic acid, *Biochem. Biophys. Res. Commun.*, 78, 136, 1977.
200. **Fernandez-Pol, J. A.**, Isolation and characterization of a siderophore-like growth factor from mutants of SV 40-transformed cells adapted to picolinic acid, *Cell*, 14, 489, 1978.
201. **Rosenberg, B. and Van Camp, L.**, The successful regression of large solid sarcoma 180 tumors by platinum compounds, *Cancer Res.*, 30, 1799, 1970.
202. **Howle, J. A. and Gale, G. R.**, Cis-dichlorodiammine platinum II: persistant and selective inhibition of DNA synthesis, *in vivo, Biochem. Pharmacol.*, 14, 2757, 1970.
203. **Roberts, J. J. and Pascoe, J. M.**, Cross-linking of complementary strands of DNA in mammalian cells by antitumour platinum compound, *Nature (London)*, 235, 282, 1972.
204. **Muchausen, L. L.**, The chemical and biological effects of cis-dichloridimmine platinum II, an antitumor agent, on DNA, *Proc. Natl. Acad. Sci. U.S.A.*, 71, 4519, 1974.
205. **Roberts, J. J.**, Bacterial, viral, and tissue cultures studies on neutral platinum complexes, *Recent Results Cancer Res.*, 48, 79, 1974.
206. **Stone, P. J., Kelman, A. D., and Sinex, F. M.**, Specific binding of antitumor drug cis-$Pt(NH_3)_2Cl_2$ to DNA rich in guanine and cytosine, *Nature (London)*, 251, 736, 1974.
207. **Aggarwal, S. K. and Rosenberg, B.**, Cell surface-associated nucleic acid in tumorigenic cells made visible with platinum-pyrimidine complexes by electron microscopy, *Proc. Natl. Acad. Sci. U.S.A.*, 72, 928, 1975.
208. **Rosenberg, B.**, On the mechanism of action of platinum complexes as anticancer agents, *J. Clin. Hematol. Oncol.*, 7, 817, 1977.
209. **Stone, P. J., Kelman, A. D., and Sinex, F. M.**, Resolution of α,β and γ DNA of *Saccharomyces cerevisiae* with the antitumor drug cis-$Pt(NH_3)_2$ Cl_2. Evidence for preferential drug binding by GpG sequences of DNA, *J. Mol. Biol.*, 104, 793, 1976.
210. **Meyne, J. and Lockhart, H. L.**, Cytogenetic effects of cis-platinum II diammine dichloride on human lymphocytes cultures, *Mutat. Res.*, 58, 87, 1978.
211. **Erck, A., Rainen, L., Whileyman, J., Chang, I. M., Kimball, A. P., and Bear, J.**, Studies of rhodium (II) carboxylates as potential antitumor agents, *Proc. Soc. Exp. Biol. Med.*, 145, 1278, 1974.
212. **Nagai, K., Yamaki, H., Suzuki, H., Tanaka, N., and Umezawa, H.**, The combined effects of bleomycin and sulfhydryl compounds on the thermal denaturation of DNA, *Biochim. Biophys. Acta*, 179, 165, 1969.
213. **Müller, W. E. G., Yamazaki, Z., Breter, H. J., and Zahn, R. K.**, Action of bleomycin on DNA and RNA, *Eur. J. Biochem.*, 31, 518, 1972.
214. **Haidle, C. W., Weiss, K. K., and Kuo, M. T.**, Release of free bases from DNA after reaction with bleomycin, *Mol. Pharmacol.*, 8, 531, 1972.
215. **Umezawa, H.**, Chemistry and mechanism of action of bleomycin, *Fed. Proc. Fed. Am. Soc. Exp. Biol.*, 33, 2296, 1974.
216. **Miyaki, M., Ono, T., Horis, S., and Umezawa, H.**, Binding of bleomycin to DNA in bleomycin-sensitive and -resistant rat ascites hepatoma cells, *Cancer Res.*, 35, 2015, 1975.
217. **Sausville, E. A., Stein, R. W., Peisach, J., and Horwitz, S. B.**, Properties and products of the degradation of DNA by bleomycin and iron (II), *Biochemistry*, 17, 1746, 1978.
218. **Sausville, E. A., Peisach, J., and Horwitz, S. B.**, Effect of chelating agents and metal ions on the degradation of DNA by bleomycin, *Biochemistry*, 17, 2740, 1978.
219. **D'Andrea, A. D. and Haseltine, W. A.**, Sequence specific cleavage of DNA by the antitumor antibiotics neocarzinostatin and bleomycin, *Proc. Natl. Acad. Sci. U.S.A.*, 75, 3608, 1978.
220. **Falaschi, A. and Kornberg, A.**, Antimetabolites affecting protein or nucleic acid synthesis. Phleomycin, an inhibitor of DNA polymerase, *Fed. Proc. Fed. Am. Soc. Exp. Biol.*, 23, 940, 1964.
221. **Stern, R., Rose, J. A., and Friedman, R. M.**, Phleomycin-induced cleavage of DNA, *Biochemistry*, 13, 307, 1974.
222. **Beerman, T. A. and Goldberg, I. H.**, DNA strand scission by the antitumor protein neocarzinostatin, *Biochem. Biophys. Res. Commun.*, 59, 1254, 1974.
223. **Poon, R., Beerman, T. A., and Goldberg, I. H.**, Characterization of DNA strand breakage *in vitro* by the antitumor protein neocarzinostatin, *Biochemistry*, 16, 486, 1977.
224. **Kappen, L. S. and Goldberg, I. H.**, Effect of neocarzinostatin-induced strand scission on the template activity of DNA for DNA polymerase I, *Biochemistry*, 16, 479, 1977.
225. **Hatayama, T., Goldberg, I. H., Takeshita, M., and Grollman, A. P.**, Nucleotide specificity in DNA scission by neocarzinostatin, *Proc. Natl. Acad. Sci. U.S.A.*, 75, 3603, 1978.
226. **Calendi, E., Di Marco, A., Reggiani, M., Scarpinato, B., and Valentini, L.**, On physico-chemical interactions between daunomycin and nucleic acids, *Biochim. Biophys. Acta*, 103, 25, 1965.

227. **Kersten, W., Kersten, H., and Szybalski, W.**, Physicochemical properties of complexes between DNA and antibiotics which affect RNA synthesis (actinomycin, daunomycin, cinerubin, nogalomycin, chromomycin, mithramycin, and olivomycin), *Biochemistry,* 5, 236, 1966.
228. **Fishman, M. M. and Schwartz, I.**, Effect of divalent cations on the daunomycin-DNA complex, *Biochem. Pharmacol.,* 23, 2147, 1974.
229. **Mizuno, N. S., Zakis, B., and Decker, R. W.**, Binding of daunomycin to DNA and the inhibition of RNA and DNA synthesis, *Cancer Res.,* 35, 1542, 1975.
230. **Ward, D. C., Reich, E., and Goldberg, I. H.**, Base specificity in the interaction of polynucleotides with antibiotic drugs, *Science,* 149, 1259, 1965.
231. **Behr, W., Honikel, K., and Hartmann, G.**, Interaction of the RNA polymerase inhibitor chromomycin with DNA, *Eur. J. Biochem.,* 9, 82, 1969.
232. **Cone, R., Hasan, S. K., Lown, J. W., and Morgan, A. R.**, The mechanism of the degradation of DNA by steptonigrin, *Can. J. Biochem.,* 54, 219, 1976.
233. **Lown, J. W. and Sim, S. K.**, Studies related to antitumor antibiotics. Part VIII. Cleavage of DNA by streptonigrin analogs and the relationship to antineoplastic activity, *Can. J. Biochem.,* 54, 446, 1976.
234. **Lown, J. W. and Sim, S. K.**, Studies related to antitumor antibiotics. VII. Synthesis of streptonigrin analogues and their single strand scission of DNA, *Can. J. Biochem.,* 54, 2563, 1976.
235. **White, J. R.**, Steptonigrin-transition metal complexes: binding to DNA and biological activity, *Biochem. Biophys. Res. Commun.,* 77, 387, 1977.
236. **Reich, E. and Goldberg, I. H.**, Actinomycin D and nucleic acid function, *Progr. Nucleic Acid Res. Mol. Biol.,* 3, 184, 1964.
237. **Mukherjee, K. L. and Heidelberger, C.**, Studies on fluorinated pyrimidines. IX. The degradation of 5-fluorouracil-6-^{14}C, *J. Biol. Chem.,* 235, 433, 1960.
238. **Heidelberger, C., Chaudhuri, N. K., Danneberg, P., Mooren, D., and Griesbach, L.**, Fluorinated pyrimidines. A new class of tumour-inhibitory compounds, *Nature (London),* 179, 663, 1957.
239. **Chandra, P., Zunino, F., Zaccara, A., Wacker, A., and Götz, A.**, Influence of tirolone hydrochloride on the secondary structure and template activity of DNA, *FEBS. Lett.,* 23, 145, 1972.
240. **Chandra, P. and Wright, G. J.**, Tirolone hydrochloride: the drug profile, *Top. Curr. Chemistry,* 72, 126, 1977.
241. **Kashiguchi, H. and Takahashi, H.**, Inhibition of two copper-containing enzymes, tyrosinase and dopamine β-hydroxylase by L-mimosine, *Mol. Pharmacol.,* 13, 362, 1977.
242. **Guillé, E., Grisvard, J., Tuffet, A., Villaudy, J., and Goldé, A.**, Variations of the association of copper with the DNA isolated from synchronized chick embryo fibroblasts not infected or infected with RSV, in *Avian Tumor Viruses,* Barlati, S. and De Giuli-Morghen, Eds., Piccin Medical Books, Turin, 1978, 21.
243. **Sartorelli, A. C. and Creasy, W. A.**, Cancer chemotherapy, *Annu. Rev. Pharmacol.,* 9, 91, 1969.
244. **Lindebaum, A.**, A survey of naturally occuring chelating ligands, in *Metal Ions in Biological Systems,* Dhar, S. K., Ed., Plenum Press, New York, 1975, 67.
245. **Jacobs, A.**, Low molecular weight intracellular iron transport compounds, *Blood,* 50, 433, 1977.
246. **Hurley, L. S., Duncan, J. R., Sloan, M. V., and Eckhert, C. D.**, Zinc binding ligands in milk and intestine: a role in neonatal nutrition?, *Proc. Natl. Acad. Sci. U.S.A.,* 74, 3547, 1977.
247. **Perrin, D. D.and Agrawal, R. P.**, in *Metal Ions in Biological Systems,* Vol. 2, Sigel, H., Ed., Marcel Dekker, New York, 1973, 167.
248. **Freeman, H. C.**, Crystal structures of metal-peptide complexes, in *Advances in Protein Chemistry,* Vol. 22, Anfinsen, C. B., Jr., Anson, M. L., Edsall, J. T., Eds., and Richards, F. M., Academic Press, New York, 1967, 257.
249. **Itabashi, H.**, Reaction of sex hormones and related compounds with metals in ethalonic solutions by spectrophotometry, *Endocrinol. Jpn.,* 7, 284, 1960.
250. **Heath, O. V. S. and Clark, J. E.**, Chelating agents as plant-growth substances due to action of auxin, *Nature (London),* 177, 1118, 1956.
251. **Oota, Y. and Tsudzuki, T.**, Resemblance of growth substances to metal chelators with respect to their actions on duckweed growth, *Plant Cell Physiol.,* 12, 619, 1971.
252. **Brown, S. W.**, Heterochromatin, *Science,* 151, 417, 1966.
253. **Lima de Faria, A. and Jaworska, H.**, Late DNA synthesis in heterochromatin, *Nature (London),* 217, 138, 1968.
254. **Williams, C. A. and Ockey, C. H.**, Distribution of DNA replicator sites in mammalian nuclei after different methods of cell synchronization, *Exp. Cell Res.,* 63, 365, 1975.
255. **Guillé, E. and Quêtier, F.**, Heterochromatic, redundant and metabolic DNAs: a new hypothesis about their structure and function, *Prog. Biophys. Mol. Biol.,* 27, 121, 1973.
256. **Comings, D. E.**, The role of heterochromatin, in *Proc. 4th Int. Conf.,* Motulsky, A. G. and Lutz, W., Eds., Excerpta Medica, Amsterdam, 1973, 44.

257. **Rieger, R., Michaelis, A., Schubert, I., and Kaina, B.**, Effects of chromosome repatterning in *Vicia faba* L. II Aberration clustering after treatment with chemical mutagens and X-rays as affected by segment transposition, *Biol. Zentralbl.*, 96, 161, 1977.
258. **Cooper, J. E. K.**, Chromosomes and DNA of *Microtus*. III Heterochromatin rearrangements in *M. agrestis*. Bone marrow clones, *Chromosoma*, 62, 269, 1977.
259. **Clark, R. J. and Felsenfeld, G.**, Association of arginine-rich histones with GC-rich regions of DNA in chromatin, *Nature (London) New Biology*, 240, 226, 1972.
260. **Dina, D., Crippa, M., and Beccari, E.**, Hybridization properties and sequence arrangement in a population of mRNAs, *Nature (London) New Biology*, 242, 101, 1973.
261. **Evenson, D. P., Mego, W. A., and Taylor, J. H.**, Subunits of chromosomal DNA. I. Electron microscopic analysis of partially denatured DNA, *Chromosoma*, 39, 225, 1972.
262. **Sekiya, T., Van Ormondt, H., and Khorana, H. G.**, The nucleotide sequence in the promoter region of the gene for an *Escherichia coli* tyrosine transfer RNA, *J. Biol. Chem.*, 250, 1087, 1975.
263. **Wells, R. D., Blakesley, R. W., Hardies, S. C., Horn, G. T., Larson, J. E., Selsing, E., Burd, J. F., Chan, H. W., Dadgson, J. B., Jensen, K. F., Nes, I. F. and Wartell, R. M.**, The role of DNA structure in genetic regulation, *CRC Crit. Rev. Biochem.*, 4, 305, 1977.
264. **Nakanishi, S., Adhya, S., Gottesman, M., and Pastan, I.**, Activation of transcription at specific promoters by glycerol, *J. Biol. Chem.*, 249, 4050, 1974.
265. **Crepin, M., Cukier-Kahn, R., and Gros, F.**, Effect of low molecular weight DNA-binding protein, H factor, on the *in vitro* transcription of the lactose operon in *E. coli*, *Proc. Natl. Acad. Sci. U.S.A.*, 72, 333, 1975.
266. **Pfahl, M.**, Effect of DNA denaturants on the lac repressor-operator interaction, *Biochim. Biophys. Acta*, 520, 285, 1978.
267. **Lima de Faria, A.**, The relation between chromomeres, replicons, operons, transcription units, genes, viruses and palindromes, *Hereditas*, 81, 249, 1975.
268. **Weisenberg, R. C.**, Microtubule formation *in vitro* in solutions containing low calcium concentrations, *Science*, 177, 1104, 1972.
269. **Pogo, A. O., Littan, V. C., Allfrey, V. G., and Mirsky, A. E.**, Modification of RNA synthesis in nuclei isolated from normal and regenerating liver. Some effects of salt and specific divalent cations, *Proc. Natl. Acad. Sci. U.S.A.*, 57, 743, 1967.
270. **Falchuk, K. H., Krishan, A., and Vallee, B. L.**, DNA distribution in the cell cycle of *Euglena gracilis*. Cytofluorometry of zinc-deficient cells, *Biochemistry*, 14, 3439, 1975.
271. **Falchuk, K. H., Fawcett, D., and Vallee, B. L.**, Role of zinc in cell division of *Euglena gracilis*, *J. Cell Sci.*, 17, 57, 1975.
272. **Falchuk, K. H., Fawcett, D. W., and Vallee, B. L.**, Competitive antagonism of cadmium and zinc in the morphology and cell division of *Euglena gracilis*, *J. Submicr. Cytol.*, 7, 139, 1975.
273. **Falchuk, K. H., Hardy, C., Ulpino, L., and Vallee, B. L.**, RNA polymerase, manganese and RNA metabolism of zinc sufficient and deficient, *E. gracilis*, *Biochem. Biophys. Res. Commun.*, 77, 314, 1977.
274. **Krishan, A.**, Cytochalasin, B: time-lapse cinematographic studies on its effects on cytokinesis, *J. Cell Biol.*, 54, 567, 1972.
275. **Desai, L. S., Krishan, A., and Foley, G. E.**, Effects of bleomycin on cells in culture. Quantitative cytochemical study, *Cancer*, 34, 1873, 1974.
276. **Sunderman, F. W., Jr., Kasprzak, K. S., Lau, T. J., Minghetti, P. P., Maenza, R. M., Becker, N., Oukelinx, C., and Goldblatt, P. J.**, Effects of manganese on carcinogenicity and metabolism of nickel subsulfide, *Cancer Res.*, 36, 1790, 1976.
277. **Kroeger, H. and Müller, G.**, Control of puffing activity in three chromosomal segments of explanted salivary gland cells of *Chironomus thummi* by variation in extracellular Na^+, K^+ and Mg^{2+}. *Exp. Cell Res.*, 82, 89, 1973.
278. **Kroeger, H. and Trösch, W.**, Influence of the explanation medium on intranuclear (Na) (K) and (Mg) of *Chironomus thummi* salivary gland cells, *J. Cell. Physiol.*, 83, 19, 1974.
279. **Cone, C. D., Jr.**, Variation of the transmembrane potential level as a basic mechanism of mitosis control, *Oncology*, 24, 438, 1970.
280. **Cone, C. D., Jr. and Tongier, M. Jr.**, Contact inhibition of division: involvement of the electrical transmembrane potential, *J. Cell. Physiol.*, 82, 373, 1973.
281. **Cone, C. D. and Cone, C. M.**, Induction of mitosis in mature neurons in central nervous system by sustained depolarization, *Science*, 192, 155, 1976.
282. **Cone, C. D.**, Unified theory on the basic mechanism of normal mitotic control and oncogenesis, *J. Theor. Biol.*, 30, 151, 1971.
283. **Parenti, R., Guillé, E., Grisvard, J., Durante, M., Giorgi, L. and Buiatti, M.**, Presence of transient satellite in dedifferentiating pith tissue of *Nicotiana glauca*, *Nature (London), New Biology*, 246, 237, 1973.

284. **Durante, M., Geri, C., Nuti-Ronchi, V., Martini, G., Guillé, E., Grisvard, J., Giorgi, L., Parenti, R., and Buiatti, M.**, Inhibition of *Nicotiana glauca* pith tissue proliferation through incorporation of 5-BrdU into DNA, *Cell Differ.*, 6, 53, 1977.
285. **Epstein, S. M., Benedetti, E. L., Shinozuka, H., Bartus, B., and Farber, E.**, Altered and distorted DNA from a premalignant liver lesion induced by 2-fluorenylacetamide, *Chem. Biol. Interact.*, 1, 113, 1969/1970.
286. **Poried, W. J., Henzel, J. H., and Rob, C. A.**, Acceleration of wound healing in man with zinc sulfate given by mouth, *Lancet*, 1, 121, 1967.
287. **Cohen, C.**, Zinc sulfate and bedsores, *Br. Med. J.*, 2, 561, 1968.
288. **Haddow, A.**, Molecular repair, wound healing, and carcinogenesis: tumor production a possible overhealing ? in *Advances in Cancer Research*, Vol. 16, Klein, G. and Weinhouse, S., Eds., Academic Press, New York, 1972, 181.
289. **Flynn, A., Pories, W. J., Strain, W. H., and Weiland, F. L.**, Manipulation of blead zinc by progesterone, *Naturwissenchaften*, 60, 162, 1973.
290. **Hurley, L. S.**, Interaction of genes and metals in development, *Fed. Proc. Fed. Am. Soc. Exp. Biol.*, 35, 2272, 1976.
291. **Andronikashvili, E. L., Mosulishvili, L. M., Belokobil'skiy, A. I., Kharabadze, N. E., Shonia, N. I., Desai, L. S., and Foley, G. E.**, Human leukaemic cells determination of trace elements in nucleic acids and histones by neutron activation analysis, *Biochem. J.*, 157, 529, 1967.
292. **Andronikashvili, E. L., Mosulishvili, L. M., Manjgaladze, V. P., Belokobil'skiy, A. I., Kharabadze, N. E., and Efremova, E. Y.**, D. Binding of trace amounts of certain elements by the nucleic acids of malignant tumors, *Dokl. Akad. Nauk S.S.S.R.*, 195, 359, 1970.
293. **Andronikashvili, E. L., Mosulishvili, L. M., Belokobil'skiy, A. I., Kharabadze, N. E., Terzieva, T. K. and Efremova, E. Y.** Content of some trace elements in Sarcoma M1 DNA in dynamics of malignant growth, *Cancer Res.*, 34, 271, 1974.
294. **Apelgot, S., Coppey, J., Sissöeff, I., Grisvard, J., and Guillé, E.**, Studies on DNA-metal complexes in eukaryotic cells systems. I. Distribution of ^{64}Cu in mice bearing Krebs ascites tumors and in their normal counterparts, in press.
295. **Mc Quitty, Jr., J. T., Dewys, W. D., Monaco, L., Strain, W. H., Rob, C. G., Apgar, J., and Pories, W. J.**, Inhibition of tumor growth by dietary zinc deficiency, *Cancer Res.*, 30, 1387, 1970.
296. **Duncan, J. R., Dreosti I. E., and Albrecht C. F.**, Zinc intake and growth of a transplanted hepotoma induced by 3′-methyl-4-dimethyl amino-azobenzene in rats, *J. Natl. Cancer Inst.*, 53, 277, 1974.
297. **Odell, W. D. and Wolfsen, A.**, Ectopic hormone secretion by tumors, in *Cancer*, Vol. 3., Becker, F. F., Ed., Plenum Press, New York, 1975, 81.
298. **Gospodarowicz, D. and Moran, J. S.**, Growth factors in mammalian cells in culture, *Annu. Rev. Biochem.*, 45, 531, 1976.
299. **Alpert, E., Coston, R. L., and Drysdale, J. W.**, Carcino-foetal human liver ferritins, *Nature (London)*, 242, 194, 1973.
300. **Hazand, J. T. and Drysdale, J. W.**, Ferritinaemia in cancer, *Nature (London)*, 265, 755, 1977.
301. **Holley, R.**, Control of growth of mammalian cells in cell culture, *Nature (London)*, 258, 487, 1975.
302. **Dietez, P., Lipton, A., and Klinger, I.**, Serum factor requirements of normal and SV 40-transformed 3T3 mouse fibroblasts, *Proc. Natl. Acad. Sci., U.S.A.*, 68, 645, 1971.
303. **Rudland, P. S., Eckhart, W., Gospodarowicz, D., and Seifert, W.**, Cell transformation mutants are not susceptible to growth activation by fibroblast growth factor at permissive temperatures, *Nature (London)*, 250, 337, 1974.
304. **Shen, S. S., Hamamoto, S. T., Bezn, H. A., and Steinhardt, R. A.**, Alteration of sodium transport in mouse mammary epithelium associated with neoplastic transformation, *Cancer Res.*, 38, 1356, 1978.
305. **Smith, N. R., Sparks, R. L., Pool, T. B., and Cameron, I. L.**, Differences in the intracellular concentration of elements in normal and cancerous liver cells as determined by X-ray microanalysis, *Cancer Res.*, 38, 1952.
306. **Hitotsumachi, S., Rabinowitz, Z., and Sachs, L.**, Chromosomal control of reversion in transformed cells, *Nature (London)*, 231, 511, 1971.
307. **Mitelman, F., Marks, J., Levan, G., and Levan, A.**, Tumor etiology and chromosome pattern, *Science*, 176, 1340, 1972.
308. **Fahmy, O. G. and Fahmy, M. J.**, Genetic deletions at specific loci by polycyclic hydrocarbons in relation to carcinogensis, *Int. J. Cancer*, 6, 250, 1970.
309. **Pero, R. W., Bryngelsson, T., Mitelman, F., and Levan, G.**, Changes in the deoxyadenylate regions of rat DNA in sarcomas induced by 7,12-dimethylbenz (a) anthracene and Rous sarcoma virus, *Hereditas*, 80, 153, 1975.

310. **Pero, R. W., Bryngelsson, T., Norgren, A., and Deutsch, A.,** Alterations in the deoxyadenylate regions of the DNA from four human breast tumours, *Eur. J. Cancer,* 11, 861, 1975.
311. **Pero, R. W., Bryngelsson, T., Deutsch, A., and Norden, A.,** Changes in the deoxyadenylate regions in 5 cases of chronic lymphocytic leukemic DNA, *Eur. J. Cancer,* 12, 357, 1976.
312. **Ts'o, P.O.P.,** Some aspects of the basic mechanisms of chemical carcinogenesis, *J. Toxicol. Environ. Health,* 2, 1305, 1977.
313. **Markert, C. L.,** Neoplasia: a disease of cell differentiation, *Cancer Res.,* 28, 1908, 1968.
314. **Kubinski, H., Morin, N. R., and Zeldin, P. E.,** Increased attachment of nuclei acids to eukaryotic and prokaryotic cells induced by chemical and physical carcinogens and mutagens, *Cancer Res.,* 36, 3025, 3033.
315. **Kubinski, H., Andersen, P. R., and Kellicutt, L. M.,** Association of nucleic acids and cellular membranes induced by a carcinogen, β-propiolactone, *Chem. Biol. Interact.,* 5, 279, 1972.
316. **Andersen, P. R., Gibbs, P., and Kubinski, H.,** Effects of neuropharmacological agents on *in vitro* formation of complex between nucleic acids and proteins, *Neuropharmacology,* 13, 11, 1974.
317. **Kubinski, H. and Szybalski. W.,** Intermolecular linking and fragmentation of DNA by β-propiolactone, a monoalkylating carcinogen, *Chem. Biol. Interact.,* 10, 41, 1975.
318. **Zeldin, P. E., Bhattacharya, P. K., Kubinski, H., and Nietert, W. C.,** Macromolecular complexes produced by 1,3-propanesultone, *Cancer Res.,* 35, 1445, 1975.
319. **Fishman, W. H.,** Activation of developmental genes in neoplastic transformation, *Cancer Res.,* 36, 3423, 1976.

Chapter 5

SYNERGISM AND ANTAGONISM IN METAL TOXICOLOGY

L. Magos

TABLE OF CONTENTS

I. QUALITATIVE AND QUANTITATIVE CHANGES IN METAL TOXICITIES

A. Introduction

Knowledge on different factors which influence the distribution, metabolism, and toxicology of metals is important for different reasons. First, it has been realized that the dose-effect and dose-response relationship of one metal can be changed by the presence of another metal or metalloid. Second, the accumulation of data on environmental contamination revealed that the ideal condition of single exposure is nonexistent, and we are continuously exposed to more than one toxic compound and more than one toxic metal.[1,2] Third, it was shown that in the tissues of aquatic mammals[3] or in the brain of those men who for a long time before their death had been exposed occupationally to mercury vapor,[4] the mercury and selenium concentrations were constantly near to equimolar, indicating an adjustment and most likely an association between these elements. Fourth, the presence of methylmercury and selenium in sea fish raised the possibility that selenium protects against the toxicity of the organomercurial.[5] These considerations prompted the Scientific Committee on the Toxicology of Metals of the Permanent Commission and Interactional Association on Occupational Health to organize a workshop and publish its concensus report on *Factors Influencing Metabolism and Toxicity of Metals* in 1977.[6]

B. Synergism and Antagonism

The combined action of two metals or one metal and another foreign compound can be additive, synergistic, or antagonistic. The difference between an additive effect and synergism is that while in the case of additive effects the two compounds have the same target and the same effect in that target, in the case of synergism or potentiation the two compounds or metals might have different toxic properties and different target organs.[1] Potentiation might render the target role to an otherwise unaffected organ or change the sequence of organ sensitivities in relation to dose. Conversely, a target organ is protected by an antagonistic effect. Synergism and antagonism are always the consequence of some interaction between the toxic metal and the synergist or antagonist.

Interaction is likely to change the organ distribution of the toxic metal, though concentration increases or decreases in sensitive organs do not always correspond to a synergistic or antagonistic effect. Selenite given simultaneously with $HgCl_2$ or pretreatment with $CdCl_2$ protects against the nephrotoxic effect of $HgCl_2$, but the selenite decreases[7,8] and cadmium increases the uptake of Hg^{2+} by the kidneys.[9]

Both in synergism and antagonism the target organs of the two interacting toxicants can be identical or different as illustrated by the following examples.

1. Synergism

Identical targets — Both cadmium and lead can damage the testes, but their combined effect is more than the algebraic summation of individual effects.[10]

Different targets — Cysteine increases the renotoxicity of cadmium.[11]

2. Antagonism

Identical targets — Cadmium given in moderately renotoxic doses protects the kidneys against the renotoxic effect of $HgCl_2$ given a few days later.[9]

Different targets — Cysteine given with $CdCl_2$ protects the testes against the necrotizing effect of Cd^{2+}.[11]

C. Shift in Dose-Effects and Dose-Response Relationships

Synergistic and antagonistic effects can be measured by a shift in the dose-effect and dose-response relationships of toxic compounds. They will show that the same intensity of an effect or the same frequency of a selected effect is produced by smaller or larger doses.

According to Zielhuis[12] the dose-effect relationship shows the relationship between the dose and the intensity of a graded response in an individual. In a broader sense, dose-effect relationships include all responses (signs and symptoms) which occur one after the other when the dose is increased from the minimum toxic dose, to a highly toxic or lethal exposure level. Some of these responses can be graded on an ordinal or rank scale (e.g., ataxia); some can be measured on an interval scale (like ALA in urine). Due to individual differences in sensitivity to toxic metals or other toxic compounds, the establishment of synergistic or antagonistic interactions usually requires epidemiological studies or experimentation followed by statistical evaluation. This implies that not only individuals but also populations have their own dose-effect relationships. The intensity of a graded response of different individuals of the same population plotted against the log of the dose (or exposure) will give a regression line which represents the dose-effect relationship for this population. The confidence interval for a dose-effect relationship in a given population depends on the homogeneity of this population: the more homogeneous the population, the more narrow is the confidence interval.

Examples of dose-effect relationships were given by Goyer and Rhyne[13] for lead in rats, and for methylmercury in man by Bakir et al.[14] Thus 400 ppm lead in the drinking water given for 10 weeks caused, in rats, intranuclear inclusion bodies in the kidneys; 1200 ppm decreased weight gain; 4000 ppm caused aminoacidurea, renal oedema, reticulocytosis, and increased urinary ALA excretion; and 10,000 ppm caused anemia. In man exposed to methylmecrcury, the frequency of parestesia started to increase at 25 to 40 mg body burden of mercury, ataxia at 50 to 78 mg, and death at 200 to 312 mg body burden.

It is implicit in a dose-effect relationship constructed only from a graded response, e.g., from the urinary ALA-excretion in the case of lead exposure, that there is a definite relationship between the selected response and the clinical stages of intoxication. However, interaction can change this rule. For example, testicular damage caused by $CdCl_2$ can be diagnosed from the decrease in live sperm counts[15] and this graded response will indicate the severity of intoxication. When $CdCl_2$ is given with cysteine, instead of testicular damage, renal damage is caused[11] and the live sperm numbers show no correlation with the severity of intoxication.

A dose-response relationship is a relationship between dose and the frequency of individuals with a specified intensity of response,[12] like death, cancer, or urinary ALA-excretion above a certain level. Intercepts of several dose-response curves of the same toxic compound on the background frequency of responses give the threshold doses for these responses. When the response has zero frequency in an unexposed population, the intercept is naturally on the abscissa (log of the dose). Interaction might change the position of intercepts without changing their sequence, or as in the case of $CdCl_2$ when live sperm numbers in the semen and the number of casts in the urine are used in dose response curves, the sequence of intercepts is changed by cysteine.

D. Interaction between Toxic Metals and Essential Elements

Synergism and antagonism between two foreign compounds are not the only mechanisms which affects toxicity. Thus hypophysectomy diminished sublimate-induced nephrocalcinosis,[16] a low protein diet increased the toxicity of cadmium[17] and lead,[18] and interaction between toxic metals and essential elements might have a variety of consequences as reviewed by Levander:[19]

1. Deficiency of an essential element may increase toxicity, e.g., lead poisoning is aggravated by diets low in calcium and iron,[18] cadmium intoxication is aggravated by diets low in calcium[20] and copper.[21]
2. Excess intake of an essential element may decrease toxicity e.g., excess calcium decreases the toxicity of lead,[22] excess iron protects agains cadmium,[23] excess selenium against $HgCl_2$.[24]
3. Secondary deficiency may be caused by a toxic metal when the supply of the interacting essential element is just adequate, e.g., iron, zinc, and copper deficiency by Cd^{2+}.[19]

II. THE BIOCHEMICAL MECHANISMS OF INTERACTIONS

A. Introduction

The biochemical mechanism for synergistic and antagonistic effects is always some kind of interaction between two chemicals. The interaction either changes their distribution between target and nontarget organs, between critical and noncritical binding sites (pharmacokinetic interaction), or affects the ability of the organism to maintain organs, cells, and biochemical processes in a viable condition (pharmacodynamic interaction). In the case of metal-metal interactions the distinction between pharmacokinetic and pharmacodynamic interactions is not always possible, but all the available evidence suggests that change in kinetics is a sensitive indicator for the presencc of an interaction. Thus, selenium might protect against methylmercury by the stabilization of membranes[25] or mopping up free radicals,[26] but the pharmacokinetic effect of selenite on the distribution of methylmercury is very significant after their single dose administration.[27]

The possible mechanisms of metal-metal interactions, as reviewed by Magos and Webb,[28] can be extended to include the interactions of metals with nonmetallic compounds.

B. Interaction by Association

1. Chelation

The simplest form of interaction is association. Metal chelation therapy is based on the association of a metal with a chelator followed by increased excretion. However, in some cases chelation does not increase excretion and may increase the concentration of toxic metal in the sensitive organ. The effects of the following metal chelating agents are summarized below: 2,3-dimercaptopropanol (BAL), dimercaptosuccinic acid (DMSA), dimercaptopropane sulfonate (DMPS or unithiol), pencillamine, and calcium ethylene diamine tetra-acetate (EDTA).

Dimercaptopropanol — This is an effective antidote against trivalent arsenic, less effective against pentavalent arsenic, and ineffective against arsine.[29] Given shortly after the administration of Hg^{2+}, BAL protects the kidneys against nephrotoxic effects.[30,31] It protects against acute doses of antimony, chromium, bismuth, and nickel,[32] but not against lead, selenium,[32] or $MeHg^{+}$.[33] BAL actually increases the toxicity of lead and selenium.[32] The effect of BAL against $HgCl_2$ conforms with the decreased kidney concentration and increased urinary excretion of Hg^{2+}.[33,34] The increase in urinary excretion results in the depletion of Hg^{2+} from other organs, like the liver. The urinary excretion of arsenic,[29] lead,[31,35] and gold[36] are also increased. The kidney concentrations of gold[36] and cadmium[37] are increased by BAL, but the urinary excretion of cadmium is unaffected.[37] The brain uptake of methylmercury is accelerated by BAL.[38] BAL has no similar effect after a single dose of Hg^{2+},[29] but when the chelator was given with daily doses of mercury it increased the brain concentration of Hg^{2+}.[40]

Dimercaptosuccinic acid — This water soluble derivative of BAL is an effective antidote against both Hg^{2+} and $MeHg^{+}$ intoxication.[41,42] It increases the urinary elimination of mercury after $MeHg^{+}$ administration more than after that of Hg^{2+},[41] and decreases the concentration of mercury both in the kidneys and the brain[41-44] including the brains of neonatal pups of treated mothers.[45] DMSA given with lead prevents the accumulation of porphyrins in red blood cells,[46] increases urinary excretion, and decreases the concentration of lead in many tissues, including the kidneys and bone.[46,47]

Dimercaptopropane sulfonate (DMPS) — This water-soluble derivative of BAL has proved to be an effective antidote against acute cobalt and Hg^{2+} intoxication.[48,49] It increases the urinary excretion and decreases the kidney content of Hg^{2+} after the administration of $HgCl_2$.[50,51] After the administration of methylmercury, DMPS increases the urinary excretion, decreases the body burden, and the accumulation of $MeHg^{2+}$ in the brain, although it does not seem to increase the elimination of methylmercury from that organ.[52] On the molar bases it is less effective than DMSA in decreasing the body burden of mercury.[44]

D-Pencillamine — It is effective against the acute effects of $HgCl_2$[53,54] and proved to be useful in the treatment of workers exposed to lead.[55] Lead mobilization by D-penicillamine can increase temporarily the concentration of lead in soft tissues.[56] In animal experiments D-pencillamine decreased the excretion of ALA, without improving anemia.[57] D-pencillamine is able to increase the urinary excretion of lead[55-57] cadmium,[37] gold,[36] and $MeHg^{2+}$.[58] It shortened the biological half-time of Au^{2+} and Hg^{2+}[59] but the effect on Hg^{2+} is not high enough to cause an appreciable increase in urinary inorganic mercury excretion.[51,60,61]

However one of its derivatives, acetyl-D,L-penicillamine, was shown to mobilize mercury in human cases of sublimate[62] or chronic mercury vapor intoxications.[63,64] D-Penicillamine decreases the brain concentration of $MeHg^{2+}$[58] and increases the kidney content of cadmium.[37]

EDTA — EDTA is able to relieve the effects of acute and chronic lead intoxication[65] and it is an antidote against acute cadmium intoxication.[66] It is not effective against acute $HgCl_2$ or organomercurial intoxication.[33] EDTA is a nephrotic agent for older male rats,[67] and in some cases of lead intoxication EDTA therapy and lead have a synergistic effect on the renal tissue as shown by the development of nephrosis.[68-70] After prolonged exposure to cadmium it also potentiates the nephrotoxic effect of cadmium.[66] The urinary excretion of lead is increased by EDTA more than by unithiol or D-pencillamine;[57] it increases the urinary excretion of cadmium,[37,66] but not mercury after $HgCl_2$, phenylmercury, or methylmercury administration.[33] After a single dose, it prevents the accumulation of cadmium in the liver;[37,66] after prolonged administration the loss of cadmium by excretion is not sufficient to significantly alter organ concentrations.[66]

2. Interaction by the Association of Metals and Metalloids

Association between two metals or metalloids is the simplest form of metal-metal interaction. When 0.5 μmol/kg selenite per 200-g rat was injected simultaneously with an equimolar dose of Hg^{2+} or Cd^{2+} at the same subcutaneous injection site, the retention of ions increased only in the case of selenium and mercury.[71] As mercury selenite precipitates at a lower pH than cadmium selenite,[72] the probability of the formation of mercury selenite precipitate at the site of the injection is higher. It was suggested that the formation of a complex between lead and selenite ($PbSeO_3$) might be responsible for the protective effect of selenite against lead.[73] The association of selenium with mercury and cadmium is the main form of interaction only after the metabolic conversion of selenite to selenide. The biochemical mechanism of the formation of CdSe colloid was investigated and described in detail by Gasiewicz and Smith[74,75] and the process is most likely valid for HgSe formation. In rat plasma there is a protein component which binds selenium and cadmium or selenium and mercury in 1:1 atomic ratio.[74,75] In rats given Hg^{2+} and selenite black particles that contain mercury and selenium in a 1:1 ratio occur in macrophages and intranuclearly in the renal tubular cells.[76] Tellurite has a weaker effect than selenite on the distribution of Hg^{2+} but tellurium does not form particles with mercury, and if present, prevents the formation of intranuclear bodies at all.[76] The formation of mercury selenide supposes not only the formation of selenide, but also the concurrent presence of mercuric ion. After the administration of $MeHg^{2+}$ and selenite, the retention of mercury in the plasma is not increased, but decreased[27] and with in vitro experiments, blood in the presence of selenite is able to convert a large proportion of methylmercury to a form which can be extracted with benzene without acidification.[77] The extractable methylmercury cannot be attached to thiol compounds or have any charge. It cannot be MeHgCl either, as this form is incompatible with the presence of GSH and other thiol groups. This form has not been identified so far, but it might be bis-methylmercury selenide (MeHgSe-HgMe).

C. Competition for Carriers

Absorption of a metal from the site of administration depends on the available transport ligands. Thus mercurycysteinate is more rapidly absorbed from the subcutaneous injection site than $HgCl_2$[78] and the absorption of Hg^{2+} in proportion to the dose of $HgCl_2$ decreases with increasing doses.[79] Similar competetion for the available transfer ligands might explain why 0.5 μmol Zn^{2+} or Cd^{2+} slowed the absorption of 0.5 μmol Hg^{2+} from the subcutaneous injection site.[79]

Competition for transport ligands in the gastrointestinal tract or intestinal wall can influence the absorption of many metals. Thus, the proportion of zinc taken up from the lumen and transferred to the body was greater from lower than from higher doses, and the addition of cadmium or iron inhibited transport.[80] Zinc interferes with the absorption of iron, cadmium,[80] and copper.[81] Cadmium is able to disrupt the absorption of iron[82] and zinc[83] and iron shares transfer binding sites with cobalt[84] and manganese.[85]

Interaction between metals at the level of intestinal absorption is more complex than competition for a simple carrier as shown by the mutual interference between zinc and copper on their intestinal absorption.[86,87] It has been shown that the elution profile of orally administered zinc and copper is different, both in the intestinal lumen and in the epithelial cell supernatant, and on this basis Evans and Hahn[88] suggested that the elements may be transported from the lumen to the blood by two separate metal binding components. While the Zn^{2+} binding protein may be thionein,[89] the Cu^{2+} binding protein differs in amino acid composition from thionein and chelatin.[90] However, the synthesis of intestinal metallothionein is inversely related to the absorption of zinc[89] and the synthesis of the copper binding protein to the transport of copper.[90] Thus it is possible that though zinc and copper are bound to different binding proteins in the intestine, the same small molecular weight carriers compete with these two metal binding proteins for their respective metals.

In the blood both small molecular weight compounds and plasma proteins act as carriers. Thus albumin has a common binding site for Cu^{2+} and Zn^{2+}.[88] When the same chelator can transport two metals, the competition between the two metals depends on their relative affinities to physiological binding sites and to the chelator.

D. Interaction by Metabolic Interference

Metabolism of metals can involve different chemical transformations: change in the oxidation state, alkylation, or the cleavage of metal to carbon bonds. A less direct interference is change in the binding affinity or capacity of proteins for a metal or induction of metal binding proteins.

Atomic mercury is oxidized in red blood cells or organs to Hg^{2+} by catalase.[91] Catalase inhibitors, like ethanol[92,93] or aminotriazole[94] inhibit the oxidation of Hg^0 and by this mechanism decrease the retention and increase the exhalation of mercury. As a small proportion of Hg^{2+} is continuously reduced to Hg^0 and reoxidized to Hg^{2+} ethanol increases the loss of mercury even from Hg^{2+} deposits.[95,96] Ascorbic acid increases the retention of mercury at the subcutaneous injection site[78] probably by the conversion of mercuric mercury to mercurous mercury.

The effects of dietary selenite[97] on the distribution of Hg in rats exposed to mercury vapor show some similarities with the effects of aminotriazole:[94] decrease in lung and heart and increase in blood and liver mercury concentrations. As selenium increases the activity of GSH peroxidase,[99] the effect of selenite on the catalase mediated oxidation of Hg^0 to Hg^{2+} may be through competition with catalase for H_2O_2.

Many other metals change oxidation state in vivo. Selenate and selenite are converted to selenide[98,100] which is either bound to plasma proteins[100] or converted to volatile dimethylselenide[99] and/or water-soluble trimethylselenonium.[101] The simultaneous administration of Hg^{2+} or Cd^{2+} markedly decreases the respiratory excretion of dimethylselenide[102] and the toxicity of selenite.[103] The toxicity of selenium is also decreased by the simultaneous administration of arsenic,[104] copper,[103] lead,[73] and silver.[105] However the toxicity of selenite is increased when Hg^{2+},[8] or arsenic[106] are given 1 hr after selenite. The signs of intoxication are similar to those of dimethylselenide intoxication.[8] The toxicity of dimethylselenide is also increased by arsenic[106] and $HgCl_2$[8] though not by $CdCl_2$[8]. As male rats are more sensitive to dimethylselenide than females, and they also retain more selenium from this volatile selenide compound than females,[8] it is most likely that arsenic and Hg^{2+} increase the conversion of dimethylselenide to a highly toxic nonvolatile selenium species. However, the protective effect of Hg^{2+} and Cd^{2+} given before or with selenite administration is most likely caused by the inhibition of dimethylselenide formation.

Dietary selenite is able to increase the activity of the liver enzyme which cleaves the carbon to mercury bond of phenylmercury, but it has only a slight effect on the cleavage of ethylmercury and no effect at all on the cleavage of methylmercury.[107] Another effect of selenite is a reduction in the binding capacity of serum proteins to Hg^{2+} and phenylmercury with an increase in binding affinities.[108] Selenium given alternately with $HgCl_2$ eliminates the stimulation of renal metallothionein synthesis by Hg^{2+}.[109] Selenium pretreatment also diverts cadmium from kidney metallothionein to higher molecular weight proteins.[110] A similar shift from a protein of 30,000 molecular weight to a protein having a molecular weight of 115,000 is present in the testes of selenium pretreated rats. Probably this shift is responsible for the prevention of cadmium-induced testicular damage and might account for the significantly increased cadmium uptake by the testes.[110]

Metallothionein is a low molecular weight protein. Its characteristics according to Webb[111] are inducibility by Cd^{2+}, Zn^{2+}, Cu^{2+}, and Hg^{2+} and the ability to bind these cations and Ag^+ and Sn^{2+}. There are some reports that thionein in the liver can be induced besides the above-mentioned metals, by large doses of Pb^{2+}, Mn^{2+}, In^{3+}, Cr^{3+}, Sn^{4+},[112] Co^{2+}, Fe,$^{3+}$ and Bi^{3+}.[113] It is possible that some of these metals are not genuine thionein inducers but act through zinc mobilization. Nevertheless, even if one considers Cd^{2+}, Zn^{2+}, Cu^{2+}, and Hg^{2+} as thionein inducers, the role of metallothionein in metal-metal interactions is important.

Cadmium pretreatment protects against the renotoxic effect of Hg^{2+},[9] and increases in the kidneys the concentration of mercury bound to metallothionein.[114] This increase is caused partly because Hg^{2+} is able to replace some Cd^{2+} on metallothionein and partly because after Cd^{2+} pretreatment Hg^{2+} is able to induce metallothionein without the usual lag period.[114] However, the increase in Cd^{2+}-bound mercury is less than the increase in Hg^{2+} uptake by the whole kidney or by the kidney-soluble fraction, and therefore thionein alone cannot be responsible for the protective effect. Pretreatment with a low dose of Cd^{2+} gives protection against a subsequent lethal dose of the same cation. The protective effect is maximal 1 to 3 days after pretreatment, though the increased tissue thionein concentrations are maintained for a longer time.[115] The lack of correlation between preinduced thionein concentration and protection makes doubtful the dominant role of thionein in this protective mechanism. It might be that tolerance is linked at least partly to the morphological and biochemical characteristics of the regenerating tubular cells. These have less organelles and microvilli and incorporate necrotic cell debris by phagocytosis.[116]

Zinc also gives protection against cadmium toxicity. Cadmium replaces Zn^{2+} from the preinduced thionein and the synthesis of new thionein after the injection of Cd^{2+} starts without a lag phase.[117] Accumulation of cadmium is associated with increased zinc concentration in the liver and increased copper concentrations in the kidneys. The increase in the concentrations of zinc and copper is correlated with increases in liver and kidney metallothioneins induced by cadmium.[118] Iron, mainly from ferritin, is lost from the liver of cadmium-fed animals.[118] Simultaneous administration of high doses of Cd^{2+} and Cu^{2+} salts results in increase in the toxicity of Cd^{2+} accompanied by a failure of Cd sequestration.[119]

Competition of toxic metals with essential elements for intestinal transport can lead to mineral deficiencies.[19] Deficiency in calcium, zinc, and magnesium can depress the activities of microsomal mixed function oxidases.[120] It is possible that the depression of these enzymes results in an additional detrimental effect on the toxicity of some metals by the same mechanism which makes beneficial the effect of phenobarbitone, an inducer of these enzymes. Thus phenobarbitone increases the biliary excretion of methylmercury[121] and Cd^{2+}.[122] Phenobarbitone also protects against the nephrotoxic effect of methylmercury[123] and the lethal effect of a single dose of Cd^{2+}, Pb^{2+}, and Hg^{2+}.[124] Another inducer, spirinolactone, protects against $HgCl_2$[125] and methoxyethylmercury.[126] When given before the mercury compound, it increases the urinary excretion of Hg^{2+},[127] $MeHg^{2+}$,[128] and copper.[129]

The biliary excretion of $MeHg^{2+}$ can be increased temporarily by cysteine and GSH[130] and decreased by diethylmaleate, a depleter of liver GSH.[131] However there is no straightforward relationship between free GSH or cysteine concentrations in blood or liver, liver $MeHg^{2+}$ concentration, and biliary excretion.[130] It was suggested that not free GSH, but ligandin-bound GSH might be responsible for the transport of $MeHg^{2+}$ through the hepatocyte to the site of secretion.[130] This hypothesis is in agreement with the effects of phenobarbitone which is an inducer of ligandin[132] and increases the biliary excretion of methylmercury. Sodium dehydrocholate, which increases bile flow more than phenobarbitone, has no such effect.[133] The role of biliary excretion is an important fact in the antagonism of arsenic against selenium. Arsenic markedly decreases the retention of selenium in the liver[134] and increases the biliary excretion of selenium.[135]

The identification of the essential mechanism in metabolic interactions between metals or a metal and a nonmetallic compound is in an exploratory stage like research on other factors which influence toxicity. Some of the more important factors within a species are sex and age. Thus the susceptibility of the heme biosynthetic pathway to lead, as reflected by increased free erythrocyte porphyrin, is in the order children >women>men.[136] The acute oral toxicities of many metals shows a low sensitivity in rats between 3 and 6 weeks of age.[137] The immune status of the individual is an additional variation in the metabolism and toxicity of beryllium, chromium, gold, mercury, nickel, and platinum. Genetically controlled immune responses have been demonstrated with mercury, beryllium, and chromium.[138]

III. CONCLUSIONS

The shift in the dose-effect and dose-response relationships caused by interacting compounds has very definite practical consequences: the harmfulness of exposure to a toxic metal depends not only on its dose, but on the presence of synergistic or antagonistic compounds, which can be metals, metalloids, or organic. Synergism or antagonism sometimes depends on the simultaneous uptake of two interacting compounds; in other cases interaction requires a definite sequence of exposure. Change in the sequence might result in an increase in the toxicity of one metal and a decrease in the toxicity of the other. However, in every case the threshold limit or threshold exposure necessary to produce a harmful effect is altered. This means that the combined exposure to a toxic metal with an interacting compound will increase or decrease the safety margin compared with the condition of exposure to the toxic metal only.

Another practical problem is the minimal requirement for the interacting compound to exert a synergistic or antagonistic effect on a toxic metal. When an essential element competes with a toxic metal for intestinal absorption, the absorption of the toxic metal is continuously decreased when the supply of the essential element is increased from deficiency to excess. However, postabsorptive interaction might require the essential element to be supplied above a certain threshold dose. Certainly synthesis of metal binding proteins (thus the induction of renal metallothionein synthesis) requires a threshold dose or a renal threshold concentration of Cd^{2+}.[139]

The synergistic and antagonistic effects of one metal on the toxicity of another metal become modified by the presence of other elements transported by the same system or are bound by the same metal binding protein. This multiple interaction means a shift in the basic synergistic or antagonistic response.

REFERENCES

1. **Magos, L.**, Problems of simultaneous exposure to two or more foreign compounds, in *Comparative Studies of Food and Environmental Contamination,* International Atomic Energy Agency, Proceedings Series, Vienna, 1974, 505.
2. **Magos L.**, The role of synergism and antagonism in the toxicology of metals, in *Effects and Dose-Response Relationships of Toxic Metals,* Nordberg, G., Ed., Elsevier, Amsterdam, 1976, 491.
3. **Koeman, J. H., Ven, W. S. M., Goeij, J. J. M., Tjioe, P. S., and Haaften, J. L.**, Mercury and selenium in marine mammals and birds, *Sci. Total Environ.,* 3, 279, 1975.
4. **Kosta, L., Bryne, A. R., and Zelenko, V.**, Correlation between selenium and mercury in man following exposure to inorganic mercury, *Nature (London),* 254, 238, 1975.
5. **Ganther, H. E., Goudi, C., Sunde, M. L., Kopecky, M. J., Wagner, P., Oh, S. -H., and Hoekstra, W. G.**, Selenium: relation to decreased toxicity of methylmercury added to diets containing tuna, *Science,* 175, 1122, 1972.

6. Task Group on Metal Interaction, Factors influencing metabolism and toxicity of metals: a concensus report, *Environ. Health Perspect.*, 25, 3, 1978.
7. **Eybl, V., Sykora, J., and Mertl, F.**, Einfluss von Natriumselenit, Natriumtellurit and Natriumsulfit auf Retention und Verteilung von Quecksilber bei Mäusen, *Arch. Toxikol.*, 25, 196, 1969.
8. **Parizek, J., Ostadalova, I., Kalouskova, J., Babicky, A., and Benes, J.**, The detoxifying effects of selenium interrelations between compounds of selenium and certain metals, in *Newer Trace Elements in Nutrition*, Mertz, W. and Cornatzer, W. E., Eds., Marcel Dekker, New York, 1971, 85.
9. **Magos, L., Webb, M., and Butler, W. H.**, The effect of cadmium pretreatment on the nephrotoxic action and kidney uptake of mercury in male and female rats, *Br. J. Exp. Pathol.*, 55, 589, 1974.
10. **Der, R., Fahim, Z., Yousef, M., and Fahim, M.**, Environmental interaction of lead and cadmium on reproduction and metabolism of male rats, *Res. Commun. Chem. Pathol. Pharmacol.*, 14, 689, 1976.
11. **Gunn, S. A., Gould, T. C., and Anderson, W. A. D.**, Selectivity of organ response to cadmium injury and various protective measures, *J. Pathol. Bacteriol.*, 96, 89, 1968.
12. **Zielhuis, R. L.**, Dose-response relationships for inorganic lead. I. Biochemical and haematological responses, *Int. Arch. Occup. Health*, 35, 1, 1975.
13. **Goyer, R. A. and Rhyne, B. C.**, Pathological effects of lead, *Int. Rev. Exp. Pathol.*, 12, 1, 1973.
14. **Bakir, F., Damluji, S. F., Amin-Zaki, L., Murtadha, M., Khalidi, A., Al Rawi, N. Y., Tikriti, S., Dhahir, H. I., Clarkson, T. W., Smith, J. C., and Doherty. R. A.**, Methylmercury poisoning in Iraq, *Science*, 181, 230, 1973.
15. **Paufler, S. K. and Foote, R. H.**, Effect of triethylenemelamine (TEM) and cadmium chloride on spermatogenesis, *J. Reprod. Fertil.*, 19, 309, 1969.
16. **Szabo, S. and Selye, H.**, Inhibition of hypophysectomy of nephrocalcinosis produced by mercuric chloride, *Urol. Int.*, 26, 39, 1971.
17. **Fitzhugh, O. G. and Meiller, F. H.**, The chronic toxicity of cadmium, *J. Pharmacol. Exp. Ther.*, 72, 15, 1941.
18. **Goyer, R. A. and Mahaffey, K. R.**, Susceptibility to lead toxicity, *Environ. Health Perspect.*, 2, 73, 1972.
19. **Levander, O. A.**, Nutritional factors in relation to heavy metal toxicants, *Fed. Proc. Fed. Am. Soc. Exp. Biol.*, 36, 1683, 1977.
20. **Larsson, S. E. and Piscator, M.**, Effects of cadmium on skeletal tissue in normal and calcium-deficient rats, *Isr. J. Med. Sci.*, 7, 495, 1971.
21. **Hill, C. H., Matrone, G., Payne, W. L., and Barber, C. W.**, In vivo interactions of cadmium with copper, zinc and iron, *J. Nutr.*, 80, 227, 1963.
22. **Hsu, F. S., Krook, L., Pond, W. G., and Duncan, J. R.**, Interactions of dietary cadmium with toxic levels of lead and zinc in pigs, *J. Nutr.* 105, 112, 1975.
23. **Banis, R. J., Pond, W. G., Walker, E. F., and O'Connor, J. R.**, Dietary cadmium, iron and zinc interactions in the growing rat, *Proc. Soc. Exp. Biol. Med.*, 130, 802, 1969.
24. **Parizek, J. and Ostadalova, I.**, The protective effect of small amounts of selenite in sublimate intoxication, *Experientia*, 23, 142, 1967.
25. **Kasuya M.**, Effect of selenium on the toxicity of methylmercury on nervous tissue culture, *Toxicol. Appl. Pharmacol.*, 35, 11, 1976.
26. **Ganther, H. E.**, Modification of methylmercury toxicity and metabolism by selenium and vitamin E: possible mechanisms, *Environ. Health Perspect.*, 25, 71, 1978.
27. **Magos, L. and Webb, M.**, Effect of selenium on the brain uptake of methylmercury , *Arch. Toxicol.*, 38, 201, 1977.
28. **Magos, L. and Webb, M.**, Theoretical and practical considerations on the problem of metal-metal interactions, *Environ. Health Perspect.*, 25, 151, 1978.
29. **Brugsch, H. G.**, Recent advances in chelation therapy, *J. Occup. Med.*, 7, 394, 1965.
30. **Gilman, A., Allen, R. P., Philips, F. S., and St. John, E.**, Clinical use of 2,3-dimercaptopropanol (BAL) X. The treatment of acute systemic mercury poisoning in experimental animals with BAL, thiosorbitol and BAL glucoside, *J. Clin. Invest.*, 25, 549, 1946.
31. **Longcope, W. T. and Luetscher, J. A., Jr.**, The use of BAL (British Anti-Lewisite) in the treatment of the injurious effects of arsenic, mercury and other metallic poisons, *Ann. Intern. Med.*, 31, 545, 1949.
32. **Braun, H., Lusky, L. M., and Calvery, H. O.**, The efficacy of 2,3-dimercaptopropanol (BAL) in the therapy of poisoning by compounds of antimony, bismuth, chromium, mercury and nickel, *J. Pharmacol. Exp. Ther.*, 87, 119, 1946.
33. **Swensson, A. and Ulfvarson, U.**, Experiments with different antidotes in acute poisoning by different mercury compounds. Effects on survival and on distribution and excretion of mercury, *Int. Arch. Gewerbepathol. Gewerbehyg.*, 24, 12, 1967.

34. **Adam, K. R.**, The effects of dithiols on the distribution of mercury in rabbits, *Br. J. Pharmacol.*, 6, 483, 1951.
35. **Aronson, A. L. and Hammond, P. B.**, Effect of two chelating agents on the distribution and excretion of lead, *J. Pharmacol. Exp. Ther.*, 146, 241, 1964.
36. **Rubin, M., Sliwinski, A., Photias, M., Feldman, M. and Zuaifler, N.**, Influence of chelation on gold metabolism in rats, *Proc. Soc. Exp. Biol. Med.*, 124, 290, 1967.
37. **Dequidt, J., Haguenoer, J.-M., and Fromont, B.**, Action de quelques detoxicants dans l'intoxication experimentale per le cadmium, *Arch. Mal. Prof. Med. Trav. Secur. Soc.* 34, 427, 1973.
38. **Berlin, M., Jerksell, L. G., and Nordberg, G.**, Accelerated uptake of mercury by brain caused by 2,3-dimercaptopropanol (BAL) after injection into the mouse of a methylmercuric compound, *Acta Pharmacol. Toxicol.*, 23, 312, 1965.
39. **Magos, L.**, Effect of 2,3-dimercaptopropanol (BAL) on urinary excretion and brain content of mercury, *Br. J. Ind. Med.*, 25, 152, 1968.
40. **Berlin, M. and Lewander, T.**, Increased brain uptake of mercury caused by 2,3-dimercaptopropanol (BAL) in mice given mercuric chloride, *Acta Pharmacol. Toxicol.*, 22, 1, 1965.
41. **Magos, L.**, The effects of dimercaptosuccinic acid on the excretion and distribution of mercury in rats and mice treated with mercuric chloride and methylmercury chloride, *Br. J. Pharmacol.*, 56, 479, 1976.
42. **Magos, L., Peristianis, G. C., and Snowden, R. T.**, Postexposure preventive treatment of methylmercury intoxication in rats with dimercaptosuccinic acid, *Toxicol. Appl. Pharmacol.*, 45, 463, 1978.
43. **Friedheim, E. and Corvi, C.**, Meso-dimercaptosuccinic acid, a chelating agent for the treatment of mercury poisoning, *J. Pharm. Pharmacol.*, 27, 624, 1975.
44. **Aaseth, J. and Friedheim, E. A.**, Treatment of methylmercury poisoning in mice with 2,3-dimercaptosuccinic acid and other complexing thiols, *Acta Pharmacol. Toxicol.*, 42, 248, 1978.
45. **Hughes, J. A. and Sparber, S. B.**, Reduction of methylmercury concentration in neonatal rat brains after administration of dimercaptosuccinic acid to dams while pregnant, *Res. Commun. Chem. Pathol. Pharmacol.*, 22, 357, 1978.
46. **Graziano, J. H., Leong, J. K., and Friedheim, E.**, 2,3-dimercaptosuccinic acid: a new agent for the treatment of lead poisoning, *J. Pharmacol. Exp. Ther.*, 206, 696, 1978.
47. **Friedheim, E., Corvi, C., and Wakker, C. H.**, Meso-dimercaptosuccinic acid a chelating agent for the treatment of mercury and lead poisoning, *J. Pharm. Pharmacol.*, 28, 711, 1976.
48. **Cherkes, A. I. and Braver-Chernobulskaia, B. S.**, Unithiol — a cobalt antidote, *Pharmacol. Toxicol.*, 21, No. 3, 264, 1959.
49. **Belonozshko, G. A.**, Unithiol therapy of inorganic mercury poisoning, *Pharmacol. Toxicol.*, 21 (No. 3), 280, 1959.
50. **Dutkiewicz, T. and Oginski, M.**, Dislokation und Ausscheidung des Quecksilbers bei Ratten nach Applikation von Unitiol, *Int. Arch. Gewerbepathol. Gewerbehyg.*, 28, 201, 1967.
51. **Gabard, B.**, The excretion and distribution of inorganic mercury in the rat as influenced by several chelating agents, *Arch. Toxicol.*, 35, 15, 1976.
52. **Gabard, B.**, Treatment of methylmercury poisoning in the rat with sodium 2,3-dimercaptopropanol-sulfonate: influence of dose and mode of administration, *Toxicol. Appl. Pharmacol.*, 38, 415, 1976.
53. **Aposhian, H. V.**, Protection by D-penicillamine against the lethal effects of mercuric chloride, *Science*, 128, 93, 1958.
54. **Aposhian, H. V. and Aposhian, M. M.**, N-acetyl-DL-penicillamine, a new oral protective agent against the lethal effects of mercuric chloride, *J. Pharmacol. Exp. Ther.*, 126, 131, 1959.
55. **Cramer, K. and Selander, S.**, Penicillamine in lead poisoning, *Postgrad. Med. J.*, 45 (Suppl. Penicillamine), 45, 1968.
56. **Hammond, P. B.**, The effects of D-penicillamine on the tissue distribution and excretion of lead, *Toxicol. Appl. Pharmacol.*, 26, 241, 1973.
57. **Hofmann, U. and Segewitz, G.**, Influence of chelation therapy on acute lead intoxication in rats, *Arch. Toxicol.*, 34, 213, 1975.
58. **Magos, L. and Clarkson, T. W.**, The effect of oral doses of a polythiol resin on the excretion of methlymercury in mice treated with cysteine, D-penicillamine or phenobarbitone, *Chem. Biol. Interact.*, 14, 325, 1976.
59. **Silva A. J., Fleishman, D. G., and Shore, B.**, The effects of pencillamine on the body burdens of several heavy metals, *Health Phys.*, 24, 535, 1973.
60. **Teisinger, J. and Srbova, J.**, Vliv D-penicilaminie na vylucovani stuti a clova v moci (effect of D-penicillamine on the urinary excretion of mercury and lead), *Prackt. Lek.*, 16, 433, 1964.
61. **Magos, L. and Stoytchev, T.**, Combined effect of sodium maleate and some thiol compounds on mercury excretion and redistribution in rats, *Br. J. Pharmacol.*, 35, 121, 1969.
62. **Aronow, R. and Fleischmann, L. E.**, Mercury poisoning in children. The value of N-acetyl, D, L-pencillamine in combined therapeutic approach, *Clin. Pediatr. Bologna*, 15, 936, 1976.

63. **Kark, R. A., Poskanzer, D. C., Bullock, J. D., and Boylen, G.**, Mercury poisoning and its treatment with N-acetyl-D, L-pencillamine, *N. Engl. J. Med.*, 285, 11, 1971.
64. **Clarkson, T. W.**, Recent advances in the toxicology of mercury with emphasis on the alkymercurials, *CRC Crit. Rev. Toxicol.*, 1, 203, 1972.
65. **Chenoweth, M. B.**, Clinical use of metal-binding drugs, *Clin. Pharmacol. Ther.*, 9, 365, 1968.
66. **Friberg, L.**, Edathamil calcium-disodium in cadmium poisoning, *Arch. Ind. Health*, 13, 18, 1956.
67. **Reuber, M. D.**, Severe nephrosis in older male rats given calcium disodium edetate, *Arch. Environ. Health*, 15, 141, 1967.
68. **Moeschlin, S.**, Zur Klinik und Therapie der Bleivergiftunge mit Bericht über eine todliche toxische Nephrose durch Ca EDTA (Calcium Versenat), *Schweiz. Med. Wochenschr.*, 87, 1091, 1957.
69. **Reuber, M. D. and Bradley, J. E.**, Acute versenate nephrosis: occurring as a result of treatment for lead intoxication, *JAMA*, 174, 263, 1960.
70. **Yver, L., Marechaud, R., Picaud, D., Touchard, G., Talin d'Eyzac, A., Matuchansky, C., and Patte, D.**, Insuffisance renale aigue au cours d'un saturnisme professionnel. Responsibilité du traitment chelateur, *Nouv. Presse Med.* 7, 1541, 1978.
71. **Magos, L. and Webb, M.**, Differences in distribution and excretion of selenium and cadmium or mercury after their simultaneous administration subcutaneously in equimolar doses, *Arch. Toxicol.*, 36, 63, 1976.
72. **Neville, G. H.**, Selenium, in *Thorpe's Dictionary of Applied Chemistry*, Vol. 10, 4th ed., Thorpe, J. F. and Whitley, M. A., Eds., Longmans, Green, London, 1950, 23.
73. **Rastogi, S. C., Clausen, J., and Srivastava, K. C.**, Selenium and lead: mutual detoxifying effects., *Toxicology*, 6, 377, 1976.
74. **Gasiewicz, T. A. and Smith, J. C.**, Interactions of cadmium and selenium in rat plasma in vivo and in vitro, *Biochim. Biophys. Acta*, 428, 113, 1976.
75. **Gasiewicz, T. A. and Smith, J. C.**, Properties of the cadmium and selenium complex formed in rat plasma in vivo and in vitro, *Chem. Biol. Interact.*, 23, 171, 1978.
76. **Groth, D. H., Stettler, L., and Mackay, G.**, Interactions of mercury, cadmium, tellurium, arsenic and beryllium, in *Effects of Dose-Response Relationships of Toxic Metals*, Nordberg, G. F., Ed., Elsevier, Amsterdam, 1976, 527.
77. **Sumino, K., Yamamoto, R., and Kitamura, S.**, A role of selenium against methylmercury toxicity, *Nature (London)*, 268, 73, 1977.
78. **Magos, L.**, Factors affecting the uptake and retention of mercury by kidneys in rats, in *Mercury, Mercurials and Mercaptans*, Miller, M. W. and Clarkson, T. W., Eds., Charles C Thomas, Springfield, Ill., 1973, 167.
79. **Magos, L. and Webb, M.**, The interaction between cadmium, mercury and zinc administered subcutaneously in a single injection, *Arch. Toxicol.*, 36, 53, 1976.
80. **Hamilton, D. L., Bellamy, J. E. C., Valberg, J. D., and Valberg, L. S.**, Zinc, cadmium, and iron interactions during intestinal absorption in iron-deficient mice, *Can. J. Physiol. Pharmacol.*, 56, 384, 1978.
81. **Hahn, C. J. and Evans, G. W.**, Absorption of trace metals in the zinc deficient rat, *Am. J. Physiol.*, 228, 1020, 1975.
82. **Hamilton, D. L. and Valberg, L. S.**, Relationship between cadmium and iron absorption, *Am. J. Physiol.*, 227, 1033, 1974.
83. **Evans, G. W., Grace, C. I., and Hahns, C. H.**, The effect of copper and cadmium on ^{65}Zn absorption in zinc deficient and zinc supplemented rats, *Bioinorg. Chem.*, 3, 115, 1974.
84. **Thompson, A. B. R., Valberg, L. S., and Sinclair, D. G.**, Competitive nature of the intestinal transport mechanism for cobalt and iron in the rat, *J. Clin. Invest.*, 50, 2384, 1971.
85. **Thompson, A. B. R., Olatanbosun, D., Valberg, L. S., and Ludwig, J.**, Inter-relationship of intestinal transport system for manganese and iron, *J. Lab. Clin. Med.*, 78, 642, 1971.
86. **Van Campen, D. R. and Scaife, P. V.**, Zinc interference with copper absorption in rats, *J. Nutr.*, 91, 473, 1967.
87. **Van Campen, D. R.**, Copper interference with the intestinal absorption of zinc-65 by rats, *J. Nutr.*, 97, 104, 1970.
88. **Evans, G. W. and Hahn, C. J.**, Copper- and zinc-binding component in rat intestine, *Adv. Exp. Med. Biol.*, 48, 285, 1974.
89. **Richards, M. P. and Cousins, R. J.**, Mammalian zinc homeostasis: requirement for RNA and metallothionein synthesis, *Biochem. Biophys. Res. Commun.*, 64, 1215, 1975.
90. **Evans, G. W. and LeBlanc, F. N.**, Copper-binding protein in rat intestine: amino acid composition and function, *Nutr. Rep. Int.* 14, 281, 1976.
91. **Magos, L., Halbach, S., and Clarkson, T. W.**, Role of catalase in the oxidation of mercury vapor, *Biochem. Pharmacol.*, 27, 1373, 1978.

92. **Nielsen Kudsk, F.**, The influence of ethyl alcohol on the absorption of mercury vapour from lungs in man, *Acta Pharmacol. Toxicol.*, 23, 263, 1965.
93. **Magos L., Clarkson, T. W., and Greenwood, M. R.**, The depression of pulmonary retention of mercury vapor by ethanol: identification of the site of action, *Toxicol. Appl. Pharmacol.*, 26, 180, 1973.
94. **Magos, L., Sugata, Y., and Clarkson, T. W.**, Effects of 3-amino-1,2,4-triazole on mercury uptake by in vitro human blood samples and by whole rats, *Toxicol. Appl. Pharmacol.*, 28, 367, 1974.
95. **Hursh, J. B., Clarkson, T. W., Cherian, T. W., Vostal, J. J., and Vander Mallie, R.**, Clearance of mercury (Hg-197, Hg-203) vapor inhaled by human subjects, *Arch. Environ. Health*, 31, 302, 1976.
96. **Dunn, J. D., Clarkson, T. W., and Magos, L.**, Ethanol-increased exhalation of mercury in mice, *Br. J. Ind. Med.*, 35, 241, 1978.
97. **Nygaard, S.-P. and Hansen, J. C.**, Mercury selenium interaction at contentrations of selenium and of mercury vapours as prevalent in nature, *Bull. Environ. Contam. Toxicol.*, 20, 20, 1978.
98. **Hoekstra, W. G.**, Biochemical function of selenium and its relation to vitamin E, *Fed. Proc. Fed. Am. Soc. Exp. Biol.*, 34, 2083, 1975.
99. **Ganther, H. E. and Hsieh, H. S.**, Mechanism for the conversion of selenite to selenides in mammalian tissues, in *Trace Element Metabolism in Animals*, Vol. 2, Hoekstra, W. G., Suttie, J. W., Ganther, H. E., and Mertz., H. E., Eds., University Park Press, Baltimore, 1974, 339.
100. **Gasiewicz, T. A. and Smith, J. C.**, The metabolism of selenite by intact erythrocytes in vitro, *Chem. Biol. Interact.*, 21, 299, 1978.
101. **Palmer, I. S., Gansalus, R. P., Halverson, A. W., and Olson, O. E.**, Trimethylselenonium ion as a general excretory product from selenium metabolism in the rat, *Biochim. Biophys. Acta*, 208, 260, 1970.
102. **Parizek, J., Benes, I., Babicky, A., Benes, J., Prochazkova, V., and Lener, J.**, Metabolic interrelations of trace elements, the effect of mercury and cadmium on the respiratory excretion of volatile selenium compounds, *Physiol. Bohemoslov.*, 18, 105, 1969.
103. **Hill, C. H.**, Reversal of selenium toxicity in chicks by mercury, copper and cadmium, *J. Nutr.*, 104, 593, 1974.
104. **Moxon, A. L.**, The effect of arsenic on the toxicity of seleniferous grain, *Science*, 88, 81, 1938.
105. **Jensen, L. S.**, Modification of a selenium toxicity in chicks by dietary silver and copper, *J. Nutr.*, 105, 769, 1975.
106. **Palmer, I. S. and Halverson, A. V.**, Potentiation of acute selenium toxicity by arsenic, *Fed. Proc. Fed. Am. Soc. Exp. Biol.*, 33, 694, 1974.
107. **Fang, S. C.**, Induction of C-Hg cleavage enzymes in rat liver by dietary selenite, *Res. Commun. Pathol. Pharmacol.*, 9, 579, 1974.
108. **Fang, S. C., Chen, R. W., and Fallin, E.**, Influence of dietary selenite on the binding characteristics of rat serum proteins to mercurial compounds, *Chem. Biol. Interact.*, 15, 51, 1976.
109. **Chmielnicka, J. and Brzeznicka, E. A.**, The influence of selenium on the level of mercury and metallothionein in prolonged exposure to different mercury compounds, *Bull. Environ. Contam. Toxicol.*, 19, 183, 1978.
110. **Chen, R. W., Whanger, P. D., and Weswig, P. H.**, Selenium-induced redistribution of cadmium binding to tissue proteins: a possible mechanism of protection against cadmium toxicity , *Bioinorg. Chem.*, 4, 125, 1975.
111. **Webb, M.**, The metallothioneins, *Biochem. Soc. Trans.*, 3, 631, 1974.
112. **Suzuki, Y. and Yoshikawa, H.**, Induction of hepatic zinc-binding proteins of rats by various metals, *Ind. Health*, 14, 25, 1976.
113. **Piotrowski, J. K. and Szymanska, A.**, Influence of certain metals on the level of metallothionein-like proteins in the liver and kidneys of rats, *J. Toxicol. Environ. Health*, 1, 991, 1976.
114. **Webb, M. and Magos, L.**, Cadmium-thionein and the protection by cadmium against the nephrotoxicity of mercury, *Chem. Biol. Interact.*, 14, 357, 1976.
115. **Webb, M. and Verschoyle, R. D.**, An investigation of the role of metallothioneins in protection against the acute toxicity of the cadmium ion, *Biochem. Pharmacol.*, 25, 673, 1976.
116. **Siegel, F. L. and Bulger, R. E.**, Scanning and transmission EM of rat kidney following low dose of mercuric chloride administration, *Beitr. Pathol.*, 156, 313, 1975.
117. **Webb, M.**, Protection by zinc against cadmium toxicity, *Biochem. Pharmacol.*, 21, 2767, 1972.
118. **Stonard, M. D. and Webb, M.**, Influence of dietary cadmium on the distribution of the essential metals copper, zinc and iron in tissues of the rat, *Chem. Biol. Interact.*, 15, 349, 1976.
119. **Irons, R. D. and Smith, J. C.**, Prevention by copper of cadmium sequestration by metallothionein in liver, *Chem. Biol. Interact.*, 15, 289, 1976.
120. **Basu, K. B. and Dickerson, J. W. T.**, Inter-relationships of nutrition and the metabolism of drugs, *Chem. Biol. Interact.*, 8, 193, 1974.

121. **Magos, L. and Clarkson, T. W.**, Effect of phenobarbitone on the biliary excretion of methylmercury in rats and mice, *Nature (London)*, 246, 123, 1973.
122. **Ohsawa, M. and Fukuda, K.**, Enhancement by phenobarbital of the biliary excretion of methylmercury and cadmium in rats, *Ind. Health*, 14, 7, 1976.
123. **Fowler, B. A., Lucier, G. W., Folsom, M. D., and Brown, H. W.**, Phenobarbital protection against methylmercury nephrotoxicity, *Environ. Health Perspect.*, 4, 100, 1973.
124. **Yoshikawa, H. and Ohsawa, M.**, Protective effect of phenobarbital on cadmium toxicity in mice, *Toxicol. Appl. Pharmacol.*, 34, 517, 1975.
125. **Selye, H.**, Mercury poisoning: prevention by spirinolactone, *Science*, 169, 775, 1970.
126. **Lehotzky, K.**, Protection by spirinolactone and different antidotes against acute organic mercury poisoning of rats, *Int. Arch. Arbeitsmed.*, 33, 329, 1974.
127. **Kitani, K., Morita, Y., and Kanai, S.**, The effects of spirinolactone on the biliary excretion of mercury, cadmium, zinc and cerium in rats, *Biochem. Pharmacol.*, 26, 279, 1977.
128. **Klaassen, C. D.**, Biliary excretion of mercury compounds, *Toxicol. Appl. Pharmacol.*, 33, 356, 1975.
129. **Haddow, J. E. and Lester, R.**, Biliary copper excretion in the rat is enhanced by spirinolactone, *Drug Metab. Dispos.*, 4, 499, 1976.
130. **Magos, L., Clarkson T. W., and Allen, J.**, The inter-relationship between non-protein bound thiols and the biliary excretion of methylmercury, *Biochem. Pharmacol.*, 27, 2203, 1978.
131. **Refsik, T.**, Excretion of methylmercury in rat bile: the effect of diethylmaleate, cyclohexene oxide and acrylamide, *Acta Pharmacol. Toxicol.*, 42, 135, 1978.
132. **Arias, I. M., Fleischner, G., Kirsch, R., Mishkin, S., and Gatmaitan, Z.**, On the structure, regulation, and function of ligandin, in *Glutathione: Metabolism and Function*, Arias, I. M. and Jakoby, W. B., Eds., Raven Press, New York, 1976.
133. **Magos, L., McGregor, J. T., and Clarkson, T. W.**, The effect of phenobarbital and sodium dehydrocholate on the biliary excretion of methylmercury in the rat, *Toxicol. Appl. Pharmacol.*, 30, 1, 1974.
134. **Ganther, H. E. and Baumann, C. A.**, Selenium metabolism. I. Effects of diet, arsenic and cadmium, *J. Nutr.*, 77, 2210, 1962.
135. **Levander, O. A.**, Metabolic inter-relationships between arsenic and selenium, *Environ. Health Perspect.*, 1977.
136. **Rocls, H. A., Buchet, J.-P., Bernadr, A., Hubermont, G., Lauwerys, R. R., and Masson, P.**, Investigations of factors influencing exposure and response to lead, mercury and cadmium in man and in animals, *Environ. Health Perspect.*, 25, 91, 1978.
137. **Kostial, K., Kello, D., Jugo, S., Rabar, I., and Maljkovic, T.**, Influence of age on metal metabolism and toxicity, *Environ. Health Perspect.* 25, 81, 1978.
138. **Kazantzis, G.**, The role of hypersensitivity and the immune response in influencing susceptibility to metal toxicity, *Environ. Health Perspect.* 25, 111, 1978.
139. **Cempel, M. and Webb, M.**, The time-course of cadmium-thionein synthesis in the rat, *Biochem. Pharmacol.*, 25, 2067, 1976.

Chapter 6

HAIR AS A BIOPSY MATERIAL

Robert A. Jacob

TABLE OF CONTENTS

I. INTRODUCTION

Metallic ions are released systemically and locally from implanted medical devices as covered in detail in other parts of this series. Determinations of metal in hair have been useful to nutritionists and toxicologists in establishing body deficiencies or excesses of a variety of trace metals. Since most of the metals comprising implant devices are normally present in body compartments in only trace or ultratrace amounts, hair metal determinations might be useful for monitoring excess body burdens of these metals. This application has been suggested by Owen et al.[1] who determined hair chromium in patients with stainless steel hip joint implants and found no evidence for systemic chromium accumulation. This review will cover the utility of hair as a biopsy material for assessing body burden of trace metals, with particular emphasis on those metals commonly found in implant materials, i.e., aluminum, chromium, cobalt, copper, manganese, mercury, molybdenum, nickel, silver, titanium, and vanadium.

The use of hair as a specimen for diagnosing mineral status has distinct advantages over specimens such as blood, urine, or tissue. Hair is collected easily and without trauma. It is stable over long periods of time without refrigeration (as evidenced by Swift's inferences about Jane Austens habits from microscopic analysis of her approximately 150 year old lock of hair[2] and mineral analysis of ancient Peruvian hair).[3] Once minerals are incorporated into the hair shaft during synthesis in the follicle, they are immune to factors known to affect metal levels in blood and urine such as circadian, postprandial, infectious, and hormonal states.

The longitudinal profile of metals along the shaft can provide a history of metal body burden or environmental exposure. The concentrations of trace metals in hair are usually more than ten times as great as those in blood or urine samples. This is particularly important for elements such as chromium that appear in blood and urine at ultratrace or part per billion (ppb) levels. The much higher levels of about 1 part per million (ppm) of chromium in hair eliminate problems of analytical sensitivity and contamination that are associated with ultratrace analysis. Analytically, assessment of chromium status is much more feasible via hair than via blood or urine samples.

Hair as a biopsy material also has some unique problems, chief of which is its susceptibility to environmental contamination. The metal in hair can have an endogenous source (through the follicle via ingestion, inhalation, implantation) or an exogenous source (direct contamination of hair filaments via air, dust, cosmetic treatments, and so on). Nutritionists and toxicologists who wish to assess body metal burden are interested in measuring endogenous hair metal, whereas environmentalists may be more interested in exogenous hair metal, as a measure of environmental contamination.[4] Forensic scientists have studied the profile of trace metals along hair shafts as possible fingerprints for identification. For this purpose, the reproducibility of the trace metal pattern in hair is more important than the origins of the hair metals. Hence, washing procedures, developed to remove oils and exogenous metal contamination on hair surfaces but to leave most if not all of the endogenous metal, are important in the use of hair metal as a measure of body metal status. This includes use of hair metal levels as a measure of systemic metal ion toxicity from implanted medical devices. The question of procedures for washing hair samples is covered in detail in the methods section.

As Klevay suggested,[5] the possibility that exogenous contamination might obscure real differences is probably greater for nutritionists who test for subnormal levels related to trace metal deficiency than for toxiciologists who test for supranormal levels. Analytical sensitivity is often less of a problem in the latter situation.

The principles and practices of hair metal determination, therefore, depend on the purpose for which the hair analysis is being considered. Books and reviews have dealt with the utility of hair metal determination in general[6-9] and with regard to heavy metal toxicity[10-12] or environmental exposure[13] in particular. Pertinent to an interest in the longitudinal pattern of hair metal as a record of past exposure are publications on the dynamics of hair growth.[14-20] Where the true growth rate was not measured, Smith[21] suggested an approximate rate of 1.3 cm/month, Hambidge et al.[22] used 1 cm/month, and Sexton et al.[23] assumed 1.0 to 2.0 cm/month. Condensed overviews of the utility of hair as a biopsy material were recently published by Klevay[5] and Maugh.[24]

II. METHODS

A. Sample Collection and Preparation

1. Sampling

Contamination should be avoided by precautions that are common to all trace metal work throughout sampling and analysis procedures. Plastic and acid washed glassware are generally the best laboratory materials; stainless steel is adequate for the determination of most metals but is suspect for nickel, chromium, and vanadium contamination. Materials and procedures suitable for trace metal analysis of environmental and biological samples have been described.[25,26]

Hair samples are usually taken with stainless steel or surgical scissors. Use of such scissors has not been shown to significantly contaminate hair samples with trace metals such as nickel, chromium, and vanadium; however, glass or quartz knives, laser beams,[25] and a titanium knife[27] have also been used to sample biological materials. A more imporatnt sampling consideration has to do with the longitudinal section of the hair filament taken for analysis.

Heterogeneity of hair metal concentration is well known.[22,28-39] Although the pattern of metal distribution along the hair shaft is not always consistent,[22,40] metal levels usually tend to increase from proximal to distal sections. This trend has been reported for copper and lead by Renshaw et al.[33] The pattern has also been shown by Hambidge[28] for copper and by Bate and Dyer[32] for zinc; they have suggested that the elevated levels of metal found in the outer hair segments were of exogenous origin, i.e., due to external contamination. In addition to the above mechanism, Renshaw et al.[33] suggested that evaporating perspiration, traveling along the hair shaft by capillary action, might also contribute to elevation in trace element concentrations along the length of the hair shaft. Hambidge[28] has suggested that nutritionists analyzing hair for evidence of essential trace metal deficiency take only the most proximal section of the hair shafts for analysis. Variations in sampling may have contributed to reported differences in normal values and differences ascribed to sex[21,41-48] or age.[42-46] Comparison of normal values for hair copper among investigators (Table 1) supports the suggestion that analysis of the whole hair, or of distal sections, resulted in higher values for females than for males. When the proximal 3 cm of hair was analyzed in our laboratory, values for hair copper of males and females were virtually identical. Our values were also similar to Hambidge's[28] values for proximal sections but were lower than his values for distal sections of hair.

Patterns of longitudinal metal heterogeneity are not always of increasing metal, proximal to distal. Smith[21] analyzed a single hair from a subject after about 5 months of chronic arsenic poisoning. The results showed an arsenic level that was about 4 times as great in the proximal 4 cm as in the distal 2 cm; the trend of longitudinal metal distribution was attributed to excess endogenous arsenic excretion. Endogenous metal ion release from implant materials may also result in a specific longitudinal hair metal pattern. Hambidge et al.[22] found that large variations in chromium along the hair shafts of normal subjects did not show a consistent pattern and hence reflected

Table 1
REPORTED NORMAL VALUES FOR HAIR COPPER

Mean hair copper (μg/g)				
Males		**Females**	**Section analyzed**	**Investigator**
16			Whole hair	Schroeder and Nason[41]
		56	Distal end	
22		31	Whole hair	Klevay[42]
18		51	Whole hair	Ryan et al.[34]
	12[a]		Proximal sections	Hambidge[28]
	21[a]		Distal sections	
15			Whole hair	Harrison et al.[53]
14		25	Distal end	Creason et al.[106]
23		47	Whole hair	Deeming and Weber[43]
13		14	Proximal 3 cm	Jacob

[a] Mean of 8 males and 19 females.

changes in the past chromium nutritional status of the individual rather than differences in exogenous contamination.

Although the use of proximal segments is not necessary for all purposes of hair metal analysis, investigators should sample consistently and be aware that differences in hair lengths or the longitudinal sections analyzed may contribute markedly to results. Renshaw et al.[33] strongly stated this often overlooked caution and also discussed the relevance of mechanisms by which trace metals enter the hair to the distribution of the metals along the shaft. The analyst also should remember when sampling hair that different longitudinal sections represent the body burden of metal at different times. Since hair grows slowly, distal segments may reflect body burdens many months in the past, while proximal segments reflect the most recent status.

The impact of sampling variations is apparently less among different locations on the scalp than along the length of the hair shafts. This has been shown by Hambidge et al.[22] for chromium and by Perkons and Jervis[49] for other elements. Other studies regarding this question have been published.[29] Again, sampling consistently from the same anatomical region is advisable. This author, as does Klevay,[5] takes hair samples from the occipitonuchal region. This area is low in the back of the head, a less exposed area where the hair loss is less visible and, perhaps, the hair less contaminated.

2. *Washing*

Most investigators who are interested in determining endogenous hair metal agree that hair should be washed before analysis to remove oil and exogenous metal contamination. Many studies on the effects of various washing procedures have been reported.[32-34,41,50-56] The usual components of washing procedures, either alone or in combination, include water, organic solvents, and detergent. Investigators, however, do not agree on the utility of washing procedures to remove exogenous contamination that might otherwise confound results. Nordlund et al.[57] found that the green hair of two young women was caused by contamination with copper from shampoo water. They stated that standard hair washing procedures do not remove such exogenously absorbed metal and, therefore, the results of hair trace metal determinations should be interpreted with caution. Smith[21] has stated that hair samples should not be washed at all. Hilderbrand and White[50] and Assarian and Oberleas,[51] who studied the effect of washing procedures on the trace metal content of hair, found that values for some metals depend on the particular washing procedure. Hilderbrand and White found that EDTA-containing washes removed considerable endogenous hair metal and advised against the use of such chelate-type agents. Hilderbrand and White[50] and Assarian and Oberleas[51] concluded that hair is an unreliable diagnostic sample because results can vary among washing techniques. Hambidge et al.,[54] however, have shown that organic solvents or detergent washes are similarly effective in removing a large percentage of exogenous hair chromium contamination without loss of tightly bound endogenous chromium. This author feels it is better to overwash than underwash hair samples as a loss of some endogenous metal is unlikely to affect important correlations whereas remaining exogenous metal will certainly be confounding. Researchers in our laboratory use the vigorous washing procedure of Klevay,[5,42,58-60] consisting of successive washes with acetone, ether, detergent, and deionized water. Tatro[61] found that this washing procedure removed essentially no endogenous zinc or copper from replicate hair samples. Harrison et al.[53] also stated that levels of zinc and copper in hair are not influenced by the type of wash.

This author agrees with Klevay[5] and Shapcott[52] that the possible effect of washing techniques on results does not negate the utility of hair metal determinations. Investigators should compare data for hair metal only when washing techniques are identical. Use of consistent sampling and washing techniques will increase chances of obtaining biologically and clinically important results. Many such reported results of hair metal determinations will be presented in the Section III.

B. Metal Determination

1. Instrumental Methods

The two most common methods for determining minerals in hair are atomic absorption spectrophotometry (AAS)[30,33,53,62-68] and neutron activation analysis.[36,40,49,69-73] X-ray,[31,35,74-76] spectrophotometric, and emission spectroscopic[77] methods have also been reported.

Use of AAS methods is probably most practical as the equipment is commonly available, the procedures are simple, and the technique is versatile and sensitive. Sunderman[62,78] has authored reviews on the application of AAS, both flame and electrothermal, to the determination of trace metals in a variety of biological materials. As the sensitivity of flame atomization methods are limited, electrothermal atomization is appropriate for studies where greater sensitivity is needed, for example, studies of the longitudinal hair metal pattern[30,33,63] where centimeter or millimeter lengths of hair need to be analyzed, and studies in humans where only a small amount of hair sample can be taken with proximal sections cut for analysis. Preparation of hair samples prior to AAS metal determination usually involves digestion of the organic hair material. An authoritative compendium of methods for wet or dry ashing of organic material has been published in the *Analyst.*[79] We have found that hair samples wet digest easily and cleanly in nitric acid or nitric/perchloric acids. The possibility of loss of volatile forms of chromium, mercury, or silver should be recognized, particularly in dry ashing procedures.[63] Hambidge et al.[54] used low temperature plasma ashing to prepare samples for chromium analysis. Prior ashing is avoided and single strands of hair or sections of hair filament can be analyzed by placing the hair sample directly into graphite atomizers.[30,33] Data from analyses of single hairs should be interpreted with caution because the samples taken may not be representative. In addition, Renshaw[80] has found that concentrations of lead contaminating the surface of single hairs can vary appreciably due to differences in the hair diameters.

Table 2
PREPARATION OF A HOMOGENEOUS HAIR POOL FOR QUALITY CONTROL OF TRACE METAL DETERMINATIONS

1. Collect several grams of hair from one or more people.
2. Cut finely as possible with stainless steel scissors in beaker with acetone (2—4 mm sections).
3. Mix vigorously on wrist-shaker for 30 min in erlenmeyer flask.
4. Decant acetone, shake with ether, then acetone, 15 min each.
5. Wrist-shake slowly with 5% sodium lauryl sulfate, 30 min, decant, filter.
6. Rinse with ultrapure H_2O, acetone, air-dry.
7. Store in desiccator.

Neutron activation analysis has also been widely used for hair metal determinations. The equipment needed is complex, but the method offers excellent sensitivity and simultaneous multi-element analysis.[34] Samples are irradiated and counted as supplied, hence prior ashing procedures are not required, eliminating a potential contaminating step. Neutron activation analysis also allows the analysis of single hairs or even millimeter sections of hairs.[21]

2. Quality Control

Because standard hair reference material is not available, several laboratories have prepared homogeneous hair pools[50,81,82] that can be used to assess precision and accuracy via standard metal recovery tests. In this laboratory, hair pools were prepared according to the procedure shown in Table 2. The reproducibility (CV%) of sampling 50 mg of this hair pool has been: zinc, 2.7%; copper, 6.6%; cadmium, 8.2% and chromium, 13.3%. Alder et al.[83] have investigated the use of silk and animal hairs as a standard for hair analysis. Other investigators[34] have used National Bureau of Standards, Standard Reference Materials, such as Bovine Liver and Orchard Leaves as standards that are representative of organic matrices.

III. CLINICALLY RELEVANT STUDIES

Many types of studies have been conducted regarding the effects of individual, demographic, and sample-preparation variables on hair metal concentrations. Although these studies may elucidate factors that should be appreciated, studies demonstrating the relationship between levels of metal in hair and in various body compartments are basic to an understanding of the utility of hair metal as a measure of body burden. Efforts of toxicologists to monitor body heavy metal intoxication have led to several such demonstrations. Norberg and Nishiyama[84] injected mice intravenously with radioactive cadmium chloride and counted the whole body and hair cadmium concentrations at regular intervals for up to 112 days. By sacrificing animals at 44 and 112 days, cadmium concentrations on whole blood and other body tissues were also obtained. The concentration of cadmium in newly grown hair was found to correlate highly ($r = +0.99$) with whole body cadmium levels. The cadmium concentrations in liver and heart seemed to decrease at approximately the same rate as that in newly grown hair and the whole body, while the concentrations in kidney and pancreas did not decrease at all. The concentration of cadmium in whole blood decreased much faster than that in whole body or hair and hence was not a good indicator of body accumulation. The authors concluded that for mice, and possibly for man, hair provides a better index of cadmium accumulation in the body than either blood or urine. Kello and Kostial[85] administered radioactive lead and cadmium to two different age groups of rats and found that hair lead was a good indicator of the lead body burden in both age groups. Hair cadmium, while a good indicator of cadmium body burden in adult rats, led to overestimation of body burden in the young rats. Brancato et al.[86] studied rats fed different amounts of cadmium in the drinking water and found that hair cadmium correlated with liver and kidney cadmium but blood levels did not. In humans, Oleru[87] found positive correlations of hair cadmium with liver, kidney, and lung cadmium, respectively. Gross et al.,[88] however, found poor correlation between hair cadmium and liver and kidney cadmium on humans from fetal through old age.

Jacob et al. measured zinc and copper in hair and in whole liver and subcellular liver fractions in rats fed different amounts of the metals in their drinking water.[89] Hair copper correlated directly with liver copper and copper in the nuclear and cytosol subcellular liver fractions. Hair copper also paralleled dietary intakes, but hair zinc did not. Neither did hair zinc show any correlation with liver zinc. Bellanger and Roth[90] determined copper in blood, liver, and hair of cattle at slaughter and found significant correlations of copper in hair with liver copper and with blood plasma copper for plasma copper levels <70 μg/dℓ. Studies of the concentrations of copper, iron, manganese, zinc, and cadmium in human hair, liver, muscle, and lung autopsy samples showed no significant correlations between metal concentrations in hair and other tissues.[91,92] The investigators, therefore, disputed the use of trace metal concentrations in hair as indicators of metal concentration in other tissues. In normal subjects, Deeming and Weber[43] found that females had higher serum and hair copper levels than did males; however hair copper levels did not correlate well with either serum or diet concentrations.

In an accidental exposure of humans to nickel carbonyl, the levels of nickel in urine and hair were in some cases, well above normal.[93] Scheiner et al.[94] determined nickel in hair of guinea pigs fed excess nickel in drinking water and found no significant difference from control animals receiving no supplementary nickel. It was concluded that hair is not a valid biopsy specimen for assessment of nickel ingestion. Spruit and Bongaarts[95,96] measured nickel in plasma, urine, and hair in subjects occupationally exposed to nickel, nickel hypersensitive subjects, and normal subjects. Concentrations of nickel did not differ between hypersensitive and normal subjects although nickel levels in plasma, urine, and hair of occupationally exposed subjects were about tenfold higher than in the controls. Li and Furst[97] injected nickel powder, intramuscularly, into shaved rats and found no subsequent changes in nickel content of the growing hair but the hair zinc and copper levels changed markedly.

Gruhn and Ludke[98] found that feeding chrome leather hydrolyzate waste to swine slightly increased the chromium content of hair, liver, and kidney, but not of meat. Hambidge and Baum[99] found that hair chromium levels in infants and young children paralleled reported changes in chromium levels in other tissues throughout the first 2 years of life. They concluded that hair chromium could be a useful index of body chromium status. Taylor et al.[100] studied small mammals adjacent to cooling towers, who were chronically exposed to increased chromium through both ingestion and inhalation pathways. Concentrations of chromium in hair, pelt, and bone were significantly higher than tissues from nonexposed control animals. Reported normal values for hair chromium and other metals commonly found in implant alloys are summarized in Table 3.

Deeming and Weber[101] fed rats graded levels of zinc in order to evaluate hair zinc as a measure of zinc status. Hair zinc did not correlate with plasma or liver zinc, but correlations of hair zinc with zinc intake and zinc in bone and testes were found both below and above the zinc requirement. This indicates that hair zinc reflected both

Table 3
REPORTED NORMAL VALUES OF HAIR METALS COMMONLY FOUND IN MEDICAL IMPLANT MATERIALS

	μg/g			
	Mean	Range[a]	Method[b]	Ref.
Al	6.5	1—9	AAS	63
	5	1—17	NA	34
Co	0.31	DL—8	AAS	63
	0.21	DL—0.7	AAS	41
Cr	0.33	DL—35	AAS	63
	0.62[c]	0.06—5.30	ES	106
	0.71	DL—2.5	AAS	41
	0.23	0.09—0.52	ES	54
	0.31	0.06—0.54	AAS	107
Hg	20.6	5—100	UV	82
	0.77[c]	0.05—14.0	AAS	106
Mn	0.18	DL—30	AAS	63
	0.95[c]	0.07—11.0	AAS	106
	0.60	0.03—3.7	NA	34
Ni	2.0	DL—28	AAS	63
	0.22	0.13—0.51	AAS	108
	0.74[c]	0.05—11.0	ES	106
	1.69	DL—12.0	AAS	41
Ag	0.3	DL—2.6	AAS	63
	0.17[c]	0.01—4.3	ES	106
Ti		0.03—0.06	ES	109
V	0.18[c]	0.01—2.2	ES	106
	0.06	0.01—0.56	NA	34
	0.04	0.01—0.09	NA	27

[a] DL = detection limit.
[b] AAS = atomic absorption spectrophotometry, ES = emission spectrosocpy, NA = neutron activation, UV = ultraviolet.
[c] Geometric mean.

subnormal and supranormal body zinc states. Because of the inportance of zinc as an essential nutrient many studies have reported human hair and plasma zinc values as indices of zinc status. Usually the two indices did not correlate.[102] Henley et al.[75] have used protein-induced X-ray emission spectroscopy to analyze 2-mm lengths of the root ends of human hair for a variety of minerals. Except for calcium no correlations of root end minerals with blood minerals were found.

Banta and Markesbary[103] found that a patient with suspected manganese intoxication had abnormally high levels of managanese in serum, urine, feces, cerebrum, and scalp, axillary, and pubic hair. Forseca and Lang[104] showed that the manganese content of hair from dairy cows was directly related to the manganese content of the forage.

Mercury in hair and red cells of Japanese islanders was directly related to the amount of fish consumed.[105] A good demonstration of the unique potential of hair to provide a history of excess body burden was Sexton's[23] recent report of mercury vapor poisoning in people living in a trailer park. Hair samples were taken 6 and 9 months after mercury exposure and analyzed in 1-cm segments. Blood and hair mercury levels were abnormally high and fell after termination of mercury exposure. The pattern of mercury in hair taken at 6 months (June) and 9 months (September) after exposure is

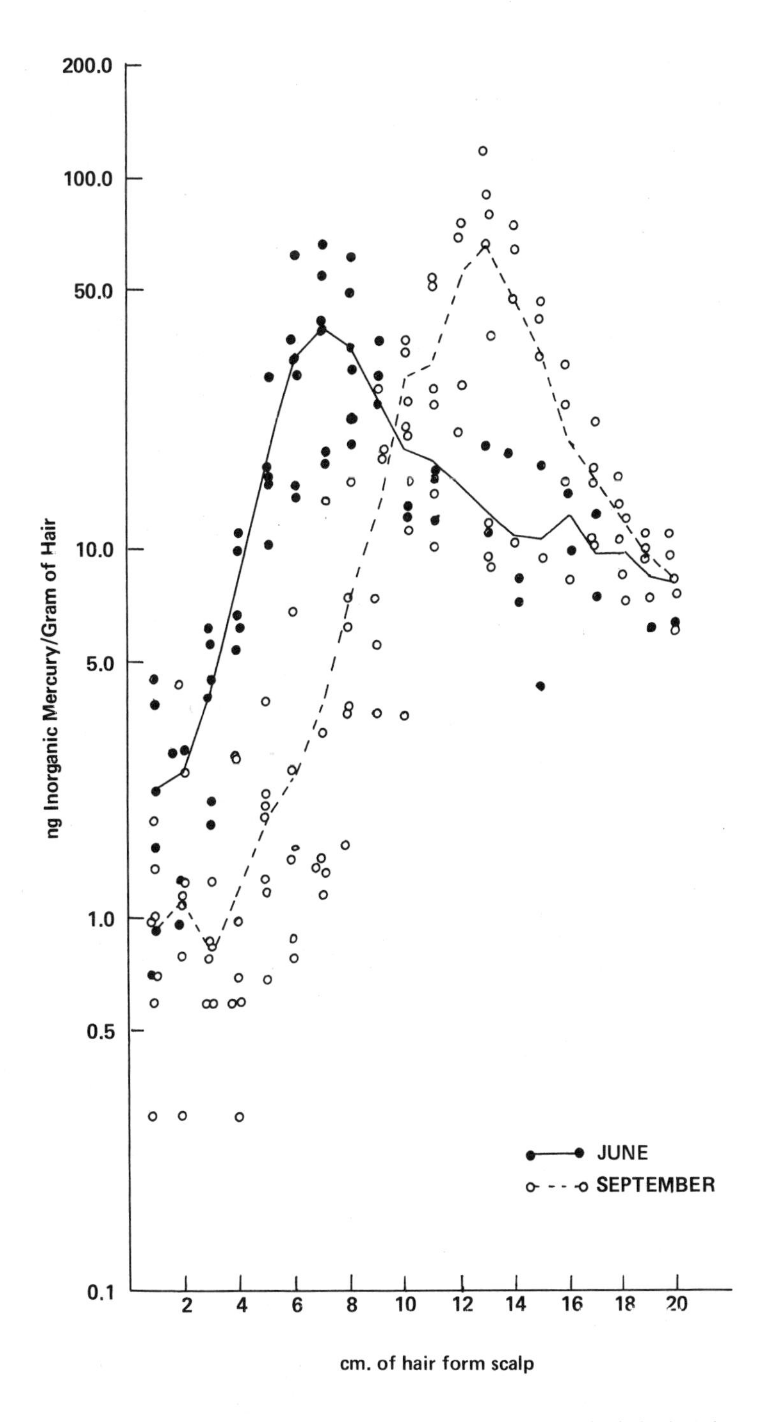

FIGURE 1. The pattern of mercury concentration along the hair shaft in human hair samples taken at 6 months (June) and 9 months (September) after termination of exposure to inorganic mercury vapor. (From Sexton, D. J., Powell, K. E., Liddle, J., Smrek, A., Smith, J. C., and Clarkson, T. W., *Arch. Environ. Health,* 33, 186, 1978. With permission.)

shown in Figure 1. Over this time period, the peak mercury concentration in the hair shifted from 7 cm to 13 cm from the scalp. The mercury concentration of the newly grown hair decreased to one quarter the peak level over this same period. The authors estimated that hair farther than 12 cm from the scalp of the June hair sample was formed before the mercury exposure. Hence, the high mercury level in these hair sections probably was due to exogenous contamination, but the peak was due primarily to endogenous mercury excretion.

In summary, reported studies have both supported and refuted the utility of hair metal determinations as a measure of body burden. Correlations with blood levels are generally not seen while correlations with organ levels have been demonstrated, although not consistently. Correlations with liver levels are significant since the liver is important in both intermediary metabolism and homeostasis of trace metals. Apparently hair metal determinations are less useful for discriminating among either normal subjects or subjects in marginal deficiency or excess than among subjects in more severe states. Hence, the use of hair as a tool for diagnosing systemic metal ion release from implant materials should have clinical utility, particularly where metal body burden becomes much greater than normal. The relationship between metals in hair and in whole bodies or organs should be further evaluated with careful attention to important variables of sample collection and preparation.

REFERENCES

1. **Owen, R., Meachim, G., and Williams, D. F.,** Hair sampling for chromium content following Charnley hip arthroplasty, *J. Biomed. Mater. Res.,* 10, 91, 1976.
2. **Swift, J. A.,** Scanning electron microscope study of Jane Austen's hair, *Nature (London),* 238, 161, 1972.
3. **Benfer, R. A., Typpo, J. T., Graf, V. B., and Pickett, E. E.,** Mineral analysis of ancient Peruvian hair, *Am. J. Phys. Anthropol.,* 48, 277, 1978.
4. **Gordus, A. A., Maher, C. C., and Bird, G. C.,** Human hair as an indicator of trace metal environmental exposure, in *Proc. Annu. NSF Trace Contam. Conf.,* Fulkerson, W., Shults, W. D., and Van Hook, R. I., Eds., National Technical Information Service, Springfield, Va., 1974, 463.
5. **Klevay, L. M.,** Hair as a biopsy material. Progress and prospects, *Arch. Intern. Med.,* 138, 1127, 1978.
6. **Hopps, H. C.,** The biologic bases for using hair and nail for analyses of trace elements, *Trace Subst. Environ. Health,* 8, 59, 1974.
7. **Brown, A. C.,** Blood, urine, and hair, *Hum. Hair Symp.,* 1, 1, 1974.
8. **Brown, A. C., Ed.,** *The First Human Hair Symposium,* Medcom, New York, 1974.
9. **Valkovic, V.,** *Trace Elements in Human Hair,* Garland STPM Press, New York, 1978.
10. **Clarkson, T. W.,** Metal concentrations in blood, urine, hair, and other tissues as indicators of metal accumulation in the body, in *Int. Conf. Environ. Sensing Assessment Proc.,* Institute of Electrical and Electronics Engineers, New York, 1976, 4.
11. **Lockeretz, W.,** Lead content of human hair, *Science,* 180 (4090), 1080, 1973.
12. **Takeuchi, T.,** Relation between mercury concentration in hair and the onset of minamata disease, in *Environ. Mercury Contam. Int. Conf.,* Ann Arbor Science, Ann Arbor, Mich., 1972, 302.
13. **Glubrecht, H. and Ryabukhin, Y.,** Elemental composition of human hair determined by activation analysis as an indicator of environmental contamination in man, in *Sci. Better Environ. Proc. Int. Congr. Human Environ.,* Pergamon Press, Oxford, 1977, 540.
14. **Johnson, E.,** The control of hair growth, *Physiol. Pathophysiol. Skin,* 4, 1351, 1977.
15. **Moretti, G., Rampini, E., and Rebora, A.,** Hormones and hair growth in man, *Hautarzt,* 28, 619, 1977.
16. **Rook, A. J., Ed.,** *An Introduction to the Biology of the Skin,* Blackwell, Oxford, 1970, 167.
17. **Moretti, G., Rampini, E., and Rebora, A.,** The hair cycle reevaluated, *Int. J. Dermatol.,* 15, 277, 1976.
18. **Ebling, F. J., Hair,** *J. Invest. Dermatol.,* 67, 98, 1976.

19. **Pelfini, C., Garimele, D., and Pisnau, G.,** Aging of the skin and hair growth in man, in *Advances in Biology of Skin,* Pergamon Press New York, 1969, 153.
20. **Demis, J. D., Crounse, R. G., Dobson, R., and McGuire, J.,** *Clinical Dermatology, I,* Harper & Row, New York, 1972, 3.
21. **Smith, H.,** The interpretation of the arsenic content of human hair, *J. Forensic Soc.,* 4, 192, 1964.
22. **Hambidge, K. M., Franklin, M. L., and Jacobs, M. A.,** Changes in hair chromium concentrations with increasing distances from hair roots, *Am. J. Clin. Nutr.,* 25, 380, 1972.
23. **Sexton, D. J., Powell, K. E., Liddle, J., Smrek, A., Smith, J., and Clarkson, T. W.,** A non-occupational outbreak of inorganic mercury vapor poisoning, *Arch. Environ. Health,* 33, 186, 1978.
24. **Maugh, T. H., II,** Hair: A diagnostic tool to complement blood serum and urine, *Science,* 202 (4374), 1271, 1978.
25. **Maienthal, E. J. and Becker, D. A.,** A survey on current literature on sampling, sample handling, for environmental materials and long term storage, *Interface,* 5, 49, 1975.
26. **Moody, J. R. and Lindstrom, R. M.,** Selection and cleaning of plastic containers for storage of trace element samples, *Anal. Chem.,* 49, 2264, 1977.
27. **Byrne, A. R. and Kosta, L.,** Vanadium in foods and in human body fluids and tissues, *Sci. Total Environ.,* 10, 17, 1978.
28. **Hambidge, K. M.,** Increase in hair copper concentration with increasing distance from the scalp, *Am. J. Clin. Nutr.,* 26, 1212, 1973.
29. **Alder, J. F., Samuel, A. J., and West, T. S.,** The anatomical and longitudinal variation of trace element contents in human hair, *Anal. Chim. Acta,* 92, 217, 1977.
30. **Lech, J., Delles, F., and Culver, B.,** Hair — the body's trace metal diary, *Var. Instrum. Appl.,* 8, 8, 1974.
31. **Zeitz, L., Lee, R., and Rothschild, E. O.,** Element analysis in hair by X-ray fluorescence, *Anal. Biochem.,* 31, 123, 1969.
32. **Bate, L. C. and Dyer, F. E.,** Trace elements in human hair, *Nucleonics,* 23, 74, 1965.
33. **Renshaw, G. D., Pounds, C. A., and Pearson, E. F.,** Determination of lead and copper in hair by non-flame atomic absorption spectrophotometry, *J. Forensic Sci.,* 18, 143, 1973.
34. **Ryan, D. E., Holzbecher, J., and Stuart, D. C.,** Trace elements in scalp-hair of persons with multiple sclerosis and of normal individuals, *Clin. Chem. (Winston-Salem, N.C.,)*24, 1996, 1978.
35. **Valkovic, V., Miljanic, U., Wheeler, R. M., Liebert, R. B., Zabel, T., and Phillips, G. C.,** Variation in trace metal concentrations along single hairs as measured by protein induced X-ray emission photometry, *Nature (London),* 243, 543, 1973.
36. **Obrusnik, I., Gislason, J., Maes, D., McMillan, D. K., D'auria, J., and Pate, B. D.,** Variation of trace element concentrations in single human head hairs, *J. Radional. Chem.,* 15, 115, 1973.
37. **Maes, D. and Pate, B. D.,** The spatial distribution of copper in individual human hairs, *J. Forensic Sci.,* 21, 127, 1976.
38. **Cookson, J. A. and Pilling, F. D.,** Trace element distributions across the diameter of human hair, *Phys. Med. Biol.,* 20, 1015, 1975.
39. **Maes, D. and Pate, B. D.,** The spatial distribution of zinc and cobalt in single human head hairs, *J. Forensic Sci.,* 22, 75, 1977.
40. **Schlesinger, H. L., Lukens, H. R., and Settle, D. M.,** Examination of Actual Case Evidence Samples by Neutron Activation Analysis, Special Publ. No. 312, U.S. National Bureau of Standards, 1969, 265.
41. **Schroeder, H. A. and Nason, A. P.,** Trace metals in human hair, *J. Invest. Dermatol.,* 53, 71, 1969.
42. **Klevay, L. M.,** Hair as a biopsy material. II. Assessment of copper nutriture, *Am. J. Clin. Nutr.,* 23, 1194, 1970.
43. **Deeming, S. B. and Weber, C. W.,** Hair analysis of trace metals in human subjects as influenced by age, sex, and contraceptive drugs, *Am. J. Clin. Nutr.,* 31, 1175, 1978.
44. **Petering, H. G., Yeager, D. W., and Witherup, S. O.,** Trace metal content of hair. II. Cadmium and lead of human hair in relation to age and sex, *Arch. Environ. Health,* 27, 327, 1973.
45. **Petering, H. G., Yeager, D. W., and Witherup, S. O.,** Trace metal content of hair. I. Zinc and copper content of human hair in relation to age and sex, *Arch. Environ. Health,* 23, 202, 1971.
46. **Reeves, R. D., Jolley, K. W., and Buckley, P. D.,** Lead in human hair. Relation to age, sex, and environmental factors, *Bull. Environ. Contam. Toxicol.,* 14, 579, 1975.
47. **Verghese, G. C., Koshore, R., and Gurin, V. P.,** Differences in trace element concentrations in hair between males and females, *J. Radioanal. Chem.,* 15, 329, 1973.
48. **Katz, S. A., Bowen, H. F. M., Comaish, J. S., and Samitz, M. H.,** Tissue nickel levels and nickel dermatitis. I. Nickel in hair, *Br. J. Dermatol.,* 93, 187, 1975.
49. **Perkons, A. K. and Jervis, R. E.,** Application of radioactivation analysis in forensic investigations, *J. Forensic Sci.,* 7, 449, 1962.

50. **Hilderbrand, D. C. and White, D. H.**, Trace-element analysis in hair: An evaluation, *Clin. Chem.*, 20, 148, 1974.
51. **Assarian, G. S. and Oberleas, D.**, Effect of washing procedures on trace-element content of hair, *Clin. Chem.*, 23, 1771, 1977.
52. **Shapcott, D.**, More on use of hair in trace-metal analysis, *Clin. Chem.*, 24, 391, 1978.
53. **Harrison, W. W., Yurachek, J. P., and Benson, C. A.**, The determination of trace elements in human hair by atomic absorption spectroscopy, *Clin. Chim. Acta*, 23, 83, 1969.
54. **Hambidge, K. M., Franklin, M. L., and Jacobs, M. A.**, Hair chromium concentration: effects of sample washing and external environment, *Am. J. Clin. Nutr.*, 25, 384, 1972.
55. **Bate, L. C.** The use of activation analysis in procedures for the removal and characterization of the surface contaminants of hair, *J. Forensic Sci.*, 10, 60, 1965.
56. **Scoble, H. A. and Litman, R.**, Preparation of hair and nail samples for trace element analysis, *Anal. Lett.*, B11, 183, 1978.
57. **Nordlund, J. J., Hartley, C., and Fister, J.**, On the cause of green hair, *Arch. Dermatol.*, 113, 1700, 1977.
58. **Klevay, L. M.**, Hair as a biopsy material. I. Assessment of zinc nutriture, *Am. J. Clin. Nutr.*, 23, 284, 1970.
59. **Klevay, L. M.**, Hair as a biopsy material. III. Assessment of environmental lead exposure, *Arch. Environ. Health*, 26, 169, 1973.
60. **Klevay, L. M.**, Hair as a biopsy material: Part IV. Geographic variations in the concentration of zinc, *Nutr. Rep. Int.*, 10, 181, 1974.
61. **Tatro, M.**, unpublished data, 1976.
62. **Sunderman, F. W., Jr.**, Atomic absorption spectrometry of trace metals in clinical pathology, *Human Pathol.*, 4, 549, 1973.
63. **Alder, J. F., Samuel, A. J., and West, T. S.**, The single element determination of trace metals in hair by carbon-furnace atomic absorption spectrometry, *Anal. Chim. Acta*, 87, 313, 1976.
64. **Alder, J. F., Alger, D., Samule, A. J., and West, T. S.**, The design and development of a multichannel atomic absorption spectrometer for the simultaneous determination of trace metals in hair, *Anal. Chim. Acta*, 87, 301, 1976.
65. **Renshaw, G. D., Pounds, C. A., and Pearson, E. F.**, Variation in lead concentration along single hairs as measured by non-flame atomic absorption spectrophotometry, *Nature (London)*, 238, 162, 1972.
66. **Sorenson, J. R., Melby, E. G., Nord, P. F., and Petering, H. G.**, Inferences in the determination of metallic elements in human hair. Evaluation of zinc, copper, lead, and cadmium using atomic absorption spectrophotometry, *Arch. Environ. Health*, 27, 36, 1973.
67. **Lau, H. K. and Ashmead, H.**, Trace minerals in human hair by atomic absorption spectrophotometry. Recovery study, *Anal. Lett.*, 8, 815, 1975.
68. **Murthy, L., Menden, E. E., Eller, P. M., and Petering, H. G.**, Atomic absorption determination of zinc, copper, cadmium, and lead in tissues solubilized by aqueous tetramethylammonium hydroxide, *Anal. Biochem.*, 53, 365, 1973.
69. **Quittner, P., Szabo, E., Perneczki, G., and Major, A.**, Determination of short-lived radionuclides in neutron activated human head-hair samples, *J. Radioanal. Chem.*, 5, 133, 1970.
70. **Bowen, H. J.**, Determination of trace elements in hair samples from normal and protein deficient children by activation analysis, *Sci. Total Environ.*, 1, 75, 1972.
71. **Menke, H., Leszczynski, C., and Weber, M.**, Comparative study of X-ray and gamma-ray spectrometry applied to neutron activation analysis of human head hair, *Radiochem. Radioanal. Lett.*, 14, 217, 1973.
72. **Coleman, R. F., Cripps, F. H., Stimson, A., and Scott, H. D.**, The Determination of Trace Elements in Human Hair by Neutron Activation and the Application of Forensic Science, Atomic Weapons Research Establishment, Report 0-86/66, 1967.
73. **Turkstra, J., Beukes, P. J. L., Brits, R.J. N., and Hambleton-Jones, B. B.**, Multi-element characterization of human hair by neutron activation analysis, *S. Afr. J. Sci.*, 74, 182, 1978.
74. **Biegert, E. K.**, Trace element analysis of biological material using PIXE, *Proc. Annu. Conf. Microbeam Anal. Soc.*, 12, 93A, 1977.
75. **Henley, E. C., Kassoury, M. E., and Nelson, J. W.**, Proton-induced X-ray emission analysis of single human hair roots, *Science*, 197, 277, 1977.
76. **Rendic, D., Holjevic, S., Valkovic, V., Zabel, T. H., and Phillips, G. C.**, Trace-element concentrations in human hair measured by proton-induced X-ray emission, *J. Invest. Dermatol.*, 66, 371, 1976.
77. **Hambidge, K. M.**, Use of static argon atmosphere in emission spectrochemical determination of chromium in biological materials, *Anal. Chem.*, 43, 103, 1971.
78. **Sunderman, F. W., Jr.**, Electrothermal atomic absorption spectrometry of trace metals in biological fluids, *Ann. Clin. Lab. Sci.*, 5, 421, 1975.

79. Analytical Methods Committee, Methods for the destruction of organic matter, *Analyst*, 85, 643, 1960.
80. **Renshaw, G. D.**, The distribution of trace elements in human hair and its possible effect on reported elemental concentration levels, *Med. Sci. Law*, 16, 37, 1976.
81. **Jacob, R. A.**, Use of a hair pool for quality control of trace metal determinations in hair, paper presented at 173rd Am. Chem. Soc. Natl. Meet. Anal. Chem. Div., March, 1977.
82. **Nord, P. J., Kadaba, M. P., and Sorenson, J. R. J.**, Mercury in human hair, *Arch. Environ. Health*, 27, 40, 1973.
83. **Alder, J. F., Pankhurst, C. A., Samuel, A. J., and West, T. S.**, The use of silk and animal hairs as a standard for hair analysis, *Anal. Chim. Acta*, 91, 407, 1977.
84. **Norberg, G. F. and Nishiyama, K.**, Whole body and hair retention of cadmium in mice, *Arch. Environ. Health*, 24, 209, 1972.
85. **Kello, D. and Kostial, K.**, Lead and cadmium in hair as an indicator of body burden in rats of different age, *Bull. Environ. Contam. Toxicol.*, 20, 618, 1978.
86. **Brancato, D. J., Picchioni, A. L., and Chin, L.**, Cadmium levels in hair and other tissues during continuous cadmium intake, *J. Toxicol. Environ. Health*, 2, 351, 1976.
87. **Oleru, G. U.**, Kidney, liver, hair and lungs as indicators of cadmium absorption, *J. Am. Ind. Hyg. Assoc.*, 37, 617, 1976.
88. **Gross, S. B., Yeager, D. W., and Middendorf, M. S.**, Cadmium in liver, kidney, and hair of humans, fetal through old age, *J. Toxicol. Environ. Health*, 2, 153, 1976.
89. **Jacob, R. A., Klevay, L. M., and Logan, G. M., Jr.**, Hair as a biopsy material. V. Hair metal as an index of hepatic metal in rats: copper and zinc, *Am. J. Clin. Nutr.*, 31, 477, 1978.
90. **Belanger, J. and Roth, L. C.**, Relations among levels of copper in the blood, liver, and hair in 250 cattle at slaughter, *Rech. Vet.*, 1, 127, 1968.
91. **Vuori, E., Huunan-Seppala, A., and Kilpio, J. O.**, The concentrations of copper, iron, manganese, zinc, and cadmium in human hair as a possible indicator of their tissue concentrations, in *Proc. Int. Symp. Recent Adv. Assess. Health Eff. Environ. Pollut.*, a report from the Department of Public Health Science, University of Helsinki, Helsinki, 1975, 2263.
92. **Gibbs, G. W. and Bogdanovic, E.**, Trace elements in lung and hair, in *Proc. Int. Symp. Recent Adv. Asses. Health Eff. Environ. Pollut.*, a report from the Department of Epidemiol. Health, McGill University, Montreal, 1975, 2271.
93. **Hagedorn-Goetz, H., Kueppers, G., and Stoeppler, M.**, Nickel contents in urine and hair in a case of exposure to nickel carbonyl, *Arch. Toxicol.*, 38, 275, 1977.
94. **Scheiner, D. M., Katz, S. A., and Samitz, M. H.**, Nickel levels in hair and nickel ingestion in guinea pigs, *Environ. Res.*, 12, 355, 1976.
95. **Spruit, D. and Bongaarts, P. J. M.**, Nickel content of plasma, urine and hair in contact dermatitis, *Dermatologica*, 154, 291, 1977.
96. **Spruit, D. and Bongaarts, P. J. M.**, Nickel content of plasma, urine and hair in contact dermatitis, *Dev. Toxicol. Environ. Sci.*, 1, 261, 1977.
97. **Li, M. and Furst, A.**, Zinc and copper levels in rat hair following nickel administration, *Proc. West. Pharmacol. Soc.*, 20, 445, 1977.
98. **Gruhn, K. and Ludke, H.**, Use of hydrolyzed chrome leather waste in pig feeding and the content of chromium in flesh liver, kidneys, and hair, *Arch. Tiernaehr.*, 22, 113, 1972.
99. **Hambidge, K. M. and Baum, J. D.**, Hair chromium of human newborn and changes during infancy, *Am. J. Clin. Nutr.*, 25, 376, 1972.
100. **Taylor, F. G., Parr, P. D., and Dahlman, R. C.**, Distribution of Chromium in Vegetation and Small Mammals Adjacent to Cooling Towers, Report CONF-750503-6, 1975.
101. **Deeming, S. B. and Weber, C. W.**, Evaluation of hair analysis for determination of zinc status using rats, *Am. J. Clin. Nutr.*, 30, 2047, 1977.
102. **McBean, L. D., Mahloud, I. M., Reinhold, J. G., and Halsted, J. A.**, Zinc concentrations in human plasma and hair, *Am. J. Clin. Nutr.*, 24, 506, 1971.
103. **Banta, R. G. and Markesbery, W. R.**, Elevated manganese levels associated with dementia and extrapyramidal signs, *Neurology*, 27, 213, 1977.
104. **Forseca, H. and Lang, C.**, Manganese content of Orosi Valley [Costa Rica] fodder and its effect on hair concentration and reproduction in dairy cows, in *Nucl. Tech. Anim. Prod. Health, Proc. Int. Symp.*, International Atomic Energy Agency, Vienna, 1976, 171.
105. **Tsuguyoshi, T., Takemoto, T., Kashiwazaki, H., Togo M., Toyokawa, H., and Miyama, T.**, Man, fish, and mercury on small island in Jpana, *Tohuko J. Exp Med.*, 118, 181, 1976.
106. **Creason, J. P., Hinners, T. A., Bumgarner, J. E., and Pinkerton, C.**, Trace elements in hair, as related to exposure in metropolitan New York, *Clin. Chem.*, 21, 603, 1975.

107. **Mahalko, J. R. and Bennion, M.**, The effect of parity and time between pregnancies on maternal hair chromium concentration, *Am. J. Clin. Nutr.*, 29, 1069, 1976.
108. **Nechay, M. W. and Sunderman, F. W., Jr.**, Measurements of nickel in hair by atomic absorption spectrometry, *Ann. Clin. Lab. Sci.*, 3, 30, 1973.
109. **Goldblum, R. W., Derby, S., and Lerner, A. B.**, The metal content of skin, nails, and hair, *J. Invest. Dermatol.*, 20, 13, 1953.

Chapter 7

SOME BIOLOGICAL ASPECTS OF NICKEL

N. Jacobsen, A. Hensten-Pettersen, and H. Hofsøy

TABLE OF CONTENTS

I. INTRODUCTION

In this chapter an attempt is made to summarize important aspects of nickel metabolism, toxicity, carcinogenicity, and hypersensitivity induction as a brief background for readers who are primarily interested in biocompatibility problems. Several comprehensive reviews are available on these subjects.[1-7]

Nickel (sp. gr. 8.9, at. wt. 58.11) is a grayish white metal consisting of five stable isotopes. Nickel occurs normally with a valence of 0 or 2, but it can also exist in other valence states. It is ubiquitous in the earth, constituting about 0.008% of the crust and is element No. 23 in the Periodic Table. Nickel ore deposits are found as sulfide minerals containing iron or copper or as magnesium-containing silicates.[3]

About half of the world's nickel production occurs in Canada, but the U.S.S.R., Australia, and New Caledonia are important producers, together with several minor production sites.[4] Mining and refinery are not necessarily combined and may take place at widely distant places.

The refinery processes include a series of pyrometallurgical or electrometallurgical steps producing gasses, dusts, and fumes containing nickel or nickel compounds.[8] In addition, compounds of nickel are used or produced in widely different industries including the processing of a variety of nickel alloys for hard goods and electronic devices. Accompanying activities such as melting, grinding, welding, polishing, etc. increase the possibility of occupational and environmental exposure to nickel.

A. Nickel Compounds

Important nickel compounds are nickel sulfate ($NiSO_4$ $6H_2O$), nickel subsulfide, (Ni_3S_2), and the nickel oxides (NiO and Ni_2O_3), which have been demonstrated in refinery plant environments. Nickel carbonyl, $Ni(CO)_4$, is also an intermediate product, in certain refinery processes. Other nickel compounds are nickel acetates, $Ni(COOCH_3)$ and $Ni(COOCH_3)\cdot 4H_2O$, several nickel ammonium sulfates, nickel chloride and fluoride, $NiCl_2$ and NiF_2, nickel hydroxides, $Ni(OH)_2$, $Ni(OH)_3$, and $NiO\cdot nH_2O$, nickelocene, $(C_5H_5)_2Ni$, nickel nitrate, $Ni(NO_3)_2$, and nickel sulfamates, $Ni(SO_3NH_2)_2$. Some compounds are water soluble (nickel sulfates, ammoniumsulfates, sulfamates, nitrate, chloride), whereas other are practically insoluble in water (nickel subsulfide, hydroxides, carbonates, acetates, carbonyl) but may be soluble in organic solvents.[4]

B. Environmental Nickel

Small quantities of nickel are present in coal, petroleum, and seawater, but nickel is not always detected in surface or ground water, probably because of the formation of insoluble hydrolysates during the rockweathering process. The small amounts of nickel present in some drinking waters, therefore, probably derive from human activities.[3]

Nickel in air is dependent on the degree of urbanization and the temperature, indicating that airborne nickel is derived from oil and coal burning and traffic exhaust. The average nickel concentration in certain urban and nonurban sites in the U.S. have been measured to 0.021 and 0.006 $\mu g/m^3$.[9] Tobacco smoke contains nickel.[3,10]

Plant, seafood, and animal sources of food contain some nickel. The concentrations vary from 0 to 7 ppm of fresh weight. Some of the nickel in food is considered to derive from man-made sources in connection with food processing and preservation. However, the release of nickel from kitchen utensils is low, since nickel alloys are highly resistant to corrosion in acidic and caustic environments.[3] The daily intake of nickel has been calculated to about 0.3 to 0.6 μg in the U.S.[11] and about 0.7 μg in Sweden.[12]

C. Nickel Monitoring

Several methods are reported suitable for the monitoring of nickel in tissues and biologic fluids.[14-19] A few are briefly mentioned here.

For atomic absorption spectrophotometry, the preparation of samples to be measured generally includes wet digestion, alkalization, and extraction with chelating agents such as oximes[20] or ammonium pyrrolidine dithiocarbamate into an organic phase of methyl isobutyl ketone.[17] Organic-phase nickel may be determined directly by flame or flameless AAS techniques.[19] The sensitivity for nickel detection is usually in the range of submicrograms.

Recently Mikac-Davić et al. described an extremely sensitive method (0.2 μg Ni per liter) for flameless AAS determination of nickel or biologic fluids, using furil-dioxime.[15]

Neutron activation analyses for determination of nickel in organic materials is also used.[18]

II. NICKEL METABOLISM

A. Routes of Uptake

The human organism takes up nickel by various routes:

1. Inhalation of airborn gaseous nickel compounds or nickel dusts
2. Dietary intake
3. Parenteral uptake from nickel containing implants
4. Percutaneous and permucous absorption

The two last routes are probably not of quantitative significance but are clinically important in the pathogenesis of nickel dermatitis.

B. Tissue Distribution and Elimination

Old data from several sources indicate that nickel is present in the lungs, liver, and kidneys of stillborn human fetuses.[21] More recent data on nickel in various tissues and organs from adults show a relative nickel distribution of bone > lung > large intestine > small intestine > skin.[22] The concentration of nickel in biologic fluids is usually maintained within a characteristic range. Observed variations between reference values of nickel in serum and urine collected from different populations probably depend on differences in environmental concentrations of nickel.[23] Nomoto and Sunderman[17] reported nickel concentrations in blood (4.8 ± 1.3 μg/ℓ), serum (2.6 ± 0.8 μg/ℓ), and urine (2.3 ± 1.4 μg/ℓ) collected from 26 healthy subjects in central Connecticut where the environmental concentration of nickel is relatively low. Total excretion of nickel in urine was 2.4 ± 1.1 μg/day. Comparable data[23] from subjects not occupationally exposed to nickel but living in an area of nickel mining showed increased serum nickel (4.6 ± 1.4 μg/ℓ) and a significantly higher mean value of urinary excreted nickel (7.9 ± 3.7 μg/day).

Nickel concentrations of 52 ± 36 μg/ℓ have been found in sweat collected from healthy men.[24] The nickel concentration in salivary secretions appears to be comparable to that of serum.[25] Data on nickel concentrations in human milk are sparse. Animal experiments indicate that variation in nickel concentration administered orally to rats is reflected in the milk[26] and that nickel given to lactating mice by injection is transferred to the sucklings.[27] Nickel is also present in hair. Determination of the nickel concentration in human hair samples has therefore been used by several workers as a guide for nickel exposure.[13,18,28-30] However, large disparities occur between the nickel values obtained. The validity of using hair specimens for the monitoring of nickel exposure instead of urine or serum analysis is therefore uncertain.[31]

Characteristic alterations of nickel concentrations in human serum are observed in connection with certain diseases. Increased nickel levels are found in patients after acute myocardial infarction,[32-34] stroke, and severe burns.[33] The hypernickelemia observed in myocardial infarction is thought to be a secondary manifestation of leukocytosis and leukocytolysis.[6] McNeely et al.[33] reported hyponickelemia in patients with hepatic cirrhosis or uremia associated with marked hypoalbuminemia.

Experimental data from different species of rodents on the distribution of parenteral administered $^{63}Ni(II)$ show that the relative nickel distribution in various organs and tissues follows a certain pattern during the first hours after nickel injection.[35-38] The highest nickel concentrations are invariably found in the kidneys. The pituary gland,[35,36] lung, spleen, and skin also show high contents of nickel. The lowest relative nickel concentrations are generally observed in the central nervous system. Some nickel affinity to mouse calvaria and long bones,[27,39] cartilage, and various other connective tissues[40] has also been found. Retention of single dose nickel is seen in lungs and kidneys,[40] liver, calvaria, and teeth.[27] The affinity of nickel to mineralized tissues is probably localized to the soft tissue lining and is not equivalent to the bone seeking properties of Sr and Ca.

Administered Ni(II) accumulates in the visceral yolk sack and embryos of pregnant rats.[41] In contrast to the Ni(II) distribution in the mother, considerable nickel concentrations are observed in the brain of mouse fetuses.[27]

Intravenously administered Ni(II) is rapidly eliminated from the serum through urine and bile, and the major part of an administered nickel dose is recovered in the urine after 24 hr.[42]

The metabolism of any nickel that enters the body by the pulmonary route is similar to that of parenterally administered nickel.[43] An appreciable amount of nickel is excreted under conditions of profuse sweating.[24]

The major part of orally ingested nickel is excreted in the feces.[44,45] Fecal excretion of nickel is, therefore, largely dependent on the dietary intake. Horak and Sunderman found a total fecal excretion of nickel from ten adults to average 259 ± 126 μg/day.[45]

C. Nickel Binding

Three distinct forms of nickel are normally present in human and rabbit serum, the nickeloplasmin,[46] the albuminbound nickel,[47] and ultrafiltrable nickel.[48] A metalloprotein rich in nickel was fractionated from human serum by Himmelhoch et al.[49] Later Nomoto et al.[46] isolated a similar nickel-containing protein, characterized as an α_1-macroglobulin, from rabbit serum. The nickel present in this metalloprotein is not readily exchangeable with $^{63}Ni(II)$ in vivo or in vitro.[50] In normal rabbit serum the nickeloplasmin constitutes approximately 44% of total nickel present.[51]

A 9.5S α_1-glycoprotein isolated from human serum by Haupt et al.[52] shows strong binding for Ni(II). The relationship between this particular protein and nickeloplasmin is not clear.

Ni(II) affinity to some acidic salivary proteins has been shown after cultivation of human glands in a medium containing $^{63}NiCl_2$[53] and to an acidic fraction of rabbit salivary proteins after intraperitoneal injection of $^{63}NiCl_2$.[54]

Ni(II) may activate or inhibit numerous enzymatic reactions under various conditions,[3] but no animal nickel metalloenzyme is described. The discovery of Dixon et al.[55] that jack bean urease is a nickel metalloenzyme has intensified the search for nickel-dependent animal enzymes.

Several investigations have shown that albumin is the principal Ni(II)-binding protein in serum from various species.[42,47,56-58] Ni(II) competes with Cu(II) in the binding to the N-terminus of albumin through the terminal amino group, the first two peptide nitrogen atoms, and the imidazole nitrogen of the histidine residue at the third position.[57] Serum albumins containing tyrosine in the third position from the N-terminus have less affinity for nickel.[47]

Ultrafiltrable serum nickel averages 41% of total nickel in normal human serum.[59] Considerable proportional variations of ultrafiltrable serum nickel are observed in different species, reflecting differences in the nickel binding properties of serum albumins.[47] Ultrafiltrable Ni(II)-binding substances play an important role in extracellular transport of nickel,[42,59] intracellular binding of nickel,[60] and excretion of nickel in urine and bile.[48]

The identity of Ni(II)-binding ultrafiltrable ligands in serum is not established. Investigations by Asato et al.[48] demonstrated the existence of several distinct ninhydrin-positive nickel complexes in rabbit serum on thin layer chromatography. Similar experiments have shown the presence of histidine and a peptide which after hydrolysis contained glycine, aspartic acid, glutamic acid, alanine, and lysine in a thin layer ^{63}Ni-fraction from rabbit serum.[61] It is suggested that cysteine, histidine, and aspartic acid may act either singly or in combination as Ni(II) ligands.[6]

Intracellular Ni(II)-binding mechanisms are unclear. Metallothionein and related proteins do not bind Ni(II) and parenteral administration of Ni(II) seems not to influence the concentrations of such proteins in the liver and kidney of rats.[62,63]

Webb and Weinzierl[64] showed that ^{63}Ni(II) was taken up from various carriers of both high and low molecular weight by mouse dermal fibroblasts in vitro and bound in decreasing order of amount by the nuclei, cell sap, mitochondria, and microsomes.

D. Nutritional Essentiality

The essentiality of nickel has been suggested by several investigators.[65-67] Evidence for biological ill effects of experimental nickel deficiency was demonstrated in rats by Schnegg and Kirchgessner.[68-70] Nickel-deficient rats showed retarded growth[69] and developed a pronounced anemia[68] with reduction in blood hemoglobin concentrations, hematocrit values, and erythrocyte counts. The anemia caused by the nickel deprivation was shown to result from impaired intestinal absorption of iron.[70]

III. NICKEL TOXICITY

Toxic effects of nickel depend on the route of uptake, the dose, and the form of nickel. Information on this subject is based on animal experiments, epidemiological data, and in vitro observations. Many reports indicate that metallic nickel and nickel salts are relatively nontoxic, like zinc, manganese, and chromium. Dogs and cats tolerated daily oral doses of 4 to 12 mg/kg of metallic nickel for a prolonged time without ill effects.[71]

A. Experimental Data

Rats given a concentration of up to 1000 ppm of nickel salts in the food showed no toxic effects on their growth during an 8-week period.[71] Macaca monkeys fed similar amounts of nickel salts during a 24-week period also failed to exhibit ill effects in terms of growth and behavior and on hematologic criteria.[72] On the other hand, calves given nickel carbonate during 8 weeks showed growth reduction and kidney abnormalities at a concentration of 1000 ppm in the diet.[73] The acute oral LD_{50} dose of nickel salts have been found up to about 2 g/kg for rats. Toxic amounts of nickel chloride given to rats have led to diminished food uptake and effects on several blood parameters, hepatic enzyme activities, and hepatic proteins.[74]

Parenteral administration of nickel (i.v., i.p., i.m.) to experimental animals such as rats, mice, rabbits, dogs, guinea pigs, etc. has shown LD_{50} doses largely dependent on the nickel compound, the animal species, and the route of administration,[3] but far lower than the comparable oral dose. Hyperglycemia, accompanied by adaptive hyperinsulinemia, has been found on rats injected with NiCl.[6,75,76] As expected from the excretion pattern of parenterally injected nickel, pathologic changes of the kidneys have been demonstrated by rat experiments.[77] Other rat experiments point to embryotoxic hazards of nickel comparable with those of lead, cadmium, and mercury and indicate increased neonatal mortality, reduced litter size, and increased incidence of still births.[78-80] An unspecified number of malformations have also been reported in hamsters.[81]

B. Epidemiological Data

In man, inhalation of soluble nickel is the primary exposure route pertaining to renal affections, but no clinical data are available on the renal functions of workers exposed to nickel in this way. However, aerosols of $NiSO_4$ have led to severe damage of the nasal mucosa.[6,82] Reports of asthmatic lung disease among workers in the nickel plating industry also have been presented,[83] as well as reports on other hypersensitivity reactions and carcinogenesis (see later). No data concerning possible teratogenic or embryotoxic reactions to nickel in man are available.

C. Nickel Carbonyl Poisoning

Acute toxic effects of nickel or nickel compounds in man are limited to the nickel carbonyl, $Ni(CO)_4$. This is a colorless, volatile liquid miscible with organic solvents and occurring in certain refinery processes of nickel. Nickel carbonyl is considered to be 100 times more toxic than CO. Inhalation causes nausea, headache, dyspnea, and chest pain. Severe pulmonary symptoms develop a day or two after the exposure leading to extreme pulmonary malfunction and death from interstitial pneumonitis and cerebral injuries.[1,6] Experimental evidence points to damage of the membranous and agranular pneumocytes caused by nickel carbonyl that is freely passed through the lung epithelium.[84] At a later stage of the poisoning, oxidation of the carbonyl within erythrocytes and other cells leads to the formation of Ni(II) excreted by the kidneys and CO eliminated by expiration.[85] Over the years, several poisonings from industrial nickel carbonyl have taken place. The current therapy is the administration of chelating drugs such as diethyldithiocarbamate.

Experimental evidence shows that nickel carbonyl leads to ophthalmic malformations after prenatal inhalation.[86]

D. In Vitro Observations

In vitro studies showed that concentrations as low as 10 $\mu g/m\ell$ of nickel as chloride were toxic to epithelial cells in culture,[87] whereas rabbit macrophages tolerated higher doses in vitro.[88] Mouse tracheal rings in culture have shown 50% reduced ciliary activity at nickel concentrations of about 200 $\mu g/m\ell$ in the culture medium.[89]

Adverse effects of nickel in the reproductive system of male rats have been reported after administration of nickel salts.[90] Exposure of rainbow trout eggs to nickel led to shorter hatching time.[91] Recent investigations on the in vitro development of fertilized mouse eggs show a clear inhibitory effect at a concentration of about 6 $\mu g/m\ell$ of nickel as chloride.[92]

IV. NICKEL CARCINOGENESIS

As much as 80% of human cancers may be related to environmental factors of a chemical nature.[93] Epidemiologic data show that such factors include the occupational exposure to metals such as As, Cd, Cr, and Ni.[4,5,7] In addition, compounds of the metals Be, Cd, Co, Cr, Fe, Ni, Pb, Ti, and Zn have been reported to induce cancers in experimental animals.[2]

A. Occupational Aspects

For nickel, the frequency of respiratory cancer among workers employed in certain refinery processes strongly suggests a connection between the inhalation of nickel-containing dusts and compounds, and the development of nasal and lung cancers.[94] Observations to this effect are dated back to the 1930s and have been made in Wales, Canada, Norway, and other countries.[95-97] Workers in nonrefining industries in contact with nickel or nickel compounds such as welders, platers, grinders, etc. have also shown several cases of respiratory cancers, but no systematic epidemiological reports are available. In addition, excess laryngeal[97] and renal cancers[6,42] have been reported in workers exposed to nickel. In 1978, Sunderman reported a total of 447 cases of lung cancer and 143 cases of the nose and paranasal sinuses in nickel workers on the basis of published material.[7] Histological classification shows mainly squamous cell carcinomas and undifferentiated anaplastic cell carcinomas of the lungs. These types are also found in the nasal cavities in addition to pleomorphic adenomas and other types.[7] Histological changes of the nasal mucosa precede the development of invasive cancers for some time.[98]

One difficulty in the clarification of the relationship between the exposure to nickel and the development of cancer is the extremely long latent time, varying from 10 to 40 years, with means of 27 years for lung cancer and 23 years for nasal cavity cancers.

Strongly suspected forms of carcinogenic nickel are the insoluble nickel subsulfide, Ni_3S_2, the nickel oxides, Ni_2O_3 and NiO, and nickel-containing dusts. Vapor of nickel carbonyl, $Ni(CO)_4$, and soluble nickel compounds such as $NiNO_3$, $NiCl_2$, and $NiSO_4$ are also candidates for the induction of respiratory cancers.[6,7] It is difficult to arrange the nickel compounds according to their relative carcinogenic potential, but the subsulfide appears to be of specific interest.

B. Experimental Data

Nickel carcinogenesis is well documented experimentally in several animal species. Early experiments showed sarcomas in rats and rabbits after injection of nickel-containing dust.[99] Pulmonary cancers were observed after inhalation of nickel dust by guinea pigs.[100] More recent experiments have shown that injection of compounds such as Ni_3S_2, NiO, Ni_2O_3, $NiC_2H_3O_2$, and $Ni(C_5H_5)_2$ have given sarcomas on different rodents,[7,101-103] and inhalation of subsulfide, metallic nickel, and nickel carbonyl have given lung cancers.[104]

Injection of nickel subsulfide also led to testicular and renal cancer in rats.[105,106] The developing renal cancers were preceded by a rapid erythropoiesis. NiS had no such effects.[107] Another interesting aspect of the subsulfide carcinogenesis is the suppression of its effects by simultaneous injection of manganese dust.[102] On the other hand, synergistic effects were indicated between the nickel oxides NiO and Ni_3O_2 and certain polycyclic hydrocarbons.[108]

In vitro studies dealing with the mechanism of nickel carcinogenesis show that Syrian hamster cells are transformed after contact with the subsulfide, but not with the sulfide of nickel.[109] Again, this effect is counteracted by manganese. Other observations indicated inadequate pairing of synthetic polynucleotides after exposure to nickel, whereas nickel compounds have not led to increased mutagenesis in the bacterial mutation system of Ames,[7] except in combination with more complex welding fumes.[110]

Nickel carbonyl and nickelocene are considered to be able to penetrate the cell membranes directly.[6] The mechanism of uptake and intracellular transport of the less (lipid) soluble nickel carcinogens is unclear. One possibility is the formation of diffusible, small molecular nickel complexes.[111] Another possibility is the formation of nickel-protein complexes that enter the cells by endocytosis and later release ionic nickel intracellularly.[112] Endocytosis of metallic nickel has been observed in tissue macrophages.[113]

Intracellularly the nickel has been demonstrated in the nucleolar part of the cell nucleus, but the molecular mechanism of the neoplastic induction is unclear.[114] Interference with the cell division and different enzyme systems has been demonstrated.[115,116]

V. HYPERSENSITIVITY

A. Incidence

Nickel is a well-known hapten.[3]. Nickel salts such as sulfate or chloride are routinely included in the standard series of compounds tested as causative agents on patients with eczematous dermatitis. However, the importance allotted to nickel as an allergen varies from country to country. In France it ranks as number seven,[117] in the U.S. as first,[118] and as number six[119] in importance in a group of common contact allergens tested. In the U.K. and Scandinavia, nickel appears to be the most common, single cause of contact dermatitis in women, only rivalled by primula sensitivity in the U.K. and balsam sensitivity in Scandinavia.[120]

Epidemiological data on nickel allergy present the frequency of patients with positive skin reactions to nickel in the in- and out-patient clinics of dermatological hospital units. The prevalence of nickel allergies in the general population is therefore uncertain. However, comparison of the incidence of allergic contact allergens among selected patient populations shows a frequency of nickel sensitivity of about 12%.[119,121] Analysis of data from different time periods shows that the incidence of reactions to nickel (as sulfate) has remained remarkably constant over the years (12.3% in 1937, 11.2% in 1961, and 13.1% in 1968 to 70).[119,121]

There is a marked sex difference in the positive skin reactions to nickel. In one study, only women were found to be affected.[117] Other reports show that nickel sensitivity is three to five times more common in women than in men,[122-124] indicating that the incidence of nickel allergy recorded in a particular clinic will depend on the relative number of women tested.

B. Sensitizing Forms of Nickel

Like metallic cobalt, but different from metallic chromium, nickel as metal acts as a sensitizer. Most water-soluble nickel salts such as chloride and sulfate also have strong sensitizing potentials. In addition, oxides such as Ni_2O_3 and the hydroxide $Ni(OH)_2$ can elicit contact dermatitis, whereas heated NiO does not.[125,126] Detergents may contain nickel compounds and have been implicated as one reason for the higher incidence of nickel dermatitis in women than in men.[3]

Nickel containing alloys have been used as biomaterials in a variety of prostheses and implanted devices. The most commonly used alloys are stainless steel and the cobalt chromium types. Stainless steel for these purposes usually contain 10 to 14% nickel, whereas wires and probes connecting battery pacemakers with the heart muscle may contain as much as 35% nickel.[127] Some of the base metal alloys used as substitutes for gold alloys in dental work contain up to 80% nickel.[128]

The cobalt-chromium type of casting alloy contains only minor amounts of nickel (up to 2%), but this concentration has been sufficient to elicit allergic responses in certain individuals.[129,130]

C. Clinical Manifestations

Except for a few reports on asthmatic reactions to inhalation of nickel compounds,[83,131] most literature on nickel allergy is concerned with the delayed type of hypersensitivity reactions and various forms of dermatitis. The early cases of nickel dermatitis in nickel miners, smelters, nickel-plating workers, and so on were described as the "nickel itch". The eruption began as an itching or burning papular erythema of the finger webs, spreading to the fingers, wrists, and forearms.[3] Women are most often sensitized by objects worn close to the skin, such as buckles, studs, costume jewelry, and earrings.[122,132] Men are usually sensitized by occupational exposure.

In many patients, nickel dermatitis occurs as a papular or papulovesicular dermatitis with a tendency for lichenification. The eruption often has the characteristics of atopic dermatitis rather than eczematous contact allergy.[3] The topographic distribution pattern of the eruption is also characteristic of the nickel allergy, often spreading to regions distant from the primary affection.[122,133] In an analysis of 400 cases, Calnan[134] classified the pattern of nickel dermatitis into three groups: primary, indicating areas of direct contact with metal; secondary, indicating involvement of selective symmetric areas such as the elbow flexures and the flexor surfaces of the areas; and associated, indicating areas of dermatitis that have no apparent relation to the nickel exposure.

Primary nickel dermatitis of the hands is particularly common in certain occupations such as hairdressing and in the catering, tailoring, and metal industries. Persons previously sensitized by nickel-containing objects worn close to the skin have a considerably increased risk of dermatitis during employment in these industries.[126]

D. Allergic Reactions to "Internal" Nickel

A more recently recognized feature in metal allergy is the dermatitis associated with metal appliances in artifical joint prostheses, in surgical fixation of bone fractures, and in some dental prostheses.[135-140] A number of metals have been implicated, such as gold, chromium, nickel, molybdenum, and vanadium of which nickel and chromates appear to be encountered most often.

Both urticarial[130,136] and eczematous nickel dermatitis[135,136,140] have been attributed to implanted prostheses. Patch tests with nickel have been positive in the patients with excematous dermatitis, and in most cases removal of the metallic devices has resulted in complete resolution of urticaria and involution of the eczema.

These observations imply that metals such as nickel are released from metallic devices and produce the reaction in nickel-sensitive patients. This view is corroborated by the finding of an increased nickel concentration in parenchymal tissues from implantation of stainless steel rods.[141,142] Corrosion has also been shown in simulated in vivo conditions both for cobalt-chromium and stainless steel alloys. Such corrosion is enhanced by cyclic stressing, resembling the conditions encountered in the use of weight-bearing orthopedic implants.[143] Clinically, the sensitization also seems to be correlated with the wear characteristics of the prostheses, a high wear rate indicating higher probability for the development of metal sensitivity,[144] metal-to-metal hip joint replacements being of specific importance.

There is a high incidence of prosthesis failure among metal-sensitive patients. However, it is an unresolved question whether the loosening is caused by the sensitization or vice versa.[145,146] Furthermore, reactions of clinical importance do not occur in all nickel-sensitive patients. No adverse reactions were observed during 22 months after insertion of a mitral valve using a nickel-containing replacement in a patient with proven nickel sensitivity.[147] As pointed out by Pegum[148] the time factor may be an important aspect in this connection, since cases of dermatitis and suppuration around stainless steel plates have occurred as long as 3 years after the insertion.

Another important aspect may be the increased corrosion from nickel-containing devices taking place when dissimilar metals are used for implant screws and plates[149] and when nickel-plating is worn off in objects having continuing skin contact. The nickel coating may be relatively uncorrodible, but the close contact with exposed brass or copper covered by a conducting liquid, such as sweat, puts a corrosion cell in operation. The nickel component of the cell is anodic and tends to dissolve.[132,150]

Another suggestion is that corrosion may be enhanced by minute remnants of metal from burs and instruments used in orthopedic surgery, being of dissimilar composition from the implanted devices.[151]

E. Diagnosis of Hypersensitivity

The diagnosis of allergic nickel contact dermatitis is based on the case history, the character and distribution of the eruptions, and on skin testing. The routinely performed patch testing employs a concentration of 5% nickel sulfate in water.[126] Higher concentrations increase the possibility of irritative reactions,[152] and even a 5% concentration of nickel sulfate may result in undesirable irritant reactions, particularly in children.[153] However, using a 1% concentration of nickel sulfate may result in as much as 20% false negative reactions.[153] With this background, the North American Contact Dermatitis Group recommend using a 2.5% solution of nickel sulfate. If the chloride is used, an equivalent molar concentration is recommended.[3]

Intradermal tests with a 1:10,000 dilution of nickel sulfate yield results comparable to patch tests with 5% nickel sulfate.[154-156] Gottmann-Lückerath et al.[157] advocate that intracutaneous tests should not be conducted except in cases with a positive history and a negative patch test and for distinguishing specific patch test reactions from nonspecific. The background for this caution is the increased possibility of sensitization by this diagnostic procedure.

The risk of increased sensitization and, occasionally, persistence of allergic lesions may also contraindicate using patch testing. In such instances, certain in vitro tests may be advantageous in establishing the diagnosis.

F. In Vitro Tests

It has been observed that the addition of 0.1% nickel sulfate to the serum of nickel-sensitive persons resulted in a broader precipitation line than in the nonsensitive control, whereas complement fixation tests such as Rose, Waaler, Combs, etc. give negative results.[158] Such observations have prompted the use of the lymphocyte transformation test (LTT) and the macrophage inhibition tests (MIF).

The LTT, estimated morphologically or by the uptake of ^{14}C thymidine during the mitoses of lymphocytes, has given positive results when lymphocyte cultures from nickel-sensitive persons are challenged with 2.5% nickel sulfate or nickel acetate,[158,159] neither salts acting as nonspecific stimulators. Serial dilution of nickel sulfate has given two types of response with the LTT test in nickel-sensitive persons. In one group, the lymphoblast transformation was highest at the lowest nickel salt concentration, whereas in the other group higher concentration of nickel gave higher LTT values.[160]

The production of macrophage inhibition factor has been positive with nickel sulfate stimulation in some nickel-sensitive persons.[158,161] In other cases using nickel acetate it has been difficult to distinguish the stimulation from nonspecific reactions.[162] The similar leukocyte migration inhibition test (LMI) has also been used in cases of nickel dermatitis. Both patch test-positive and patch test-negative persons showed positive inhibition of leukocyte migration provided that the nickel stimulation took place with a sulfate-albumin conjugate and not only with unconjugated nickel sulfate.[163]

G. Treatment of Nickel Allergy

As in other allergies there is no cure except the avoidance of the offending allergen. Since nickel is a common part of many everyday appliances and is present in food, this poses quite a problem.

Nickel in studs and buckles can be avoided once the patient recognizes the problem. Some degree of protection has also been afforded by coating them with finger-nail polish or polyurethane.[164,165] Nickel in food is more difficult to cope with. One study showed that a recurrent nickel eczema could be improved by decreasing the nickel ingestion, compiling the diet on the basis of the literature. The dermatitis flared when a normal diet was resumed.[166] Other studies have employed diethyldithiocarbamate and tetraethylthiuramdisulfide which may complex with nickel (Ni^{2+}) entering the body and thereby inactivating it. There was some improvement with both substances.[167,168]

VI. SUMMARY

Nickel is an essential element present in food and the environment in varying concentrations and in various forms. Nickel alloys are commonly met within a series of man made objects including biomaterials. The absorption of nickel from the gastrointestinal tract is low and the fecal excretion is high. Inhaled or otherwise absorbed nickel is excreted by the kidneys and in some cases by sweating, keeping the nickel concentration within a certain range. Experimental data on rodents show retention of (single dose) nickel in lungs and kidneys, passage through the placenta and mammary glands, and some retention in brain tissue of fetuses. Nickel is present as a large molecular component (nickeloplasmin), as albumin-conjugated nickel, and as nickel joined to certain small molecular substances in serum. Nickel is also present intracellularly. The occurrence of direct toxic effects of orally ingested nickel is experimentally proven, but requires very large doses. Acute toxic effects of nickel in man are limited to poisoning by nickel carbonyl. However, nickel has other toxicological aspects in man, notably carcinogenesis and hypersensitivity reactions. The carcinogenic effects of nickel are focussed on the respiratory and possibly renal cancers observed in nickel workers exposed to nickel vapors for years. Experimental evidence indicates that both metallic nickel, nickel salts, and certain other nickel compounds are carcinogenic. Nickel subsulfide has been pointed out as specifically potent. Hypersensitive reactions to nickel are common, especially in women. The manifestations cover the range from small affections of the finger webs to generalized dermatitis with systemic symptoms. On this basis, the most important pathologic aspects of nickel-containing medical and dental biomaterials are the hypersensitive reactions. The slow dissolution of metals from such prosthetic implants and mucosal prostheses may act as sensitizing agents, and previously sensitized persons may react to the introduction of such materials into the body. In both cases undesired results may ensue. The carcinogenic implications of nickel are difficult to assess in this context. Theoretically, certain health personel such as dental technicians occupied with welding, grinding, and polishing of nickel containing alloys might run an occupational risk. Also, in view of the nonthreshold concept of chemical carcinogenesis, this point cannot be excluded from the discussion on biologic aspects of biomaterials. In addition, the total body burden of carcinogenic substances such as nickel met with in the environment and the possibility of unknown synergistic effects should be kept in mind.

REFERENCES

1. **Sunderman, F. W.**, Nickel poisoning, in *Laboratory Diagnosis of Diseases Caused by Toxic Agents,* Sunderman, F. W. and Sunderman, F. W., Jr., Eds., Warren H. Green, Inc., St. Louis, 1970, chap. 36.
2. **Sunderman, F. W., Jr.**, The current status of nickel carcinogenesis, *Ann. Clin. Lab. Sci.,* 3, 157, 1973.
3. **Sunderman, F. W., Jr., Coulston, F., Eichorn, G. L., Fellows, J. A., Mastromatteo, E., Reno, H. T., and Samitz, M. H.**, *Nickel. A Report of the Committee on Medical and Biologic Effects of Environmental Pollutants,* National Academy of Sciences, Washington, D.C., 1975.
4. International Agency for Research on Cancer, *Evaluation of the Carcinogenic Risk of Chemicals to Man,* Vol. M., World Health Organization, Lyon, 1976, 75.
5. **Sunderman, F. W., Jr.**, A review of the carcinogenicity of nickel, chromium and arsenic, *Prev. Med.,* 5, 279, 1976.
6. **Sunderman, F. W., Jr.**, A review of the metabolism and toxicity of nickel, *Ann. Clin. Lab. Sci.,* 7, 377, 1977.
7. **Sunderman, F. W., Jr.** Carcinogenic effects of metals, *Fed. Proc. Fed. Am. Soc. Exp. Biol.,* 37, 40, 1978.
8. **Gilman, J. P. W.**, Metal carcinogenesis. II. A study on the carcinogenetic activity of cobalt, copper, iron, and nickel compounds, *Cancer Res.,* 22, 158, 1962.
9. **Schroeder, H. A.**, A sensible look at air pollution by metals, *Arch. Environ. Health,* 21, 798, 1970.
10. **Hoffman, D. and Wynder, E. L.**, Smoking and occupational cancers, *Prev. Med.,* 5, 245, 1976.
11. **Schroeder, H. A., Balassa, J. J., and Tipton, I. H.**, Abnormal trace elements in man — nickel, *J. Chronic Dis.,* 15, 51, 1962.
12. **Dencker, I., Fristedt, B., and Øvrum, P.**, Järn, kopper, krom, nickel, bly och kobolt i føda, *Läkartidningen,* 68, 4027, 1971.
13. **Katz, S. A., Bowen, H. J. M., Comaish, J. S., and Samitz, M. H.**, Tissue nickel levels and nickel dermatitis. I. Nickel in hair, *Br. J. Dermatol.,* 92, 187, 1975.
14. **Lagesson, V. and Andrasko, L.**, Direct determination of nickel in blood by electrothermal atomic absorption spectrometry, *Ann. Clin. Lab. Sci.,* 8, 496, 1978.
15. **Mikac-Dević, D., Sunderman, F. W., Jr., and Nomoto, S.**, Furildioxime method for nickel analysis in serum and urine by electrothermal atomic absorption spectrometry, *Clin. Chem.,* 23, 948, 1977.
16. **Nieboer, E., Flora, C. J., Tomassini, F. D., and Cecutti, A. G.**, Pulse polarographic determination of nanogram quantities of nickel in urine and blood, *Ann. Clin. Lab. Sci.,* 8, 497, 1978.
17. **Nomoto, S. and Sunderman, F. W., Jr.**, Atomic absorption spectrometry of nickel in serum, urine and other biological materials, *Clin. Chem.,* 16, 477, 1970.
18. **Rendic, D., Holjević, S., Valković, B. Sc. V., Zabel, T. H., and Phillips, G. C.**, Trace element concentrations in human hair measured by proton-induced X-ray emission, *J. Invest. Dermatol.,* 66, 371, 1976.
19. **Sunderman, F. W., Jr.**, Atomic absorption spectrometry of trace metals in clinical pathology, *Hum. Pathol.,* 4, 549, 1973.
20. **Kazprzak, K. S., Górski, A., and Kopozynski, T.**, New oximes for extraction of nickel from biological materials, *Ann. Clin. Lab. Sci.,* 8, 497, 1978.
21. **Schroeder, H. A., Balassa, J. J., and Tipton, I. H.**, Abnormal trace metals in man — nickel, *J. Chronic Dis.,* 15, 51, 1962.
22. **Sumino, K., Hayakawa, K., Shibata, T., and Kitamura, S.**, Heavy metals in normal Japanese tissue, *Arch. Environ. Health,* 30, 487, 1975.
23. **McNeely, M. D., Nechay, M. W., and Sunderman, F. W., Jr.**, Measurements of nickel in serum and urine as indices of environmental exposure to nickel, *Clin. Chem.,* 18, 992, 1972.
24. **Hohnadel, D. C., Sunderman, F. W., Jr., Nechay, M. W., and McNeely, M. D.**, Atomic absorption spectrometry of nickel, copper, zinc, and lead in sweat collected from healthy subjects during sauna bathing, *Clin. Chem.,* 19, 1288, 1973.
25. **Catalanatto, F. A., Sunderman, F. W., Jr., and MacIntosh, T. R.**, Nickel concentrations in human parotid saliva, *Ann. Clin. Lab. Sci.,* 7, 146, 1977.
26. **Kirchgessner, M. and Schnegg, A.**, Nickel content of milk from lactating rats fed varying nickel levels, *Arch. Tierernaeh.,* 11, 774, 1976.
27. **Jacobsen, N., Alfheim, I., and Jonsen, J.**, Nickel and strontium distribution in some mouse tissues. Passage through placenta and mammary glands, *Res. Commun. Chem. Pathol. Pharmacol.,* 20, 571, 1978.
28. **Nechay, M. W. and Sunderman, F. W., Jr.**, Measurements of nickel in hair by atomic absorption spectrometry, *Ann. Clin. Lab. Sci.,* 3, 30, 1973.

29. **Schroeder, H. A. and Nason, A. P.,** Trace metals in human hair, *J. Invest. Dermatol.,* 53, 71, 1969.
30. **Spruit, D. and Bongaarts, P. J. M.,** Nickel content of plasma, urine, and hair in contact dermatitis, *Dermatol.,* 154, 291, 1977.
31. **Scheiner, D. M., Katz, S. A., and Samitz, M. H.,** Nickel levels in hair and nickel ingestion in guinea-pigs, *Environ. Res.,* 12, 355, 1976.
32. **D'Alonzo, C. A. and Pell, S.,** A study of trace metals in myocardial infarction, *Arch. Environ. Health,* 6, 381, 1963.
33. **McNeely, M. D., Sunderman, F. W., Jr., Nechay, M. W., and Levine, H.,** Abnormal concentration of nickel in serum in cases of myocardial infarction, stroke, burns, hepatic cirrhosis and uremia, *Clin. Chem.,* 17, 1123, 1971.
34. **Sunderman, F. W., Jr., Nomoto, S., Pradhan, A. M., Levine, H., Bernstein, S. H., and Hirsch, R.,** Increased concentrations of serum nickel after acute myocardial infarction, *N. Engl. J. Med.,* 283, 896, 1970.
35. **Clary, J. J.,** Nickel chloride-induced metabolic changes in the rat and guinea pig, *Toxicol. Appl. Pharmacol.,* 31, 55, 1975.
36. **Parker, K. and Sunderman, F. W., Jr.,** Distribution of ^{63}Ni in rabbit tissues following intravenous injections of $^{63}NiCl_2$, *Res. Commun. Chem. Pathol. Pharmacol.,* 7, 755, 1974.
37. **Smith, J. C. and Hackley, B.,** Distribution and excretion of nickel-63 administered intravenously to rats, *J. Nutr.,* 95, 541, 1968.
38. **Wase, A. W., Goss, D. M., and Boyd, M. J.,** The metabolism of nickel. I. Spatial and temporal distribution of ^{63}Ni in the mouse, *Arch. Biochem. Biophys.,* 51, 1, 1954.
39. **Jacobsen, N. and Jonsen, J.,** Strontium, lead and nickel incorporation into mouse calvaria in vitro, *Pathol. Eur.,* 10, 115, 1975.
40. **Oskarsson, A.,** Tissue binding $^{63}Ni(II)$ in mice studied by whole-body autoradiography, *Ann. Clin. Lab. Sci.,* 8, 499, 1978.
41. **Sunderman, F. W., Jr., Shen, S. K., Mitchell, J. M., Allpass, P. R., and Damjanov, I.,** Embryo toxicity and fetal toxicity of nickel in rats, *Toxicol. Appl. Pharmacol.,* 43, 381, 1977.
42. **Van Soestbergen, M. and Sunderman, F. W., Jr.,** ^{63}Ni-complexes in rabbit serum and urine after injection of $^{63}NiCl_2$, *Clin. Chem.,* 18, 1478, 1972.
43. **Sunderman, F. W., Jr., and Selin, C. T.,** The metabolism of nickel-63 carbonyl, *Toxicol. Appl. Pharmacol.,* 12, 207, 1968.
44. **Fedeschi, R. E. and Sunderman, F. W., Jr.,** Nickel poisoining. V. The metabolism of nickel under normal conditions and after exposure to nickel carbonyl, *A. M. A. Arch. Ind. Health,* 16, 486, 1957.
45. **Horak, E. and Sunderman, F. W., Jr.,** Fecal nickel excretion by healthy adults, *Clin. Chem.,* 19, 429, 1973.
46. **Nomoto, S., Decsy, M. I., Murphy, J. R., and Sunderman, F. W., Jr.,** Isolation of ^{63}Ni-labelled nickeloplasmin from rabbit serum, *Biochem. Med.,* 8, 171, 1973.
47. **Callan, W. M. and Sunderman, F. W., Jr.,** Species variations in binding of $^{63}NiCl_2$ by serum albumin, *Res. Commun. Chem. Pathol. Pharmacol.,* 5, 459, 1973.
48. **Asato, N., Van Soenstbergen, M., and Sunderman, F. W., Jr.,** Binding of $^{63}Ni(II)$ to ultrafiltrable constituents of rabbit serum *in vivo* and *in vitro, Clin. Chem.,* 21, 521, 1975.
49. **Himmelhoch, S. R., Sober, H. A., Vallee, B. L., Peterson, E. A., and Fuwa, K.,** Spectrographic and chromatographic resolution of metalloproteins in human serum, *Biochemistry,* 5, 2523, 1966.
50. **Decsy, M. I. and Sunderman, F. W., Jr.,** Binding of ^{63}Ni to rabbit serum α_1-macroglobulin *in vivo* and *in vitro, Bioinorgan. Chem.,* 3, 95, 1974.
51. **Nomoto, S. McNeely, M. D., and Sunderman, F. W., Jr.,** Isolactine of a nickel α_2-macroglobulin from rabbit serum, *Biochemistry,* 10, 1647, 1971.
52. **Haupt, H., Heimburger, N., Kranz, T., and Baudner, S.,** Human serum proteins with a high affinity for carboxymethyl cellulose. III. Physical-chemical and immunological characterization of a metal-binding 9.55-α_1-glycoprotein (CM-Protein III), *Z. Physiol. Chem.,* 353, 1841, 1972.
53. **Jacobsen, N., Brennhovd, I., and Jonsen, J.,** Human submandibular gland tissue in culture. II. Nickel affinity to secretory proteins, *J. Biol. Buccl.,* 5, 169, 1977.
54. **Hofsøy, H., Paulsen, G., and Jonsen, J.,** Secretion of nickel in rabbit saliva, *Ann. Clin. Lab. Sci.,* in press.
55. **Dixon, N. E., Gazzola, C., Blakeley, R. L., and Zerner, B.,** Jack bean urease (EC 3.5.1.5). A metalloenzyme. A simple biological role for nickel?, *J. Am. Chem. Soc.,* 97, 4131, 1975.
56. **Chen, J. K. M., Haro, R. T., and Furst, A.,** Excretion of nickel compounds by the rat: blood and urine levels, *Wasmann J. Biol.,* 29, 1, 1971.
57. **Peters, T., Jr.,** Serum albumin: recent progress in the understanding of its structure and biosynthesis, *Clin. Chem.,* 23, 5, 1977.
58. **Peters, T., Jr. and Blumenstock, F. A.,** Copper binding properties of bovine serum albumin and its amino-terminal peptide fragment, *J. Biol. Chem.,* 242, 1574, 1967.

59. **Hendel, R. C. and Sunderman, F. W., Jr.**, Species variations in the proportions of ultrafiltrable and protein-bound serum nickel, *Res. Commun. Chem. Pathol. Pharmacol.*, 4, 141, 1972.
60. **Sunderman, F. W., Jr., Kasprzak, K. S., Lau, T. J., Minghetti, P. P., Maenza, R. M., Becker, N., Onkelinx, C., and Goldblatt, P. J.**, Effects of manganese on carcinogenicity and metabolism of nickel subsulfide, *Cancer Res.*, 30, 1645, 1970.
61. **Paulsen, G. Hofsøy, H., and Jonsen, J.**, Isolation of a possible nickel-binding peptide in the rabbit serum, *Ann. Clin. Lab. Sci.*, 9, 500, 1979.
62. **Piotrowska, J. K. and Szymenska, J. A.**, Influence of certain metals on the level of metallothionein-like proteins in the liver and kidney of rats, *J. Toxicol. Environ. Health*, 1, 991, 1976.
63. **Sabbioni, E. and Marafante, E.**, Heavy metals in rat liver cadmium binding protein, *Environ. Physiol. Biochem.*, 5, 132, 1975.
64. **Webb, M. and Weinzierl, S. M.**, Uptake of ^{63}Ni from its complexes with proteins and other ligands by mouse dermal fibroblasts in vitro, *Br. J. Cancer*, 26, 292, 1972.
65. **Nielsen, F. H. and Ollerich, D. A.**, Nickel: A new essential trace element, *Fed. Proc. Fed. Am. Soc. Exp. Biol.*, 33, 1767, 1974.
66. **Nielsen, F. H. and Sauberlich, H. E.**, Evidence for a possible requirement for nickel by the chick, *Proc. Soc. Exp. Biol. Med.*, 134, 845, 1970.
67. **Sunderman, F. W., Jr., Nomoto, S., Morang, R., Nechay, M. W., Bruke, C. N., and Nielsen, S. W.**, Nickel deprivation in chicks, *J. Nutr.*, 102, 259, 1972.
68. **Schnegg, A. and Kirchgessner, M.**, Changes in the hemoglobin content, erythrocyte count and hematocrit in nickel deficiency, *Nutr. Metabol.*, 19, 268, 1975.
69. **Schnegg, A. and Kirchgessner, M.**, Essentiality of nickel for the growth of animals, *Z. Tierphysiol. Tierernaeh. Futtermittelkd.*, 36, 63, 1975.
70. **Schnegg, A. and Kirchgessne, M.**, Absorption and metabolic efficiency of iron during nickel deficiency, *Intern. J. Vitamin Nutr. Res.*, 46, 96, 1976.
71. **Storinger, H. E.**, Nickel, in *Industrial Hygiene and Toxicology*, Vol. 2, 2nd ed., Fassett, D. W. and Irish, D. D., Eds., Interscience, New York, 1963, 1118.
72. **Phatak, S. S. and Patwardhan, V. N.**, Toxicity of nickel, *J. Sci. Ind. Res.*, 9b(3), 70, 1950.
73. **O'Dell, G. D., Miller, W. J., King, W. A., Moore, S. L., and Blackmon, D. M.**, Nickel toxicity in the young bovine, *J. Nutr.*, 100, 1447, 1970.
74. **Schnegg, A. and Kirchgessner, M.**, Für Toxizität von alimentär herabreichtem Nickel, *Landwirtsch. Forsch.*, 29, 177, 1976.
75. **Clary, J. J.**, Nickel chloride-induced metabolic changes in the rat and guinea-pig, *Toxicol. Appl. Pharmacol.*, 31, 55, 1975.
76. **Storak, E. and Sunderman, F. W., Jr.**, Effects of Ni(II) upon plasma glucagon and glucose in rats, *Toxicol. Appl. Pharmacol.*, 33, 388, 1975.
77. **Gitlitz, P. H., Sunderman, F. W., Jr., and Goldblatt, P. J.**, Aminoaciduria and proteinuria in rats after a single intraperitoneal injection of Ni(II), *Toxicol. Appl. Pharmacol.*, 34, 430, 1975.
78. **Schroeder, H. A. and Mitchener, M.**, Toxic effects of trace elements on the reproduction of mice and rats, *Arch. Environ. Health*, 23, 102, 1971.
79. **Ambrose, A. M., Larson, P. S., Barzelleca, J. F., and Henninger, G. R., Jr.**, Long term toxicologic assessment of nickel in rats and dogs, *J. Food Sci. Technol.*, 13, 181, 1976.
80. **Sunderman, F. W., Jr., Shen, S. K., Mitchell, J. M., Allpass, P. R., and Damjanov, I.**, Embryo toxicity and fetal toxicity of nickel in rats, *Toxicol. Appl. Pharmacol.*, 43, 381, 1977.
81. **Ferm, V. H.**, The teratogenic effects of metals on mammalian embryos, *Adv. Teratol.*, 5, 51, 1972.
82. **Torjussen, J. and Solberg, L. A.**, Histological findings in the nasal mucosa of nickel workers, *Acta Oto-Laryngol.*, 82, 266, 1976.
83. **McConnell, L. H., Fink, J. N., Schlueter, D. P., and Schmidt, M. G., Jr.**, Asthma caused by nickel sensitivity, *Ann. Intern. Med.*, 78, 888, 1973.
84. **Hackett, R. L. and Sunderman, F. W., Jr.**, Pulmonary alveolar reactions to nickel carbonyl. Ultrastructural and histochemical studies, *Arch. Environ. Health*, 16, 349, 1968.
85. **Kasprzak, K. S. and Sunderman, F. W., Jr.**, The metabolism of nickel carbonyl ^{14}C, *Toxicol. Appl. Pharmacol.*, 15, 195, 1969.
86. **Sunderman, F. W., Jr., Alpass, P., Mitchell, J., and Alpart, D. M.**, Ophthalmic malformations in rats following prenatal exposure to in relation of nickel carbonyl, *Ann. Clin. Lab. Sci.*, in press.
87. **Jacobsen, N.**, Epithelial-like cells in culture derived from human gingiva: response to nickel, *Scand. J. Dent. Res.*, 85, 567, 1977.
88. **Waters, M. D., Gardner, D. E., Sranuyi, C., and Coffin, D. L.**, Metal toxicity for rabbit alveolar macrophages *in vitro*, *Environ. Res.*, 9, 32, 1975.
89. **Olsen, I. and Jonsen, J.**, Effect of cadmium acetate, copper sulfate and nickel chloride on organ cultures of mouse trachea, *Acta Pharmacol. Toxicol.*, 44, 120, 1979.

90. **Holy, M. J.**, The effect of metabolic salts on the histology and functioning of rat testis, *J. Reprod. Fertil.*, 12, 461, 1966.
91. **Shaw, T. L. and Brown, V. M.**, Heavy metals and the fertilization of rainbow trout eggs, *Nature (London)*, 230, 251, 1971.
92. **Storeng R. and Jonsen, J.**, personal communication, 1978.
93. **Searle, C. E.**, Chemical carcinogens and their significance for the chemist, *Chem. Br.*, 6, 5, 1970.
94. **Barton, R.**, Nickel carcinogenesis of the respiratory tract, *J. Oto-Laryngol.*, 6, 412, 1977.
95. **Doll, R., Morgan, L. G., and Speizer, F. E.**, Cancers of the lung and nasal sinuses in nickel workers, *Br. J. Cancer*, 24, 623, 1970.
96. **Sunderman, F. W., Jr.**, Nickel carcinogenesis epidemiology of respiratory cancer among nickel workers, *Dis. Chest.*, 54, 527, 1968.
97. **Pedersen, E. A., Høgetveit, A. C., and Andersen, A.**, Cancer of the respiratory organs among workers at a nickel refinery in Norway, *Int. J. Cancer*, 12, 32, 1973.
98. **Solberg, L. A. and Torjussen, W.**, Histological changes in nasal mucosa of workers in an electrolytic refinery, *Ann. Clin. Lab. Sci.*, 8, 503, 1978.
99. **Hueper, W. C. E.**, Experimental studies in metal carcinogenesis. IV. Cancer produced by parenterally introduced metallic nickel, *J. Natl. Cancer Inst.*, 15, 55, 1955.
100. **Hueper, W. C. E.**, Experimental studies in metal carcinogenesis. IX. Pulmonary lesions in guinea pigs and rats exposed to prolonged inhalation of powdered metallic nickel, *Arch. Pathol.*, 65, 600, 1958.
101. **Damjanov, I., Mitchell, J. M., Allpass, R. R., Bigazzi, P., and Sunderman, F. W., Jr.**, Induction of rhabdomyosarcomas and fibrosarcomas by intra testicular injection of nickel subsulfide in rats, *Proc. Am. Assoc. Cancer Res.*, 18, 52, 1977.
102. **Sunderman, F. W., Jr., Kasprzak, K. S., Lau, T. J., Minghem, P. P., Maenza, R. M., Becker, N., Onkelinx, C., and Goldblatt, P. J.**, Effects of manganese on carcinogenicity and metabolism of nickel subsulfide, *Cancer Res.*, 36, 1790, 1976.
103. **Sunderman, F. W., Jr. and Maenza, R. M.**, Comparison of carcinogenecities of nickel compounds in rats, *Res. Commun. Chem. Pathol. Pharmacol.*, 14, 319, 1976.
104. **Ottolenghi, A. D., Haseman, J. K., Payne, W. W., Falk, H. L., and MacFarland, H. N.**, Inhalation studies of nickel sulfide in pulmonary carcinogenesis of rats, *J. Natl. Cancer Inst.*, 54, 1165, 1975.
105. **Mitchell, J. M. and Allpass, P. R.**, Induction of testicular sarcomas in Fischer rats by intratesticular injection of nickel subsulfide, *Cancer Res.*, 38, 268, 1978.
106. **Jasmin, G. and Riopelle, J. L.**, Renal carcinomas and erythrocytosis in rats following intrarenal injection of nickel subsulfide, *Lab. Invest.*, 35, 71, 1976.
107. **Sunderman, F. W., Jr.**, Comparison of carcinogenecities of nickel compounds in rats, *Res. Commun. Chem. Pathol. Pharmacol.*, 2, 319, 1976.
108. **Kasperzak, K. S., Marchow, L., and Breborowitez, J.**, Pathological reactions in rat lungs following intratracheal injections of nickel subsulfide and 3,4-benzpyrene, *Res. Commun. Chem. Pathol. Pharmacol.*, 6, 237, 1973.
109. **Costa, M., Nye, J., and Sunderman, F. W., Jr.**, Morphological transformation of Syrian hamster cells induced by nickel compounds, *Ann. Clin. Lab. Sci.*, 8, 502, 1978.
110. **Stern, R. M.**, Risks in chromium and nickel exposure for stainless steel welders, *Ann. Clin. Lab. Sci.*, 8, 501, 1978.
111. **Weinzierl, S. M. and Webb, M.**, Interaction of carcinogenic metals with tissue and body fluids, *Br. J. Cancer*, 26, 279, 1972.
112. **Heath, J. C., Webb, M., and Cottsey, M.**, The interactions of carcinogenic metals with tissues and body fluids, *Br. J. Cancer*, 23, 153, 1969.
113. **Sunderman, F. W., Jr., Kaspezak, K. S., Lau, T. J., Minghetti, P. P., Maenza, R., Becker, N., Onkelinx, C., and Goldblatt, P. J.**, Effects of manganese on carcinogenesis and metabolism of nickel subsulfide, *Cancer Res.*, 36, 1790, 1976.
114. **Webb, M., Heath, J. C., and Hopkins, T.**, Intranuclear distribution of the inducing metal in primary rhabdomyosarcomata induced in the rat by nickel, cobalt and cadmium, *Br. J. Cancer*, 26, 274, 1972.
115. **Swierenga, S. H. H. and Basrur, P. K.**, Effect of nickel on cultured rat embryo muscle cells, *Lab. Invest.*, 19, 663, 1968.
116. **Sunderman, F. W., Jr. and Esfahani, M.**, Nickel carbonyl inhibition of RNA polymerase activity in hepatic nuclei, *Cancer Res.*, 28, 2565, 1968.
117. **Laugier, P., Foussereau, J., and Bulté, C. I.**, Les eczemas par allergie au nickel, *Rev. Fr. Allergol.*, 6, 1, 1966.
118. Epidemiology of contact dermatitis in North America: 1972, *Arch. Dermatol.*, 108, 537, 1973.
119. **Baer, R. L., Ramsey, D. L., and Biondi, E.**, The most common contact allergens 1968—1970, *Arch Dermatol.*, 108, 74, 1973.

120. **Magnusson, B., Blohm, S.-G., Fregerth, S., Hjorth, N., Hovding, G., Pirilä, V., and Skog, E.**, Routine patch testing. IV. Supplementary series of test substances for Scandinavian countries, *Acta Derm. Venereol.*, 48, 110, 1968.
121. **Baer, R. L., Lipkin, G., Kanof, N. B., and Biondi, E.**, Changing patterns of sensitivity to common contact allergens, *Arch. Dermatol.*, 89, 3, 1964.
122. **Calnan, C. D.** Nickel dermatitis, *Br. J. Dermatol.*, 68, 229, 1956.
123. **Marcussen, P. V.**, Specificity of patch tests with 5% nickel sulphate, *Acta Derm. Venerol.*, 39, 187, 1959.
124. **Malten, K. E. and Spruit, D.**, The relative importance of various environmental exposures to nickel in causing contact hypersensitivity, *Acta Derm. Venerol.*, 49, 14, 1969.
125. **Fregerth, S. and Rorsman, H.**, Allergy to chromium, nickel and cobalt, *Acta Derm. Venereol.* 46, 144, 1966.
126. **Hjort, N. and Fregerth, S.**, Contact dermatitis, in *Textbook of Dermatology*, Rook, A., Wilkinson, D. S., and Ebling, F. J. G., Eds., Blackwell, Oxford, 1968, chap. 14.
127. **Samitz, M. H. and Katz, S. A.**, Nickel dermatitis hazards from prostheses. In vivo and in vitro solubilization studies, *Br. J. Dermatol.*, 92, 287, 1975.
128. **Hensten-Pettersen, A., and Jacobsen, N.**, Nickel corrosion of non-precious casting alloys and the cytotoxic effect of nickel in vitro, *J. Bioeng.*, 2, 419, 1978.
129. **Wood, J. F. L.**, Mucosal reaction to cobalt-chromium alloy, *Br. Dent. J.*, 135, 423, 1974.
130. **Symeonides, P. P., Paschaloglou, C., and Papageorgiou, S.**, An allergic reaction after internal fixation of a fracture using a Vitallimum plate, *J. Allergy Clin. Immunol.* 51, 251, 1973.
131. **Sunderman, F. W. and Sunderman, F. W., Jr.**, Loffler's syndrome associated with nickel sensitivity, *Arch. Int. Med.*, 107, 405, 1961.
132. **Lunn, J. A.**, The boyfriend's diagnosis: an unusual case of nickel sensitivity, *Practitioner*, 220, 312, 1978.
133. **Marcussen, P. V.**, Spread of nickel dermatitis, *Dermatol.*, 115, 596, 1957.
134. **Calnan, C. D.**, Nickel sensitivity in women, *Int. Arch. Allergy Appl. Immunol.*, 11, 73, 1957.
135. **Foussereau, J. and Laugier, P.**, Allergic eczemas from metallic foreign bodies, *Trans. St. John's Hosp. Dermatol. Soc.*, 52, 220, 1966.
136. **Brendlinger D. L. and Tarsitano, N. J.**, Generalized dermatitis due to sensitivity to a chrome cobalt removable partial denture, *J. Am. Dent. Assoc.*, 81, 392, 1970.
137. **McKenzie, A. W., Aitken, C. V., and Ridsill-Smith, R.**, Urticaria after insertion of Smith-Petersen nail, *Br. Med. J.* 4, 36, 1967.
138. **Watt, T. L. and Baumann, R. R.**, Nickel earlobe dermatitis, *Arch. Dermatol.*, 98, 155, 1968.
139. **Benson, M. K. D., Goodwin, P. G., and Brostoff, J.**, Metal sensitivity in patients with joint replacement arthroplasties, *Br. Med. J.*, 4, 374, 1975.
140. **Barranco, V. P. and Soloman, H.**, Eczematous dermatitis from nickel, *J. Am. Med. Assoc.*, 220, 1244, 1972.
141. **Ferguson, A. B., Jr., Akahoshi, Y., Laing, P. G. and Hodge, E. S.**, Characteristics of trace ions released from embedded metal implants in the rabbit, *J. Bone Jt. Surg. Am. Vol.*, 44, 323, 1962.
142. **Mears, D. C.**, Electron probe microanalysis of tissues and cells from implant areas, *J. Bone Jt. Surg. Br. Vol.*, 45, 567, 1966.
143. **Cohen, J.**, Corrosion testing of orthopaedic implants, *J. Bone Jt. Surg. Am. Vol.*, 44, 307, 1962.
144. **Elves, M. W., Wilson, J. N., Scales, J. T., and Kemp, H. B. S.**, Incidence of metal sensitivity in patients with total joint replacements, *Br. Med. J.*, 4, 376, 1975.
145. **Deutman, R., Mulder, T. J., Brian, R., and Nater, J. P.**, Metal sensitivity before and after total hip arthroplasty, *J. Bone Jt. Surg. Am. Vol.*, 59, 862, 1977.
146. **Brown, G. C., Lockshin, M. D., Salvati, E. A., and Bullough, P. G.**, Sensitivity to metal as a possible cause of sterile loosening after cobalt-chromium total hip-replacement arthroplasty, *J. Bone Jt. Surg. Am. Vol.*, 59, 164, 1977.
147. **Lyell, A. and Bain, W. H.**, Nickel allergy and valve replacement, *Lancet*, i, 408, 1974.
148. **Pegum, J. S.**, Nickel allergy, *Lancet*, i, 674, 1974.
149. **Cramers, M. and Lucht, U.**, Metal sensitivity in patients treated for tibial fractures with plates of stainless steel, *Acta Orthop. Scand.*, 48, 245, 1977.
150. **Hegyl, E., Dolezalovä, A., Büthova, D., and Husär, I.**, On epidemiology of the contact eczema caused by nickel. Contribution to the pathogenesis, *Berufs-Dermatosen*, 22, 193, 1974.
151. **Hobkirk, J. A. and Rusiniak, K.**, Metallic contamination of bone during drilling procedures, *J. Oral Surg.*, 36, 356, 1978.
152. **Vandenberg, J. J. and Epstein, W. L.**, Experimental nickel contact sensitization in man, *J. Invest. Dermatol.*, 41, 413, 1963.
153. **Marcussen, P. V.**, Primary irritant patch-test reactions in children, *Arch. Dermatol.*, 87, 378, 1963.

154. **Epstein, S.**, Contact dermatitis due to nickel and chromate. Observations on dermal delayed (Tuberculin-type) sensitivity, *Arch Dermatol.*, 73, 236, 1956.
155. **Marcussen, P. V.**, Eczematous allergy to metals, *Acta Allergol.*, 17, 311, 1962.
156. **Fisher, A. A.**, *Contact dermatitis*, Lea & Febiger, Philadelphia, 1967, 324.
157. **Gottmann-Lückerath, I., Ehring, G., and Steigleder, G. K.**, Vergleichende Untersuchungen mit dem Epi- und Intracutantest mit den Metallsalzen von Chrom, Kobalt, Kupfer und Nickel, *Arch Dermatol. Forsch.*, 246, 159, 1973.
158. **Forman, L. and Alexander, S.**, Nickel antibodies, *Br. J. Dermatol.*, 87, 320, 1972.
159. **Hutchinson, F., Raffle, E. J., and Macleod, T. M.**, The specificity of lymphocyte transformation *in vitro* by nickel salts in nickel sensitive subjects, *J. Invest., Dermatol.* 58, 362, 1972.
160. **Gimenez-Camarasa, J. M., Garcia-Calderon, P., Asensio, J., and de Moragas, J. M.**, Lymphocyte transformation test in allergic contact nickel dermatitis, *Br. J. Dermatol.*, 92, 9, 1975.
161. **Macleod, T. M., Hutchinson, F., and Raffle, E. J.**, The leukocyte migration inhibition test in allergic nickel contact dermatitis, *Br. J. Dermatol.*, 94, 63, 1976.
162. **Pappas, A., Orfanos, C. E., and Bertram, R.**, Non-specific lymphocyte transformation *in vitro* by nickel acetate. A possible source of errors in lymphocyte transformation test (LTT), *J. Invest. Dermatol.*, 55, 198, 1970.
163. **Mirza, A. M., Perera, M. G., Maccia, C. A., Dziubynskyj, O. G., and Bernstein, I. L.**, Leukocyte migration inhibition in nickel dermatitis, *Int. Arch. Allergy Appl. Immunol.*, 49, 782, 1975.
164. **Fisher, A. A.**, Management of selected types of allergic contact dermatitis through the use of proper substitutes, *Cutis*, 3, 498, 1967.
165. **Mosely, J. C. and Allen, H. J., Jr.**, Polyurethane coating in the prevention of nickel dermatitis, *Arch. Dermatol.*, 103, 58, 1971.
166. **Kaaber, K., Veien, N. K., and Tjell, J. C.**, Chronic nickel eczema treated with a low nickel diet, *Ann. Clin. Lab. Sci*, 8, 505, 1978.
167. **Spruit, D. and Bongaarts, P. J. M.**, Diethyldithiocarbamate as a possible therapy for highly sensitive nickel contact dermatitis patients, *Ann. Clin. Lab. Sci.*, 8, 504, 1978.
168. **Menne, T. and Kaaber, K.**, Treatment of pompholyx due to nickel allergy with chelating agents, *Ann. Clin. Lab. Sci.*, 8, 504, 1978.

Chapter 8

BIOLOGICAL PROPERTIES OF COBALT

G. C. F. Clark

TABLE OF CONTENTS

I. INTRODUCTION

Cobalt has in the past been considered primarily in the context of animal nutrition and vitamin B_{12} status. However, it has for some time been used in metal alloys in surgical prostheses, and dental appliances, and consequently, the biochemistry and toxicology of the non-vitamin B_{12} cobalt has become more important. The composition of some of the alloys is shown in Table 1.

Although these alloys are selected partly because of their corrosion resistance, there will always be a finite corrosion rate as discussed in *Fundamental Aspects of Biocompatibility.* Both corrosion products and, under some circumstances ingested cobalt, may accumulate or show locally raised concentrations. The effects of this must be understood both in terms of the chemistry and biochemistry of the metal and in the context of the dynamics of the organism in which they occur.

II. INORGANIC CHEMISTRY

In considering the biochemistry of any of the transition metals it is essential to start with their inorganic chemistry. In general they all form strong complexes with inorganic and organic ligands and have variable oxidation states. They behave as strong Lewis acids and as such are very potent agents in the internal milieu of the cell. Complexation and oxidation-reduction are very closely linked, so that, for example, E^0 for the reaction

$$Co(H_2O)_6^{3+} + e^- = Co(H_2O)_6{}^{2+} \quad (1)$$

is 1.84 V, while for

$$Co(NH_3)_6^{3+} + e^- = Co(NH_3)_6{}^{2+} \quad (2)$$

it is only 0.1 V.

Due to the fact that the complexes are held together by partially covalent bonding, they are also potentially kinetically stable. This is especially marked with cobalt (III), where the rate constant for the exchange of the water molecule in $Co^{III}(NH_3)_5(H_2O)$ is $6 \times 10^{-6} sec^{-1}$, compared with $8 \times 10^9 sec^{-1}$ for $Cu^{II}(H_2O)_6$.[1]

In any biological context, the role of a transition metal ion will be a function of the stability constants of all of the possible complexes in the system, the kinetic stability of these, the redox equilibria of the metal, and the possibilities for ligand exchange or redox reactions with other components of the system, these factors having been discussed in Chapter 2.

Cobalt in aqueous systems can exist in the Co(I), Co(II), and Co(III) states. Co(I) is only found in biological systems in the form of reduced vitamin B_{12}, which we are not considering here, other possible ligands which stabilize this state being absent.

Cobalt(II) is the most common oxidation state of cobalt in aqueous systems. It forms a wide range of stable complexes, behaving very much as a typical first series transition metal. The stability constants of its complexes lie between those of Ni(II) and Fe(II) in the Irving-Williams order of complex stability.[2] The octahedral complexes, such as those with water and ammonia, are pink while tetrahedral complexes such as $Co(II)Cl_4^{2-}$ are blue. The aquo complex has a pK of about 8.9, so that at physiological pH hydrolysis is negligible and polymeric species such as those found with Fe(III) are absent.[3]

Table 1
COMPOSITION OF SOME IMPORTANT COBALT-BASED SURGICAL AND DENTAL ALLOYS

Element	Cast Co-Cr surgical alloy	Cast Co-Cr dental alloy	Wrought Co-Cr surgical alloy	Wrought Co-Ni-Cr surgical alloy
Co	64	62	54	35
Cr	28	29	20	20
Ni	1	1	10	35
Mo	6	5	—	9
W	—	—	15	—
Fe	1	3	1	—
Ti	—	—	—	1

Note: Figures given are typical weight percentages of major elements, ignoring those present at less than 1%.

The most interesting feature of cobalt(II) is its very great similarity to zinc(II). The stability constants of a wide range of their complexes are extremely close, and in many zinc-containing enzymes it is possible to remove the zinc and replace it with cobalt, retaining a reasonable level of activity. The explanation lies in their mutual capacity to form high-spin tetrahedral complexes, which other transition metals are more reluctant to form, and their very similar ionic radii.[4]

There is very little data available on the stability constants of cobalt(III) complexes because they are so kinetically stable that normal methods of measurement cannot be used. However, it appears that the complexes with nitrogen-containing ligands are very much stronger than those with oxygen ligands. This is reflected in their redox potentials, as mentioned above. The aquo complex of Co(III) is capable of oxidizing water at room temperature, while the ammonia complex of Co(II) is itself oxidized by air to Co(III). Amino acid complexes appear to be more stable in the Co(II) form. The only biologically important Co(III) complex is oxidized vitamin B_{12}.

Porphyrin complexes, probably protoporphyrin, have been reported in rats injected with cobalt chloride.[5,6] They are found in the liver and are strongly protein bound. Both Co(II) and Co(III) porphyrin complexes are known and both appear to be stable.[7] Their visible spectra are very similar, but the published spectra correspond more closely to the deoxy Co(II) protoporphyrin complex than any other form.

III. DISTRIBUTION

The levels of cobalt which have been recorded from normal human tissues are shown in Table 2. The figures shown indicate the range of values reported by different investigators. Some of the variation is explicable in terms of improvements in technique in recent years and in greater awareness of possible sources of contamination. The levels usually found are at the limits of the techniques used, especially atomic absorption spectroscopy. Many of the advances in analytical chemistry in recent years have, in fact, been prompted by the susceptibility of existing techniques to serious errors when applied to trace element determination in biological systems. The most reliable methods at the present time are flameless atomic absorption spectroscopy and atomic fluorescence spectroscopy, although it is likely that the greatest sensitivity is to be obtained from voltametric techniques. At the present time these show a tenfold improvement over flameless AAS, but they are not so suited to routine determinations.

Table 2
COBALT CONCENTRATIONS IN DIFFERENT ORGANS IN MAN

Organ	Cobalt (ppm)
Adrenals	0.2
Blood	0.099—0.0003
Bone	43.5—0.01
Brain	0.21—0.055
Stomach	0.18—0.021
Small intestine	0.08
Heart	0.19—0.01
Kidney	0.071—0.008
Liver	0.80—0.017
Muscle	0.24—0.006

Note: Ranges are the extremes of the published mean values. Data from Reference 8.

The values found for the tissues are of course total cobalt and give no indication of speciation. The total can be divided first into vitamin B_{12} cobalt and the rest, which can be assorted into the protein-bound, low-molecular weight complex and inorganic cobalt, all more or less in equilibrium with one another.

The amount of vitamin B_{12} in the body is estimated at 2 to 5 mg,[9] containing 4.4% cobalt, i.e., 0.09 to 0.22 mg. The total body cobalt content is 0.7 to 49 mg.[10] Of this, the major part is in the skeleton, with up to 0.5 mg in the liver and 0.5 mg in the blood.

As mentioned, the published values for all organs, especially bone, are very variable. Injection of radioactive cobalt into pregnant mice[11] led to preferential labeling of maternal cartilage and the fetal skeleton, implying that it is incorporated into newly formed bone, while the work of Baude et al.[12] and Comar and Davis[13] showed that pigs fed labeled cobalt at low levels initially stored it in soft tissues, but that the amount in the bone gradually increased. It is known that zinc accumulates in bone, and it may be that the cobalt is laid down by the same systems, although the solubility of cobalt phosphate is sufficiently low to suggest that cobalt hydroxyapatite will accumulate in any case.

In blood the reported levels show a 300-fold range. Erythrocytes have a concentration 1.5 to 2 times the plasma level. In the plasma it is very likely that cobalt closely mimics zinc, with at least 99% protein bound to α_2-macroglobulin (40 to 45%) and serum albumin.[14] The latter is probably the main transport form. Equilibration with the tissues is via low-molecular-weight complexes with organic acids and amino acids. With zinc, computer models indicate that the predominant forms in plasma are the complexes with cysteine and the mixed ligand complexes cysteine-citrate and cysteine-histidine.[15]

In the soft tissues the highest concentrations, up to 0.2 ppm., are found in the adrenal glands and the thyroid. The next highest are the liver and the kidney. These organs show a greater variation, from 0.01 to 0.15 ppm, probably because they are the sites of initial accumulation of ingested cobalt and its excretion. Heart and skeletal muscle are variable, but insofar as they have been determined by a single investigator, they have the same levels.

IV. ABSORPTION AND EXCRETION

Cobalt in the body has virtually no effect until toxic levels are reached. Presumably because it is not an essential element, in the free form, there are no apparent mechanisms either for controlling its uptake into or loss from the body. The labile cobalt present is thus very much a function of the amount taken in. It is not possible to give exact figures for dietary cobalt, but it may be taken as approximately 0.02 ppm for vegetables and 0.05 ppm for meat. Studies on radioactive cobalt added to the diet of volunteers indicate that 80% is excreted in the feces, and that of the absorbed material, 50% is lost in the urine within 10 days.[16] The pattern of excretion varies with the method of administration, but there is general agreement that 70 to 90% of any dose is lost within 1 week. The route of absorption is probably via the small intestine. Iron-deficient animals show enhanced uptake of cobalt, so they may share a common mucosal transport pathway. In the body, cobalt does not induce the synthesis of metallothioneins in the kidney, differing from zinc in this respect.[17] Urinary excretion is fast with clearance of raised blood levels in a matter of a few days. Studies of patient receiving therapeutic doses of cobalt chloride (50 mg/day) showed that blood levels rose rapidly following a single oral dose, but fell almost to normal within 3 days.[18] In an anephric subject in the same study, the levels remained raised for 14 days.

With the introduction of surgical implants containing large amounts of cobalt, an additional source of body cobalt is present. The alloys used are designed for maximum corrosion resistance. Nevertheless, in a physiological environment loss of material will occur, especially during the early stages of the implantation when the surface layers of a cobalt chromium alloy will become depleted in cobalt, since corrosion resistance is associated mainly with the chromium oxide surface layer. The surface area of implants can be in the region of 100 cm^2, so the loss of a 1-μm layer is equivalent to 0.1 g. In the light of the lack of reports of toxicity, the rate of loss is obviously very low, but if for any reason excessive wear develops, releasing fine particles, or if infected areas exist, with a lower pH or redox potential, then the local concentration of cobalt may rise to toxic levels.

V. BIOCHEMISTRY

There is a large amount of work in the literature on the use of cobalt in biochemistry, but a very large portion of this is devoted to its use as a probe in investigating the role of the metal in zinc-containing metalloenzymes.[19] There has been much less work on the effect of cobalt ions on other enzymes or on whole organs.

In many zinc-containing enzymes, e.g., carboxypeptidase, it is possible to remove the metal by dialysis, with loss of activity of the enzyme.[19] On adding cobalt salts, activity is restored. The cobalt enzyme is very much easier to study, as the metal has a useful optical spectrum, its magnetic moment can be measured, and since it is paramagnetic, it also perturbs the NMR relaxations of groups close to it. The relative activities of some cobalt-substituted enzymes are shown in Table 3.

Although it is likely that with very high cobalt body levels this substitution occurs in vivo, there is no evidence for any adverse effects from this cause.

Cobalt injected into rats is known to cause anemia[20] by several mechanisms. The amount of liver heme oxygenase increases, leading to enhanced breakdown of heme. δ-Amino levulinate synthase activity falls, apparently due to reduced synthesis.[21] There is diversion of δ-amino levulinic acid from heme production by the accumulation of an insoluble cobalt-protoporphyrin complex in the liver. There is no direct evidence for the occurrence of these effects in man. In contrast to these effects, cobalt can also have an erythropoietic effect due to stimulation of erythropoietin production by the kidney. Studies in rats indicate that this is caused by a respiratory alkalosis which develops following the administration of cobalt chloride.[22] This increases the affinity

Table 3
ACTIVITY OF ZINC METALLOENZYMES SUBSTITUTED BY COBALT

Enzyme	Activity
Carbonic Anhydrase	50
Carboxypeptidase	150
Alkaline phosphatase	20
Alcohol dehydrogenase	70
Neutral protease	100
Aldolase	85
Superoxide dismutase (Co-Cu)	100

Note: Activity is expressed as a percentage of that of the zinc enzyme. Data from Reference 31.

of hemoglobin for oxygen, leading to hypoxia in the kidney and consequent erythropoietin production. Of these two opposed effects the erythropoiesis is the more significant, and cobalt has been used therapeutically in treating anemia. The effects are not permanent, however, and tolerance develops so that its use is being discontinued.

Apart from these specific effects which can be observed at low levels of cobalt, the metal is also toxic at high concentrations, simply as a heavy metal. Since the transition metals are so liable to form complexes, they can seriously interfere with a wide range of enzymes which have potential ligands at their active sites. These include histidine, serine, and cysteine. For cobalt the concentrations at which these effects occur are high, but they are found around freely corroding cobalt implants, which show severe localized necrosis.

VI. TOXICOLOGY

The toxicology of metals is closely linked with their biochemistry. Cobalt is generally nontoxic, at least in the diet. Studies in sheep indicate that while doses of 3 mg/kg body weight for several weeks had no effect, 4 or 10 mg/kg caused anemia, loss of appetite, and loss of weight.[23] This is probably caused by hemolysis, and reduced heme synthesis, as noted earlier, and interference with iron uptake. In man, toxic effects have been noted in two circumstances, patients receiving cobalt to combat anemia and very heavy beer drinkers drinking beer with added cobalt.

Due to its erythopoietic effects cobalt chloride is used therapeutically in treating anemia. A patient suffering from sickle cell anemia and receiving cobalt at 2.5 mg/kg/day developed a goiter,[24] and the rate of iodine uptake fell. The effect was totally reversed by stopping the cobalt treatment. Tyrosine iodinase is completely inhibited by 1 m*M* cobalt, and although this is an extremely high concentration to be found in the body, the thyroid normally contains one of the highest concentrations of any organ in the body. It is possible that in some patients there is accumulation of cobalt in the thyroid until it reaches toxic levels. People suffering from chronic renal failure are frequently anemic, and cobalt salts have been used to control this. There are a few reports that in such cases cardiomyopathy can develop, leading to death.[25] In one case reported in detail, the myocardium showed patches of necrosis and histologically there were major changes. The muscle fibers were separated by fibrous tissue, containing vacuoles and very large nuclei. The cobalt levels were 8.9 ppm for dried tissue, which is 40 to 50 times above normal.

A very similar set of symptoms was observed in an incident in Quebec.[26] A local brewer added cobalt to his beer to stabilize the foam. Within a year cases of cardiomyopathy began to appear among heavy drinkers in the area. New cases ceased to appear a month after the addition was stopped. At least 64 patients were seen, of whom 30 died from myocardial failure. Since cobalt toxicity was not at first suspected, the tissue samples were stored in formalin, so that subsequent analyses were subject to error. The values found were, however, 10 times above control samples, which is in rough agreement with those for the patients with renal failure. The histological changes are also very similar.

It appears that cobalt can be toxic in cases such as these, either because of reduced excretion or in conjunction with coexisting pathological changes such as those due to alcohol poisoning.

One aspect of cobalt toxicology which is potentially of very great significance has been observed in rats but never in man. Heath and Webb[27] showed that if they injected a suspension of cobalt powder in horse serum into the thigh muscle of rats, they were able to consistently induce rhabdosarcomata. Nickel and cadmium also had the same effect. The tumors are transplantable for many generations and do not require the metal after induction. Induction is apparently dependent upon the presence of the serum, although this is not species specific.[28] Other workers have not been able to reproduce these results, although we ourselves have observed the induction of one melanoma remote from the implant site and one very large fluid-filled tumor at the implant site when implanting polished discs of cobalt. Cobalt powder produces necrosis, but within 6 months, the wound heals, even in the presence of undissolved particles. It is possible that the effect which has been observed by Heath and Webb is due to the rapid uptake of solubilized metal bound to foreign proteins. DNA-dependent RNA polymerase is a zinc metalloenzyme,[29] so it possible that this could be substituted with cobalt if the cobalt enters the cell by an appropriate route. This hypothesis is supported by their observation that the cobalt is concentrated in the nucleus.[30] The problem emphasized by this work is that of speciation. If the different species in which the metal exists in the body are not in equilibrium, and if some are totally harmless while others are carcinogenic, then it is essential to assess the state of the metal before any comment can be made about its toxicity.

VII. SUMMARY

The biochemical effects of cobalt are normally confined to those of vitamin B_{12}. The non-B_{12} pool can fluctuate over several orders of magnitude without causing any observable effects, and there are no specific mechanisms with the function of regulating this pool. However, cobalt can have large effects if its concentration is high enough. Most of these are reversible, but cobalt poisoning has caused some deaths in man. The dietary intake is normally so low that toxic levels are not reached, but cobalt may be supplied either therapeutically as cobalt salts or as a side effect of surgery, by corrosion of implants. Raised concentrations can then occur, either due to reduced clearance or infection, in which case toxicity may be shown.

REFERENCES

1. **Basolo, F. and Pearson, R. G.,** *Mechanisms of Inorganic Reactions,* 2nd ed., John Wiley & Sons, New York, 1967.
2. **Cotton, F. A. and Wilkinson, G.,** *Advanced Inorganic Chemistry,* 3rd ed., Interscience, New York, 1972, 596.
3. **Baes, C. F. and Mesmer, R. E.,** *The Hydrolysis of Cations,* John Wiley & Sons, New York, 1976, chap. 10.6.
4. **Ochiai, E.,** *Bioinorganic Chemistry,* Allyn and Bacon, Boston, 1977, chap. 13.
5. **Sinclair, P., Gibbs, A. H., Sinclair, J. F., and De Matteis, F.,** Formation of cobalt protoporphyrin in the liver of rats, *Biochem. J.,* 178, 528, 1979.
6. **Igarashi, J., Hayashi, N. and Kikuchi, G.,** Effects of administration of cobalt chloride and cobalt protophyrin on aminolevulinate synthase in rat liver, *J. Biochem. (Tokyo),* 84, 997, 1978.
7. **Yonetani, T., Yamamto, H., and Woodrow, G. V.,** Studies on cobalt myoglobins and hemoglobins, *J. Biol. Chem.,* 249, 682, 1974.
8. **Iyengar, G. V., Kollmer, W. E., and Bowen, H. J. M.,** *The Elemental Composition of Human Tissues and Body Fluids,* Verlag-Chemie, Weinheim, 1978.
9. **Linnell, J. C.,** The fate of cobalamins in vivo, in *Cobalamin,* Babior, B. M., Ed., John Wiley & Sons, New York, 1975, chap. 6.
10. **Ochiai, E.,** *Bioinorganic Chemistry,* Allyn and Bacon, Boston, 1977, chap. 6.
11. **Flodh, H.,** Distribution and kinetics of $^{60}CoCl_2$ and labelled vitamin B_{12} using autoradiography and impulse counting, in *Trace Element Metabolism in Animals,* Mills, C. F., Ed., Livingston, Edinburgh, 1970, 67.
12. **Braude, R., Free, A. A., Page, J. E., and Smith, E. L.,** The distribution of radioactive cobalt in pigs, *Br. J. Nutr.,* 3, 289, 1953.
13. **Comar, C. L. and Davis, G. K.,** Cobalt Metabolism studies: tissue distribution of radioactive cobalt administered to rabbits, swine and young calves, *J. Biol. Chem.* 170, 379, 1947.
14. **Underwood, E. J.,** *Trace Elements in Human and Animal Nutrition,* 4th ed., Academic Press, New York, 1977, 134.
15. **May, P. M. M.,** Computer Simulation of Metal Ion Equilibria in Biochemical Systems: Models for Blood Plasma, Ph.D. thesis, University of Cape Town, South Africa, 1975.
16. **Comar, C. L. Davies, G. K., and Taylor, R. F.,** Cobalt metabolism studies: Radioactive cobalt procedures with rats and cattle, *Arch. Biochem.,* 9, 149, 1946.
17. **Ochiai, E.,** *Bioinorganic Chemistry,* Allyn and Bacon, Boston, 1977, 481.
18. **Curtis, J. R., Goode, G. C., Herrington, J., and Urdaneta, L. E.,** Possible cobalt toxicity in maintenance hemodialysis patients after treatment with cobaltous chloride: a study of blood and tissue cobalt concentrations in normal subjects and patients with terminal renal failure, *Clin. Nephrol.,* 2, 61, 1976.
19. **Ochiai, E.,** *Bioinorganic Chemistry,* Allyn and Bacon, Boston, 1977, chap. 13.
20. **Nakamura, M., Yasukochi, Y., and Minakami, S.,** Effect of cobalt on heme biosynthesis in rat liver and spleen, *J. Biochem. (Tokyo),* 373, 737, 1975.
21. **Maines, M. D. and Sinclair, P.,** Cobalt regulation of heme synthesis and degradation in avian embryo liver cell culture, *J. Biol. Chem.,* 252, 219, 1977.
22. **Miller, M. E., Howard, D., Stohlman, F., and Flanagan, P.,** Mechanism of erythropoietin production by cobaltous chloride, *Blood,* 44, 339, 1974.
23. **Underwood, E. J.,** *Trace Elements in Human and Animal Nutrition,* 4th ed., Academic Press, New York, 1977, 153.
24. **Gross, R. T., Kriss, J. P., and Spalt, T. H.,** Haemopoietic and goitrogenic effects of cobaltous chloride in patients with sickle cell anaemia, *Pediatrics,* 15, 284, 1955.
25. **Manifold, I. H., Platts, M. M., and Kennedy, A.,** Cobalt cardiomyopathy in a patient on maintenance haemodialysis, *Br. Med. J.,* 2, 1609, 1978.
26. **Sullivan, J., Parker, M., and Carson, S. B.,** Tissue cobalt content in 'beer drinkers myocardiopathy', *J. Lab. Clin. Med.,* 71, 893, 1968.
27. **Heath, J. C.,** The production of malignant tumours by cobalt in the rat, *Br. J. Cancer,* 10, 668, 1956.
28. **Heath, J. C., Webb, M., and Caffrey, M.,** The interaction of carcinogenic metals with tissues and body fluids. Cobalt and horse serum, *Br. J. Cancer,* 23, 153, 1969.
29. **Vallee, B. L.,** Zinc biochemistry and physiology and their derangements, in *New Trends in Bioinorganic Chemistry,* Williams, R. J. P. and Da Silva, J. R. R. F., Eds., Academic Press, London, 1978, 38.

30. **Webb, M., Heath, J. C., and Hopkins, T.,** Intranuclear distribution of the inducing metal in primary rhabdosarcomata induced in the rat by nickel, cobalt and cadmium, *Br. J. Cancer,* 26, 274, 1972.
31. **Chlebowski, J. F. and Coleman, J. E.,** Zinc and its role in enzymes, in *Metal Ions in Biological Systems,* Vol. 6, Sigel, H. Ed., Marcel Dekker, New York, 1976, 22.

Chapter 9

CHROMIUM TOXICOLOGY

Sverre Langård and Arne Hensten-Pettersen

TABLE OF CONTENTS

I. PHYSICAL AND CHEMICAL PROPERTIES

Chromium as an element was isolated for the first time in 1798. The metal does not occur in nature in its pure form. It is present in trace amounts in a number of ores, but the major chromium-containing ore is chromite which may contain as much as 55 to 60% of Cr_2O_3 by weight. The metal is produced either by reduction of chromium compounds or by electrolysis after chemical treatment of high carbon ferrochromium. Ferrochromium is produced by direct reduction of the chromite ore.[1,2]

In nature, chromium occurs primarily in the trivalent state, but to some extent also in the hexavalent state. The chromium metal (0 oxidation state) possesses the property of great resistance against oxidizing acids and therefore is commonly used in corrosion-resistant alloys. Besides the naturally occurring ferrous chromite, where chromium is present in the trivalent state, chromium occurs in the hexavalent state (chromates) and in the divalent state (chromous compounds). The chromous compounds are quite unstable and are rapidly oxidized to trivalent (chromic) compounds, the most stable oxidation state of chromium. The chromates are oxidizing agents, a property which makes these compounds versatile as antioxidants. The toxicity of chromates appears to be related to the reduction from the hexavalent state to the stable trivalent state.[2]

II. METHODS AND PROBLEMS OF ANALYSIS

Surveys of the problems related to chromium analysis have recently been published elsewhere.[2,3] The analytical difficulties of chromium are much related to the problems of relevant sampling methods for biological materials, working atmosphere, food, water, and soil. The relevance of analysis is also related to the different toxicity of chromium in different valence states,[2] which is further complicated by the lack of stability of hexavalent chromium. This problem has recently been illustrated by Abell and Carlberg[4] who showed that hexavalent chromium collected on PVC filters was retained in this oxidation state for up to 2 weeks, while chromates were reduced quite quickly to the trivalent state when collected on traditional filters. Fukai[5] has demonstrated and discussed quite similar problems for the analysis of chromium in seawater, demonstrating the presence of predominantly hexavalent chromium compounds and less trivalent compounds.

Traditionally, a colorimetric method employing the violet complex of 1,5-diphenylcarbazide has been applied for the analyis of chromium in urine, but this method is subject to interference by other ions.[6] Over the last 5 to 10 years atomic absorption spectrophotometry has been the most frequently applied method for analysis of chromium both in inorganic materials and in biological samples.[7-9] Recently, gas chromatography,[10] and coupled gas chromatography-atomic absorption spectrophotometry[11] have been used for chromium determination. Radiochemical neutron activation analysis has also been used in the analysis of chromium in samples from the environment[12] and biological specimens.[13]

These methods have enabled analysts to determine very low concentrations of chromium compounds in biological samples. However, none of the methods has so far solved the problem of interaction with other ions and the risk of contamination of the sample from chromium sources outside the sample material.[9]

Since the toxicity of hexavalent and trivalent chromium is quite different, separate determination of chromium in these two valence states is desirable in many situations. When applying the 1,5-diphenlycarbazide method or the atomic absorption spectrophotometry method, including organic extraction for the concentration of the sample, only the hexavalent form is determined.[14] The analytical difficulties in determining the concentrations of chromium in both these valences states have only been partly solved.[4,5]

III. HUMAN EXPOSURE TO CHROMIUM

A. Dietary

Ever since chromium was found to be an essential requirement for the maintenance of normal activity of the glucose tolerance factor (GTF), a dietary agent required to maintain a normal glucose tolerance in the rat,[15] considerable interest has been focused on the possibility that chromium is an element essential for humans.[14] Chromium being an essential element, its occurrence in the diet and foodstuff and its availability to man is therefore of great interest. Schlettwein-Gsell and Mommsen-Straub[16] presented a survey of the chromium content of food, indicating that meat, vegetables, and unrefined sugar are among the major sources of dietary chromium, while fish and fruit contain smaller amounts.

It has been indicated that the daily urinary excretion of chromium exceeds the daily uptake in the gastrointestinal tract, with the provision that the rate of uptake of chromium present in foodstuff is equal to the uptake of chromium from inorganic chromium salts after oral administration. However, this apparent discrepancy is likely to be due to either an underestimation of the chromium absorption from foodstuff[2,17] or an overestimation of the excretion.[13] Other possible explanations of this discrepancy might be that uptake from ambient air contributes significantly to the total chromium uptake.

Comprehensive surveys on the significance of dietary chromium and the influence of the element on the metabolism of lipids and carbohydrates have been presented by Mertz[14] and Gürson.[17]

B. Environment

1. Ambient Air

Urban air concentrations of chromium in American cities have been reported from less than 10 ng/m^3 up to about 50 ng/m^3, while annual mean values in most cases are lower than 10 ng/m^3.[3] No information is available on the oxidation state of chromium in ambient air and the rates of absorption in the lungs. Schroeder et al.[18] reported increased concentrations of chromium with age in human lung tissues, contrary to a decline with age in all other tissues studied. Whether this age-related increase reflects pulmonary uptake from ambient air is not clear.

2. Working Environment

A listing of 81 different occupations where the workers may be exposed to chromium has recently been presented.[3] Among the occupations included are chromate- and chromium-pigment production, chromium platers, corrosion-inhibitor workers, furniture polishers, leather finishers, oil drillers, painters, photographers, steel workers, tanners, and welders. Although this listing is far from complete, it illustrates the wide field of applications of chromium and that a large number of workers may be exposed to chromium compounds. Since the number of applications for chromium alloyed steel is increasing, the number of welders exposed to chromium-containing welding fumes is rising and represents a large chromium exposed population.[19,20]

The prime routes of occupational exposure to chromium are inhalation and skin contact.[21,22] The most commonly encountered hazard to workers is the chromium-containing aerosol, but in the chromium-plating industry and other workplaces dealing with open baths containing chromic acid, the health hazard is related to chromium-containing mists.

Atmospheric concentrations of chromium in the hazardous industries, the measurement of which is essential in order to evaluate relationship between dose and response, have been reported only to a limited degree. A survey of such reports has been presented elsewhere.[2]

IV. METABOLISM AND STORAGE

A. Uptake and Excretion

Chromic compounds are poorly absorbed from the gastro-intestinal tract. In rats, Visek et al.[23] reported intestinal absorption of less than 0.5% of chromium in orally administered $^{51}CrCl_3$, while Donaldson and Barreras[24] found that 0.1 to 1.2% of an orally administered dose of the same compound was excreted in the urine of patients. It was also found that 0.2 to 4.4% of the radioactivity from an orally administered dose of $Na_2{}^{51}CrO_4$ appeared in the urine within 24 hr after administration, while 84.7 to 96.7% of the radioactivity was excreted in the feces of these patients.[24] When administered through a duodenal tube, approximately 94% of the radioactivity from $^{51}CrCl_3$ and less than 57% of the radioactivity from $Na_2{}^{51}CrO_4$ was recovered from the feces.

These results indicate that the passage of chromates through the upper part of the gastrointestinal tract reduces the rates of chromium absorption in the lower portion of the tract. Donaldson and Barreras[24] presented evidence that acid gastric juice was capable of reducing hexavalent chromium to the poorly absorbed chromic form. Not only the acidity of the gastric juice but also the dietary state seems to have an influence on the rates of absorption of chromium. MacKenzie et al.[25] showed that more radioactivity appeared in the tissues of fasted rats than in nonfasted rats after an orally administered dose of $Na_2{}^{51}CrO_4$. The recovery of radioactivity in the urine was twice as high in the fasted as in the nonfasted rats. Increased absorption of hexavalent chromium has also been observed in achylic patients,[24] which seems to support the view that sodium chloride is involved in the reduction of hexavalent chromium in the stomach and the upper parts of the small intestine.

These absorption data are based primarily on urinary excretion after oral administration of the compound. Since it is indicated by Hopkins[26] that 7 to 8% of an intravenously administered dose of 1.0 μg/kg^{51}Cr (as $CrCl_3$) was excreted in the rats feces within 4 days, the above data may underestimate the real gastrointestinal chromium absorption.

Recent unpublished data by Langård[27] have confirmed Hopkins'[26] suggestion that chromium excretion in the rats takes place both with the urine and the feces. These studies[27] were carried out with female Wistar rats and the urine and feces were collected separately over 24-hr collection periods during four days. Two different dose levels were chosen both for $Na_2{}^{51}CrO_4$ and $^{51}CrCl_3 \cdot 6H_2O$. As shown in Table 1, the recovery of ^{51}Cr in the feces is slightly lower than in the urine when administering a subtoxic dose of sodium chromate (0.56 mg/kg body weight), while the fecal excretion exceeds the urinary excretion when administering a presumably toxic dose (10.5 mg/kg). The table also presents the corresponding figures for excretion after intravenous administration of a low and a high dose of $^{51}CrCl_3 \cdot 6H_2O$.

Together these studies[26,27] present evidence that not only the kidneys but also the gastrointestinal tract is of importance for chromium excretion. The gastrointestinal tract seems to be of greater importance for the excretion of chromium administered as chromates than for chromic compounds.[27] The sites of gastrointestinal excretion have not been determined, but evidence has been presented that biliary excretion plays an important role.[27] These data[26,27] seem to confirm the suggestion that previous stud-

Table 1
URINARY (U) AND FECAL (F) EXCRETION OF ^{51}Cr IN FEMALE RATS AFTER I.V. ADMINISTRATION OF TWO DIFFERENT DOSE LEVELS OF $Na_2{}^{51}CrO_4$ AND $^{51}CrCl_3 \cdot 6H_2O$

	Time after administration							
	1st day		2nd day		3rd day		4th day	
Dose/kg body weight	U	F	U	F	U	F	U	F
0.56 mg $Na_2{}^{51}CrO_4$	11.2	8.6	3.9	3.7	2.3	2.3	1.7	1.2
10.5 mg $Na_2{}^{51}CrO_4$	11.5	21.6	3.4	8.2	1.8	2.7	1.7	1.8
2.5 mg $^{51}CrCl_3 \cdot 6H_2O$	12.9	6.0	2.9	2.1	1.2	1.1	0.9	0.6
101.5 mg $^{51}CrCl_3 \cdot 6H_2O$	4.1	2.9	0.9	1.6	0.4	0.8	0.3	0.6

Note: The ^{51}Cr excretion is presented as percent of the total administered ^{51}Cr dose. Each figure gives the mean excretion per rat when keeping four animals together in metabolic cages.

ies[24,25] might have underestimated the role of gastrointestinal absorption of chromates in animals.

Studies on the mechanism of chromium excretion by the kidneys indicate glomerular filtration followed by tubular reabsorption of up to about 60% of the filtered amount.[28] This study, in which $^{51}CrCl_3$ was administered intravenously to dogs, also showed that the chromium excreted with urine was dialysable, indicating that protein-bound chromium is not excreted. Compartment analysis of the metabolism of trivalent chromium in the rat has shown that three different components of the excretion curve would be identified, with half-times of 0.5, 5.9, and 83.4 days, respectively.[29] Although the results were not identical, Onkelinx[30] confirmed these results in principle. So far, no compartment analysis has been carried out with hexavalent chromium compounds.

Chromium uptake through the intact skin does not seem to be of great significance to the total body uptake, but the minute amounts taken up in the skin certainly are of great importance for the induction of allergic skin reactions.[31]

Only limited information is available on chromium absorption and excretion in the respiratory organs.[2] Gylseth et al.[19] and Tola et al.[20] have presented evidence that chromium compounds present in the fumes derived from welding of stainless steel are taken up in welder's lungs and that the level of chromium exposure might be monitored by measuring the excretion in urine. Animal inhalation studies[8] have shown that chromium administered as zinc chromate dust is quickly taken up in the blood. When using inhalation exposure, however, chromium absorption from the gastrointestinal tract also has to be considered.

B. Storage and Distribution

Schroeder et al.[18] studied the chromium concentration in human tissues in the U.S. and found very high concentrations at birth, particularly in the lungs, heart, and the kidneys and a rapid fall in these and other organs only few weeks after birth. Except for lung tissues, in which the concentration increased fourfold from the second to the fifth decade, the concentration was relatively constant in all organs from the second decade onwards. The high concentrations of chromium in the newborn was considered to be consistent with the previous suggestions[15] that chromium is an essential element in man.[18] The age-related accumulation of chromium in lung tissue is most likely due to deposition of inhaled insoluble chromium-containing particulate matter.

It is well documented in animal studies that an intravenous dose of a hexavalent chromium compound gives chromium accumulation particularly in organs belonging to the recticuloendothelial system.[23,27] The distribution of chromium after intravenous administration of $Na_2{}^{51}CrO_4$ (0.56 mg/kg body weight) to the rat is illustrated in Table 2, showing that chromium(VI) is cleared rapidly from the plasma, while chromium(VI) which is taken up in the erythrocytes is retained. Table 2 also illustrates that only small amounts of chromium(VI) are taken up in the brain. Intravenous administration of trivalent chromium compounds also gives accumulation of chromium(III) in the reticuloendothelial organs.[23,26,27,32] Intravenous administration of presumably toxic doses of $^{51}CrCl_3 \cdot 6H_2O$ and $Na_2{}^{51}CrO_4$ to rats gave rise to very high concentrations of ^{51}Cr in the lungs. Histological examination of the tissues indicated that this uptake was related to a mobilization of lung macrophages.[27]

C. Transplacental Transfer

Diab and Södermark,[33] using a whole-body radioautographic method, demonstrated radioactivity in the fetus 1 hr after intravenous injection of $^{51}CrCl_3$ to pregnant mice, but this method does not quantify the transplacental chromium transfer. Studies carried out in pregnant rats by Visek et al.[23] indicated that, regardless of the chemical

Table 2
ORGAN DISTRIBUTION OF ^{51}Cr IN FEMALE RATS AFTER I.V. ADMINISTRATION OF 0.56 mg $Na_2{}^{51}CrO_4$/ kg BODY WEIGHT

	Time after administration (hr)				
	1	7	24	100	196
	% of administered dose/g tissue				
Whole blood	314.3±32.1	411.6	366.8	303.4	219.5
Plasma	162.1±1.0	98.7	66.5	7.7	3.1
Brain	8.6±0.9	9.4	7.7	8.3	4.6
Lungs	499.8±21.9	714.4	414.2	397.9	228.0
Heart	151.6±10.2	144.7	111.3	101.9	57.4
Spleen	142.4±16.4	195.0	262.1	452.5	443.9
Liver[a]	207.8±144.2	173.9	118.8	147.6	80.8
Kidneys	923.4±82.5	660.0	456.8	421.9	313.4
Adrenals	331.7±48.2	310.3	314.0	160.4	139.9
Small intestine and contents	284.0±78.8	125.6	45.7	27.1	20.0
Sternum	90.8±1.7	94.8	66.8	57.4	52.0
Ovaries	219.5±79.6	215.0	167.6	118.8	88.2

Note: Each figure gives the mean value from two animals.
[a] SD was great throughout the experiment.

state and valence state, intravenous administration of chromium resulted in very low concentrations of chromium in the fetus. The results presented by Schroeder et al.[18] give indirect evidence that chromium is taken up in human fetus. Mertz et al.[34] have presented evidence that ^{51}Cr, incorporated into brewer's yeast, was taken up in the rats' fetus after intragastric administration to the mother. At the present time the knowledge on the mechanisms of transplacental transfer of chromium is based only on fragmentary information. Recent demonstration of embryo-toxic effects of chromium trioxide in the hamster[35] emphasizes the need for further elucidation of this aspect of chromium toxicology.

D. Intracellular Distribution

Extensive knowledge of the intracellular distribution and transport mechanisms of chromium will be needed in order to understand the physiological functions of chromium in man,[15] the significance of the association between nucleic acids and chromium,[36,37] and the mechanisms of toxicity.[38]

Since chromium isotopes have found a wide diagnostic use in the tagging of human erythrocytes[39] and platelets,[40] considerable interest has been focused on the binding of chromium in red blood cells[41] and the distribution of chromium in the platelets.[40] By tagging rabbit platelets with ^{51}Cr and using electron microscopic autoradiography, Baker et al.[40] were able to demonstrate a high proportion of radioactivity in the platelets' mitochondria and in the cytoplasm. These results indirectly confirm the findings presented by Rajam and Jackson.[42] They exposed Ehrlich mouse ascites carcinoma cells to $Na_2{}^{51}CrO_4$ and found that 65% of the radioactivity was bound to soluble proteins and the protein-free supernatant, while 12 and 17% were associated with the mitochondrial fraction and nuclear fraction, respectively.

Separately, these studies present a static picture of the intracellular chromium distribution. Langård[27] recently carried out a study on the time-related intracellular ^{51}Cr distribution in liver cells after intravenous administration of $Na_2{}^{51}CrO_4$ (0.56 mg/kg) in the rat. Fifteen minutes after administration, 64% of the radioactivity was associated with the cytosol and this percentage declined subsequently to 12% after 3 days. The corresponding figures for the microsomal fractions were 6.5 and 3.1%. During the same time period, the percentage of radioactivity associated with the nuclear fraction increased from 16 to 52% and the percentage associated with the mitochondrial fractions increased from 7 to 28% of the total radioactivity. The total cellular radioactivity stayed fairly constant throughout the experiment. Dialysis of the same fractions indicated that the ^{51}Cr which was taken up in the mitochondria is firmly bound inside the organelle. The study also gave indirect evidence that the hexavalent chromium anion passes the plasma membrane freely, both in and out of the cell, and that reduction from the hexavalent to the trivalent state takes place mainly in the mitochondria. The time-related differences in the chromium distribution inside the cell demonstrated in this study seem to give an explanation to the differences in the intracellular chromium distribution demonstrated by others.[40,42]

V. NORMAL CHROMIUM LEVELS IN TISSUES

Schroeder et al.[18] carried out a comprehensive study of the tissue distribution of chromium in man in the U.S. and in human tissues from many other countries. They demonstrated considerable geographical variations in the tissue concentrations; lung tissue concentrations varied from 700 μg/kg in New York to 110 μg/kg in thc Dallas area, while the kidney concentrations varied from 90 μg/kg in New York to 22 μg/kg in Dallas and 15 μg/kg in the Miami area. An even greater variation was found between human tissues from different countries.

No attempts have been made to establish human tissue levels of chromium which might be considered as normal. These levels apparently vary considerably from one geographical region to another and also with the occupation.

Most authors have reported normal whole blood and serum levels of chromium in humans, varying from 20 to 50 μg/ℓ.[2,14] However, recent studies applying neutron activation analysis, have reported normal serum levels as low as 0.16 μg/ℓ.[13] Although there seems to be good evidence that these new, very low figures are more realistic than previous figures, a normal level for serum chromium and whole blood chromium has not yet been established. The applicability of chromium analysis in urine, whole blood, and serum for the purpose of monitoring chromium exposure has been discussed elsewhere.[8,19,20,38]

The total urinary chromium excretion has been determined to be from 5 to 10 μg/day.[2,43] However, recent studies[10] have indicated that urinary chromium content ranges from 2 to 3 μg/ℓ urine. Provided these figures are confirmed by others, the daily urinary chromium excretion is likely to be lower than 5 μg/day.

VI. CLINICAL TOXICOLOGY

A. Acute Effects

At the present time, human cases of acute toxic effects of chromium are only rarely seen in hospitals in industrialized countries. The few cases which have been reported during the last years have been caused by accidental ingestion of hexavalent chromium compounds.[45,46] At the beginning of this century, chromates were in some cases taken for suicidal purposes.[38,47] The general picture after ingestion of large doses used to be acute diarrhea and gastrointestinal bleeding followed by nephrogenic and cardiogenic shock. After ingestion of sublethal doses of hexavalent chromium compounds the nephrotoxic and hepatotoxic effects induced by these compounds are the most predominant.[38] Fatal cases have been reported where the patients died between 1 and 9 days after ingestion of the chromate compounds.[38,48,49] Since the toxic effects of sublethal doses of chromates are primarily necrosis of the proximal tubuli and liver necrosis, the patient survives provided the symptoms due to these effects are successfully treated. Further details of the treatment of these toxic effects have been discussed elsewhere.[38]

B. Subacute Effects

The classic subacute effects associated with skin contact with chromates are the so-called chrome holes, which are ulcerations on the skin due to local deposition of chromates.[50] These painless and deep ulcerations were most frequently seen on the fingers, at the nailroots, and on the back of the hands. Since these toxic effects are well known and efforts are made to prevent their development, chrome holes are rarely seen at the present time. Perforation of the nasal septum in workers exposed to chromic acid fumes and to chromate dust is another subacute toxic effect due to chromium.[50,51] Provided the concentration of chromates in the atmosphere to which the workers are exposed is kept below the present threshold limit values, the risk of acquiring such toxic effects should be negligible at the present time.

In older literature, a raised incidence of gastric ulcers and colon ulcers in chromate exposed workers, imitating those seen in ulcerative colitis, was reported.[50,52] These findings have not been supported by modern epidemiological studies.

C. Chronic Effects

1. Carcinogenic Effects

The question of whether carcinogenicity should be considered as a chronic effect or as an acute effect will not be debated in this chapter. The word "chronic" in the present context is considered as an effect induced by exposure of long duration or a long time after exposure, and the actual "moment of cancer induction" is not taken into consideration. Therefore, when not discussing the theoretical aspects of cancer induction and promotion, cancer development caused by a carcinogen might be considered as a chronic effect of the carcinogenic agent.

At the present time there is general agreement that long-term occupational exposure to dust containing hexavalent chromium compounds is associated with an increased incidence of bronchial cancer in exposed workers.[9,21,38,53,54] A scientific discussion has been going on for more than 25 years on whether or not all hexavalent chromium compounds act as carcinogens.[21] Traditionally, chromates with poor water solubility have been considered as stronger carcinogens than the water-soluble chromates. The two latest criteria documents from the U.S. National Institute of Occupational Safety and Health (NIOSH), dealing with chromium trioxide[55] and other chromates,[21] have based their recommendations on Threshold Limit Values for industrial exposure to chromates on a differentiation between what is called "carcinogenic" chromates and "noncarcinogenic" chromates. This differentiation is based primarily on the lack of epidemiological evidence for carcinogenic properties of chromium trioxide and water-soluble chromates. However, evidence has recently been presented that chromium trioxide exposure may give rise to lung cancer.[56-59] Although all but one of these reports are case reports, there does not seem to be support for the view that chromium trioxide is to be considered as noncarcinogenic. On the contrary, there seems to be evidence that all hexavalent chromium compounds should be considered as carcinogens.[38]

A group of hexavalent chromium carcinogens of particular interest are the chromium pigments. Gross and Kölsch[60] demonstrated that chromium pigment workers have an increased risk of developing lung cancer. These results have later been supported by the findings presented by Langård and Norseth[61] and by Davies.[62] That there is a differentiation of the carcinogenic properties of the different chromium pigments is being widely discussed; however, from a theoretical point of view, there is no evidence that one chromium pigment should be less carcinogenic than the others.[38] All chromium pigments should be considered as human carcinogens.

Some evidence has also been presented that workers exposed to chromates also are under particular risk in relation to cancer in the gastrointestinal tract.[63-66] The raised incidence of cancer in these organs, however, is far from as significant as it is in the respiratory organs including the nasal sinuses.

From a clinical point of view, the tumors developing in workers with previous exposure to chromates do not differ from other tumors developing in the respiratory organs; no particular morphological type has been demonstrated in chromate workers. The association between cancer and exposure to chromates can therefore only be determined by epidemiological studies or by means of a very accurate documentation of the working history of the patient.

2. Effects on the Lungs

Letterer et al.[67] and Sluis-Cremer and du Toit[68] suggested that both chromate-containing dust and chromite dust might induce pneumoconiosis in exposed workers. However, animal experiments carried out by Swenson[69] failed to confirm this suggestion. At present it seems unlikely that pneumoconiosis is caused by pure chromate or chromite dust.

Bronchial asthma induced by sensitization to chromic acid fumes and chromate dust has been reported by different workers.[70,71] The asthmatic reaction due to chromium compounds and its treatment does not differ from asthma which is caused by other agents.

3. Skin Reactions to Chromium

Polak et al.[22] recently carried out an extensive review on the contact hypersensitivity of chromium compounds. The survey covers clinical, experimental, and genetic aspects of chromium hypersensitivity and also the treatment of chromium allergy. The interested reader is advised to study this paper. In males living in industrialized countries, chromium sensitivity appears to be the most important skin sensitizer, the skin patch test being positive in 8 to 15% of all patients who suffer from eczematous dermatitis.[72] One significant feature of chromium skin allergy is the unequal distribution between men and women,[72] which most likely is caused by the difference in the prevalence of occupational exposure to chromium compounds among men and women.[38] Another essential question, which is extensively discussed by Polak et al.,[22] is the problem of whether both hexavalent and trivalent chromium compounds or only hexavalent chromium compounds are capable of inducing sensitivity to chromium. In relation to the usage of chromium in biometals, it is also of interest to consider the possible eczematous reactions to metallic chromium. Walsh[73] demonstrated that chromium in cobalt-chromium alloys, where chromium occurs as metallic chromium, are unable to induce sensitization to chromium in man.

VII. BIOCHEMICAL MECHANISMS OF CHROMIUM TOXICITY

A. Biochemical Aspects of Acute Toxicity

Thirty to forty years ago[74] the acute toxic and carcinogenic effects of chromates were explained on the basis of the "irritation theory". It was assumed that deposition of chromate dust resulted in a chronic inflammation which was supported by repeated deposition of dust. This chronic inflammation was assumed to give rise to either cancer development or local ulceration.

Although no definitive model of the biochemical mechanisms involved in the acute cell toxicity of chromates can be presented, pieces of information from different studies can be brought together in a preliminary picture. It seems evident that the reduction from the hexavalent to the trivalent state of chromium plays an essential role in the toxicity.[38] Most metal-ion oxidants require only one-electron transfer in order to attain a lower and stable oxidation state. Chromium(VI) requires three electrons for this purpose, a situation which has been reported only in a relatively small number of instances.[75] There is good evidence that the reduction of chromium(VI) takes place intracellularly.[27,38,39] Whether this reduction takes place in one, two, or three steps is unknown; however, three electrons are needed for each atom reduced from chromium(VI) to chromium(III). Consequently, these electrons must derive from electron-rich molecules such as ascorbic acid, reduced glutathione, TPNH and DPNH, or possibly direct from the electron-transport chain. It has recently been confirmed that reduced glutathionine (GSH) and reduced nicotinamide adenine dinucleotide (NADH) are able to facilitate this reduction.[76] Koutras et al.[77,78] demonstrated that reduced nicotinamide adenine dinucleotide phosphate (NADPH) is required when this reduction is facilitated by erythrocytes. The NADPH requirement has also been observed when microsomal enzyme systems are applied in order to achieve reduction from hexavalent to trivalent chromium.[79,80]

Furthermore, the activity of glutathione reductase is inhibited by the presence of hexavalent chromium compounds.[78,81] Since the reduction to glutathione is coupled to the oxidation of NADPH to NADP,[82] the chromium(VI) inhibition of glutathione reductase might be a consequence of electron depletion needed for the reduction of chromium(VI) to chromium(III), electrons which, at least to some extent, are derived from NADPH. This hypothesis finds support in the fact that hexavalent chromium is reduced very quickly to the trivalent state[39] and that the presence of GSH in reduced form is a prerequisite for this reduction to take place.[83]

From a theoretical point of view, the consumption of electrons for the reduction of chromium(VI) might inhibit a number of metabolic processes for which these electrons are essential. Formation of free oxygen radicals, depression of intracellular levels of adenosine triphosphate, and a decrease in membranous phospholipid exchange are among the biochemical effects which may be expected. The relationship between the affected biochemical processes and cellular survival or death are not yet fully elucidated.

The site(s) of reduction from chromium(VI) to chromium(III) have been determined only to some extent. In the presence of electron donors such as NADPH, microsomal enzyme systems are able to facilitate this reduction.[79,80] The rates of reduction are enhanced in such systems when the concentration of NADPH is increased.[84] This finding was interpreted as an indication that upon entry of chromate into the cells the endoplasmic reticulum-bound enzymes will reduce chromium(VI) to chromium(III). To some extent, this is likely to be correct. However, the experimental situation was highly artificial, in that NADPH was added in great amounts. The study only seems to confirm the hypothesis, which has recently been put forward, that any intracellularly located electron donor is capable of facilitating the reduction,[85] provided that the enzymes which are required for the mobilization of electrons are present. The study carried out by Langård[27] seems to indicate that, at least in the rat liver, the predominant site of chromium(VI) reduction is located in mitochondria. In cells containing considerable amounts of cytoplasmic GSH, the cytoplasma might contribute significantly to the total intracellular chromium(VI) reduction.[81]

B. Biochemical Aspects of Carcinogenicity

Although the carcinogenic hazard seems to be related only to inhalation[2,14,21,38] and possibly to ingestion[64,66] of hexavalent chromium compounds, and unlikely to be related to inhalation of chromic compounds,[38] increasing evidence indicates that chromium(III) might be the ultimate carcinogen of chromium.[9,27,38,79,80,84] Theoretical aspects of this suggestion have been discussed elsewhere.[38] Since chromium(VI) is unable to and chromium(III) is able to bind to nucleic acids,[36,37] and the chromium carcinogenicity is likely to be associated with chromium binding to the nucleic acids at sites which are different from the physiological binding sites,[37] the chromium(VI) reduction is also essential for the ultimate carcinogenicity of chromium. From this discussion it seems evident that the site(s) of chromium(VI) reduction in the cells are just as essential for the carcinogenic effects of chromium as for the acute toxicity of the chromium compounds.

VIII. USES OF CHROMIUM IN BIOMETALS

Clinical experience over the last decades, coupled with results of many investigations on the biological behavior of metals, have led to three principal types of alloys in current usage: cobalt-chromium, stainless steel, and titanium.[86,87]

Chromium is a major alloying element in the cobalt-chromium alloys and stainless steel. In a review of the chemical and electrochemical properties of chromium,[88] it is pointed out that it behaves as if it existed in two clearly different states: the active state in which it is an extremely corrodible metal and the passive state in which it has a very low corrosion rate. The active state is produced by contact of the metal with reducing solutions (HC1 and H_2SO_4) or by cathodic polarization. The passive state is produced by contact with oxidizing solutions or by anodic polarization in solutions not containing chloride. The passive state of chromium is explained by considering this passive state to correspond to a covering of the metal with an oxide or hydroxide of very low solubility. The oxidizing action of air is often sufficient to make chromium pass from the active state to the passive state. The mechanisms involved have been discussed by Williams in *Fundamental Aspects of Biocompatibility* in this series.

Stainless steel — The addition of 18 to 20% chromium to plain carbon steel allows the formation of a tough, tenacious surface layer of oxides or hydroxides of chromium. This surface layer protects the steel from further oxidation, hence the term "stainless". However, the ability to remain stainless is related only to the particular environment which the alloy is designed for. A reducing environment may damage the surface layer and markedly decrease the corrosion resistance. In solutions containing chloride, chromium is more easily attacked by both acid and alkaline solutions because of the increased solubility of the oxides. Molybdenum is therefore often added to counteract the corrosive activity of chloride in the body.

The main advantage of stainless steel as biometal is its ready availability, low cost, and relative ease of shaping. The type normally used contains 11 to 13% nickel in addition to 16 to 18% chromium. In the cold worked state it has strength properties superior to other corrosion-resistant alloys, particularly as regards fatigue resistance. A disadvantage is the limited corrosion resistance, especially towards crevice corrosion.[89] Of 119 stainless steel implants removed from the body, 64 showed corrosion with possible consequences in relation to fracture and tissue reactions.[90]

Cobalt-chromium alloys — Cobalt-chromium alloys were introduced in dentistry under the trade name Vitallium®, and originally contained 30% chromium, 7% tungsten, and 0.5% carbon with the balance cobalt.[86] Present-day surgical Vitallium® contains about 28% chromium, 62% cobalt, and the balance manganese, iron, and molybdenum.[91] This is slightly different from the dental casting alloy which has a higher chromium content.[92] Several reviews on the biological and physical properties of Vitallium® and a host of other, similar alloys are available.[86,93-97]

It is generally accepted on the basis of electrochemical investigations and clinical experience that the corrosion resistance of the cobalt-chromium alloys in the body is very good and superior to the surgical stainless steels.

IX. ADVERSE EFFECTS OF CHROMIUM IN BIOMETALS

The elimination of a number of alloys has, in the last decades, reduced failure of devices due to gross corrosion. However, corrosion of surgical stainless steel and cobalt chromium alloys is still observed. The liberated elements and particles have been investigated with reference to the development of local toxic effects, allergies, and cancerous lesions. The corrosive release of chromium, as well as many other elements from the devices, makes it difficult to ascribe the adverse effects to chromium or to any other single element alone. In addition, the analyses of chromium have been performed either by atomic absorption spectrometry or by neutron activation analysis, which do not take into account the chemical form in which chromium is present in the tissues.[7,9] The following survey of some adverse effects observed with chromium-containing biometals must, therefore, be read with due respect to the limitations inherent in such studies.

A. Corrosive Release of Chromium

Stainless steel devices are commonly found at resection to be encapsulated by a pseudobursal membrane of the order of 1 mm which histologically appears as fibrous tissue, with the fiber axis parallel to the surface of the implant. A pseudobursal fluid may be observed within the membrane.[98] Laing et al.[99] have observed that the thickness of such a membrane appears to be proportional to the degree of metallic dissolution from the appliance. Human studies show that the tissue in direct contact with the appliances can be strikingly enriched in metals. Maximum enrichment by a factor of 380, 34, 85, and 1750 for Cr, Fe, Co, and Ni, respectively, has been observed.[100]

Animal studies on the release of elements from cobalt-chromium and stainless steel cylinders implanted in the spinal muscle of rabbits show that chromium is released from a number of surgical alloys. The chromium levels in the adjacent tissue was in the range of 100 ppm of dry-ashed tissue.[101,102] Analysis of chromium levels in other organs showed that the metal ions circulated in the body. Chromium and many other elements appeared to accumulate in the spleen.[102] However, no distinct effect of any particular alloy on the concentration of ions in vital organs was discerned. If an experimental animal showed an unusual tendency to pick up one species of metal ions, it also tended to pick up other ions derived from the embedded alloys. This indicated either an accelerated corrosion of the implants or a constitutional predisposition of the particular animal to use and store the elements involved.

By means of element correlations, it has been demonstrated that the tissues are burdened by the metals in nearly constant ratios which are not identical for human and rabbit tissues.[100] Simultaneously, the content of other elements as Zn, Se, Rb, and K appear to be less than normal in the tissues in contact with the implants. The analysis of lymphatic tissue, liver, and kidneys of rabbits with implants showed a tendency towards enrichment of Cr and Ni.[100,102]

In vivo and in vitro electrochemical studies of metals in the body show that even minute amounts (in the microgram range) could influence the properties of an implant, acting as a corrosion center. Several possible mechanisms for the effects of metallic contamination on the tissue have been proposed, including elevation of electric potentials in bone, structural bone changes, and local cytotoxic effects.[103-106] A recent study shows that surgical instruments such as bone drills transfer a few micrograms of iron to the bone, together with smaller quantities of chromium, cobalt, and tungsten[107] which leads to metal contamination at the time of surgery. It has therefore been proposed that drills and cutting instruments be made of the same materials as the implants. According to Brettle,[108] stainless steel is more susceptible to corrosive surface effects from metallic transfer than the cobalt-chromium alloys.

Experimental evidence shows that corrosion of both stainless steel and cobalt-chromium alloys is enhanced by cyclic stress, similar to that of weight-bearing orthopedic implants.[109] Another aspect is the liberation of particulate matter from such implants. Owing to the high ratio of surface area to mass in the shed particles, it is possible for even slightly soluble metals to produce local toxic levels. Rae[110-112] has shown that particulate matter from cobalt, nickel, and cobalt chromium alloys was toxic for mouse peritoneal macrophages in monolayer cultures and produced areas of localized necrosis in fetal rat knee joints grown in organ culture. In contrast, particulate pure chromium and several other metals were well tolerated under similar conditions. Further studies by Rae[113] showed that particulate cobalt, nickel, and cobalt-chromium alloys had some hemolytic action, whereas chromium did not. It thus appears that any localized cytotoxic effects of particulate matter shed from prosthetic devices is not due to chromium alone.

B. Allergic Reaction to Chromium

Allergic skin effects of chromium after environmental and occupational exposure is reviewed in Section VI.C.3 of this chapter. Another aspect is the internal allergy more recently recognized in connection with metallic dental and orthopedic devices. In such cases, the loosening and subsequent failure of total joint prostheses have been attributed to the development of sensitivity to the metal components, such as nickel, cobalt, chromium, and vanadium.[114-116] This subject has been discussed by Rae and Elves in *Fundamental Aspects of Biocompatibility,* by Mobray in *Biocompatibility of Orthopedic Implants,* and by Brown in this volume.

Of the 36 patients reported above with prosthesis failure associated with metal allergy, 6 showed positive reactions to chromium. Three reacted to chromium only, and the other three also to nickel or cobalt. Of these six, five were females whereas the sex of the last one was not reported. This is interesting, in as much as chromium allergy is three to eight times more common in men than in women, and particularly frequent in clinics where men with occupational dermatitis predominate.[72]

The occurrence of internal allergy with consequences for prosthesis failure is a controversial point. Fisher[117] states that deep tissue metal allergy is a fiction and that, for instance, stainless steel sutures can be safely used in patients with nickel allergy. Also, Brown et al.[118] in an analysis of 20 patients with sterile, loose hip replacements found none where the failure could be ascribed to hypersensitivity to either nickel, chromium, or cobalt.

The use of cobalt-chromium alloys in dentistry has generally been limited to removable partial dentures. Allergic reactions to cobalt-chromium alloys used for removable dental prostheses most often implicate nickel and cobalt, whereas only in a few instances has chromium allergy been observed (see Bergmann[96] for review). However, in the Eastern European countries, crowns and bridgework are often made of stainless steel, which apparently does not give any worse clinical tissue reaction than other dental materials. Ovrutsky and Ulyanov[119] carried out a clinical study of chromium allergy. Only 3 of 300 patients in a control group without steel dental prosthesis had positive skin reactions to chromium, whereas 12 of 83 patients, with steel prosthesis, working in acid fumes in an industrial environment, had positive reactions. Of another test group of 160 patients with steel crown and bridgework and who had oral mucosal lesions of various types, 60 (37.5%) had positive skin reactions. The incidence was highest in the group that had the greatest number of restorations (more than 9 to 13 teeth involved) and had had them in service for more than 5 years. In some patients, the subjective symptoms and generalized urticaria and eczemas occurred very quickly after the second period of prosthodontic treatment.

These patients complained of xerostomia, cheilitis, metallic taste, and local erosion of the oral mucous membranes, whereas some, in addition, had urticarial and eczematous skin reactions. The symptoms were relieved after replacement of the restorations with others of gold, other alloys, or polymeric material.

C. Carcinogenicity of Chromium-Containing Implants

Tumors have been observed in connection with implants, but the situation is by no means clear as to which factors may be operative. In addition to the chemical composition of implants and corrosion products, the physical form (sheet or powders) also is decisive (reviewed by Smith[95]). A bone sarcoma arising around a screw and plate of dissimilar metals that had been in place for 30 years has been reported.[120] In most cases it appears that tumor formation associated with corrosion of implants occurs after an interval of about 25 years in service. The number of reported cases may be few since the majority of metal implants do not function for so long.

Experimentally, it has been shown that the very fine wear particles from a cobalt-chromium alloy prosthesis produces sarcomas in rats 4 to 17 months after injection into the muscle,[121] this aspect being reviewed by Pedley et al. in *Fundamental Aspects of Biocompatibility*.

REFERENCES

1. **Morning, J. L.,** Chromium, *Miner. Yearb.,* 291, 1971.
2. **Langård, S. and Norseth, T.,** Chromium, in *Handbook on the Toxicology of Metals,* Friber, L., Nordberg, G. F., and Vouk, V. B., Eds., Elsevier, North-Holland Biomedical Press, Amsterdam, 1979, chap. 22.
3. National Research Council, Committee on Biological Effects of Atmospheric Pollutants, *Chromium,* National Academy of Science, Washington, D.C., 1974, 115.
4. **Abell, M. T. and Carlberg, J. R.,** A simple reliable method for determination of airborne hexavalent chromium, *Am. Ind. Hyg. Assoc. J.,* 35, 229, 1974.
5. **Fukai, R.,** Valency state of chromium in seawater, *Nature (London),* 213, 901, 1967.
6. **Beyermann, K.,** Das analytische Verhalten kleister Chrommengen, *Z. Anal. Chem.,* 190, 346, 1962.
7. **Guncaga, J., Lentner, C., and Haas, H. G.,** Determination of chromium in feces by atomic absorption spectrophotometry, *Clin. Chim. Acta,* 57, 77, 1974.
8. **Langård, S., Gundersen, N., Tsalev, D. L., and Glyseth, B.,** Whole blood chromium level and chromium excretion in the rat after zinc chromate inhalation, *Acta Pharmacol. Toxicol.,* 42, 142, 1978.
9. **Langård, S. and Gundersen, N., Eds.,** Nordic Symp. Chromium Analysis and Chromium Toxicology, Report HD 764/78, Nordic Council, St. Olavsgate, Oslo, 1978, 52.
10. **Ryan, T. R. and Vogt, C. R. H.,** Determination of physiological levels of Cr(III) in urine by gas chromatography, *J. Chromatogr.,* 130, 346, 1977.

11. **Wolf, W. R.**, Coupled gas chromatography-atomic absorption spectrometry, *J. Chromatogr.*, 134, 159, 1977.
12. **Gallorini, M., Greenberg, R. R., and Gills, T. E.**, Simultaneous determination of arsenic, antimony, cadmium, chromium, copper, and selenium in environmental material by radiochemical neutron activation analysis, *Anal. Chem.*, 50, 1479, 1978.
13. **Versieck, J., Hoste, J., Barbier, F., Steyaert, H., de Rudder, J., and Michels, H.**, Determination of chromium and cobalt in human serum by neutron activation analysis, *Clin. Chem. (Winston-Salem)*, 24, 303, 1978.
14. **Mertz, W.**, Chromium occurrence and function in biological systems, *Physiol. Rev.*, 49, 163, 1969.
15. **Schwartz, K. and Mertz, W.**, Chromium (III) and the glucose tolerance factor, *Arch. Biochem. Biophys.*, 85, 292, 1959.
16. **Schlettwein-Gsell, D. and Mommsen-Straub, S.**, Spurenelemente in Lebensmittelen, *Int. J. Vitam. Nutr. Res.*, Suppl. 13, 34, 1973.
17. **Gürson, C. T.**, The metabolic significance of dietary chromium, in *Advances in Nutritional Research*, Vol. 1, Draper, H. H., Ed., University of Guelph, Guelph, Ontario, 1977, 23.
18. **Schroeder, H. A., Balassa, J. J., and Tipton, I. H.**, Abnormal trace metals in man — chromium, *J. Chronic Dis.*, 15, 941, 1962.
19. **Glyseth, B., Gundersen, N., and Langård, S.**, Evaluation of chromium exposure based on a simplified method for urinary chromium determination, *Scand. J. Work Environ. Health*, 3, 28, 1977.
20. **Tola, S., Kilpiö, J., Virtamo, M., and Haapa, K.**, Urinary chromium as an indicator of the exposure of welders to chromium, *Scand. J. Work Environ. Health*, 3, 192, 1977.
21. Criteria for a Recommended Standard Occupational Exposure to Chromium (VI), National Institute for Occupational Safety and Health, U.S. Department of Health, Education, and Welfare, Washington, D.C., 1975.
22. **Polak, L., Turk, J. L., and Frey, J. R.**, Studies on contact hypersensitivity to chromium compounds, *Prog. Allergy*, 17, 145, 1973.
23. **Visek, W. J., Whitney, I. B., Kuhn, U. S. G., III, and Comar, C. L.**, Metabolism of Cr^{51} by animals as influenced by chemical state, *Proc. Soc. Exp. Biol. Med.*, 84, 610, 1953.
24. **Donaldson, R. M., Jr. and Barreras, R. F.**, Intestinal absorption of trace quantities of chromium, *J. Lab. Clin. Med.*, 68, 484, 1966.
25. **MacKenzie, R. D., Anwar, R. A., Byerrum, R. U., and Hoppert, C. A.**, Absorption and distribution of Cr^{51} in the albino rat, *Arch. Biochem. Biophys.*, 79, 200, 1959.
26. **Hopkins, L. L., Jr.**, Distribution in the rat of physiological amounts of injected Cr^{51} (III) with time, *Am. J. Physiol.*, 209, 731, 1965.
27. **Langård, S.**, The Fate of Chromium after Intravenous Administration of $Na_2{}^{51}CrO_4$ and $^{51}CrCl_3 \cdot 6H_2O$ to the Rat, Master's thesis, University of Surrey, Guildford, England, 1977.
28. **Collins, R. J., Fromm, P. O., and Collings, W. D.**, Chromium excretion in the dog, *Am. J. Physiol.*, 201, 795, 1961.
29. **Mertz, W., Roginski, E. E., and Reba, R. C.**, Biological activity and fate of trace quantities of intravenous chromium (III) in the rat, *Am. J. Physiol.*, 209, 489, 1965.
30. **Onkelinx, C.**, Compartment analysis of metabolism of chromium (III) in rats of various ages, *Am. J. Physiol.*, 232, E478, 1977.
31. **Wahlberg, J. E.**, Thresholds of sensitivity in metal contact allergy, I. Isolated and simultaneous allergy to chromium, cobalt, mercury, and/or nickel, *Berufs-Dermatosen*, 21, 22, 1973.
32. **Kraintz, L. and Talmage, R. V.**, Distribution of radioactivity following intravenous administration of trivalent chromium 51 in the rat and rabbit, *Proc. Soc. Exp. Biol. Med.*, 81, 490, 1952.
33. **Diab, M. and Söremark, R.**, Radioautographic observations on the uptake and distribution of chromium-51 in mice, *Nucl. Med.*, 11, 419, 1972.
34. **Mertz, W., Roginski, E. E., Feldman, F. J., and Thurman, E. E.**, Dependence of chromium transfer into the rat embryo on the chemical form, *J. Nutr.*, 99, 363, 1969.
35. **Gale, T. F.**, Embryotoxic effects of chromium trioxide in hamsters, *Environ. Res.*, 16, 101, 1978.
36. **Wacker, W. E. C. and Vallee, B. L.**, Nucleic acids and metals, I. Chromium, manganese, nickel, iron, and other metals in ribonucleic acid form diverse biological sources, *J. Biol. Chem.*, 234, 3257, 1959.
37. **Sissoëff, I., Grisvard, J., and Guillé, E.**, Studies on metal ions-DNA interactions: specific behaviour of reiterative DNA sequences, *Prog. Biophys. Mol. Biol.*, 31, 165, 1976.
38. **Langård, S.**, Chromium, in *Metals in the Environment*, Waldron, H. A., Ed., Academic Press, London, 1980, chap. 4.
39. **Gray, S. J. and Sterling, K.**, The tagging of red cells and plasma proteins with radioactive chromium, *J. Clin. Invest.*, 29, 1604, 1950.
40. **Baker, J. R. J., Bullock, G. R., Crawford, N., and Taylor, D. G.**, Localization in platelets of sodium ^{51}Cr-chromate, ^{125}I-antibody to whole membrane, and ^{3}H-diisopropylfluorophosphate using electron microscopic autoradiography, *Am. J. Pathol.*, 88, 277, 1977.

41. **Skrabut, E. M., Catsimpoolas, N., Crowley, J. P., and Valeri, C. R.**, Chromate incorporation into the soluble protein fraction human erythrocyte binding not associated with the hemoglobin monomeric subunit, *Biochem. Biophys. Res. Commun.*, 69, 672, 1976.
42. **Rajam, P. C. and Jackson, A. L.**, Distribution and valence state of radiochromium in intracellularly labelled Ehrlich mouse ascites carcinoma cells, *Proc. Soc. Exp. Biol. Med.*, 99, 210, 1958.
43. **Mancuso, T. F. and Hueper, W. C.**, Occupational cancer and other health hazards in a chromate plant: a medical appraisal, I. Lung cancers in chromate workers, *Ind. Med. Surg.*, 20, 358, 1951.
44. **Mitman, F. W., Wolf, W. R., Kelsay, J. I., and Prather, E. S.**, Urinary chromium levels of nine young women eating freely chosen diets, *J. Nutr.*, 105, 64, 1975.
45. **Fristedt, B., Lindqvist, B., Schütz, A., and Övrum, P.**, Survival in a case of acute oral chromic acid poisoning with acute renal failure treated by haemodialysis, *Acta Med. Scand.*, 177, 153, 1965.
46. **Kaufman, D. B., DiNicola, W., and McIntosh, R.**, Acute potassium dichromate poisoning treated by peritoneal dialysis, *Am. J. Dis. Child.*, 119, 374, 1970.
47. **Brieger, H.**, Zur Klinik der akuten Chromatvergiftung, *Z. Exp. Pathol. Ther.*, 21, 393, 1920.
48. **Reichelderfer, T. E.**, Accidental death of an infant caused by ingestion of ammonium dichromate, *South. Med. J.*, 61, 96, 1968.
49. **Schlatter, C. and Kissling U.**, Akute, tödliche Vergiftung mit Bichromat, *Beitr. Gerichtl. Med.*, 30, 382, 1973.
50. **Bloomfield, J. J.**, Health hazards in chromium plating, *Public Health Rep.*, 43, 2330, 1928.
51. **Leineberg, O.**, Om kromatskador i industriarbete, *Nord. Hyg. Tidskr.*, 36, 45, 1953.
52. **Wohlenberg, H. and Lenhart, J.**, Die Chrom-Enteropathie, Ihre differentialdiagnostische und sozialmedizinische Bedeutung, *Dtsch. Med. Wochenschr.*, 95, 1224, 1970.
53. **Mancuso, T. F.**, Occupational cancer and other health hazards in a chromate plant: a medical appraisal, II. Clinical and toxicologic aspects, *Ind. Med. Surg.*, 20, 393, 1951.
54. **Enterline, E.**, Respiratory cancer among chromate workers, *J. Occup. Med.*, 16, 523, 1974.
55. Criteria for a Recommended Standard Occupational Exposure to Chromic Acid, National Institute for Occupational Safety and Health, U.S. Department of Health, Education, and Welfare, Washington D.C., 1973.
56. **Sehnalová, H. and Barbořík, M.**, On the problem of deposition of chromium compounds in man, *Prak. Lek.*, 17, 399, 1965.
57. **Waterhouse, J. A. H.**, Cancer among chromium platers, *Br. J. Cancer*, 32, 239, 1975.
58. **Royle, H.**, Toxicity of chromic acid in the chromium plating industry (1), *Environ. Res.*, 10, 39, 1975.
59. **Michel-Briand, C. and Simonin, M.**, Cancers bronchopulmonaires survenus chez deux salariés occupés à un poste de travail dans le même de chromage électrolytique, *Arch. Mal. Prof. Med. Trav. Secur. Soc.*, 38, 1001, 1977.
60. **Gross, E. and Kölsch, F.**, Über den Lungenkrebs in der Chromfarbenindustrie, *Arch. Gewerbepathol. Gewerbehyg.*, 12, 164, 1943/44.
61. **Langård, S. and Norseth, T.**, A cohort study of bronchial carcinomas in workers producing chromate pigments, *Br. J. Ind. Med.*, 32, 62 1975.
62. **Davies, J. M.**, Lung-cancer mortality of workers making chrome pigments, *Lancet*, 1, 384, 1978.
63. **Taylor, F. H.**, The relationship of mortality and duration of employment as reflected by a cohort of chromate workers, *Am. J. Public Health*, 56, 218, 1966.
64. **Teleky, L.**, Krebs bei Chromarbeitern, *Dtsch. Med. Wochenschr.*, 62, 1353, 1936.
65. **Pokrovskaya, L. V. and Shabynina, N. K.**, Carcinogenous hazards in the production of chromium ferroalloys, *Gig. Tr. Prof. Zabol.*, 10, 23, 1973.
66. **Langård, S. and Norseth, T.**, Cancer in the gastrointestinal tract in chromate pigment workers, *Arh. Hig. Rada. Toksikol.*, suppl. 30, 301, 1979.
67. **Letterer, E., Neidhardt, K., and Klett, H.**, Chromatlungenkrebs und Chromatstaublunge, eine klinische, pathologisch-anatomische und gewerbehygieneische Studie, *Arch. Gewerbepathol. Gewerbehyg.*, 12, 323, 1944.
68. **Sluis-Cremer, G. K. and Du Toit, R. S. J.**, Pneumoconiosis in chromite miners in South Africa, *Br. J. Ind. Med.*, 25, 63, 1968.
69. **Swensson, Å.**, Experimentella undersökningar över fibrogenetiska effekten av kromit, *Arb. Hälsa*, 2, 1, 1977.
70. **Bergmann, A.**, Ueber einige Fälle von gewerblichen Asthma mit Allergie gegen einfache chemische Substanzen, *Schweiz. Med. Wochenschr.*, 63, 987, 1934.
71. **Williams, C. D.**, Asthma related to chromium compounds, *N.C. Med. J.*, 30, 482, 1969.
72. **Fregert, S., Hjorth, N., Magnusson, B., Bandmann, H.-J., Calnan, C. D., Cronin, E., Malten, K., Meheghini, C. L., Pirila, V., and Wilkinson, D. S.**, Epidemiology of contact dermatitis, *Trans. St. John's Hosp. Dermatol. Soc.*, 55, 17, 1969.
73. **Walsh, E. N.**, Chromate hazards in industry, *JAMA*, 153, 1305, 1953.

74. **Alwens, W. and Jonas, W.**, Der Chromat-lungenkrebs, *Acta Unio. Int. Cancrum,* 3, 103, 1938.
75. **McAuley, A. and Olatunji, M. A.**, Metal-ion oxidations in solution. XIX. Redox pathways in the oxidation og penicillamine and glutathione by chromium, *Can. J. Chem.*, 55, 3335, 1977.
76. **Petrilli, F. L. and Flora, S. de**, Oxidation of inactive trivalent chromium to the mutagenic hexavalent form, *Mutat. Res.*, 58, 167, 1978.
77. **Koutras, G. A., Hattori, M., Schneider, A. S., Ebaugh, F. G. Jr., and Valentine, W. N.**, Studies on chromated erythrocytes. Effect of sodium chromate on erythrocyte glutathione reductase, *J. Clin. Invest.*, 43, 323, 1964.
78. **Koutras, G. A., Schneider, A. S., and Valentine, W. N.**, Studies on chromated erythrocytes. Mechanisms of chromate inhibition of glutathione reductase, *Br. J. Haematol.*, 11, 360, 1965.
79. **deFlora, S.**, Metabolic deactivation of mutagens in the *Salmonella*-microsome test, *Nature (London),* 271, 455, 1978.
80. **Löfroth, G.**, The mutagenicity of hexavalent chromium is decreased by microsomal metabolism, *Naturwissenschaften,* 65, 207, 1978.
81. **Yawata, Y. and Tanaka, K. R.**, Red cell glutathione reductase: mechanism of action of inhibitors, *Biochim. Biophys. Acta,* 321, 72, 1973.
82. **Pigiet, V. P. and Conley, R. R.**, Purification of thioredoxin, thioredoxin reductase, and glutathione reductase by affinity chromatography, *J. Biol. Chem.*, 252, 6367, 1977.
83. **Hagenfeldt, L., Arvidsson, A., and Larsson, A.**, Glutathione and γ-glutamylcysteine in whole blood, plasma and erythrocytes, *Clin. Chim. Acta,* 85, 167, 1978.
84. **Gruber, J. E. and Jennette, K. W.**, Metabolism of the carcinogen chromate by rat liver microsomes, *Biochem. Biophys. Res. Commun.*, 82, 700, 1978.
85. **Langård, S.**, The time related subcellular distribution of chromium in the rat liver cell after intravenous administration of $Na_2{}^{51}CrO_4$, *Biol. Trace Element Res.*, 1, 45, 1979.
86. **Peters, W. J. and Jackson, R. W.**, Reaction of bone to implanted materials, in *Biology and Technology of Oral Prosthetic Implants,* Oral Science Reveiews, Vol. 5, Melcher, A. H. and Zarb, G. A., Eds., Munksgaard, Copenhagen, 1974, 56.
87. **Grenoble, D. E. and Voss, R.**, Materials and designs for implant dentistry, *Biomat. Med. Devices Artif. Organs,* 4, 133, 1976.
88. **Deltombe, E., de Zoubay, N., and Pourbaix, M.**, Chromium, in *Atlas of Electrochemical Equilibria in Aqueous Solutions,* Section 10.1, Pourbaix, M., Ed., Pergamon Press, Oxford and Cebelcar, Brussels, 1966, chap. 4.
89. **Levine, D. L. and Staehle, R. W.**, Crevice corrosion in orthopedic implant metals, *J. Biomed. Mater. Res.*, 11, 553, 1977.
90. **Williams, D. F. and Meachin, G.**, A combined metallurgical and histological study of tissue-prosthesis interactions in orthopaedic patients, in Annu. Int. Biomaterials Symp., Clemson University, Clemson, S.C., 1973.
91. **Weisman, S.**, The skeletal structure of metal implants, *Oral Implantol.*, 1, 69, 1970.
92. **Fitzpatrick, B.**, A comparative study of some implant materials, *Aust. Dent. J.*, 13, 360, 1968.
93. **Greener, E. H., Harcourt, J. K., and Lautenschlager, E. P.**, in *Materials Science in Dentistry,* Williams & Wilkins, Baltimore, 1972, 13.
94. **Phillips, R. W.**, *Skinner's Science of Dental Materials,* 7th ed., W. B. Saunders, Philadelphia, 1973, chap. 35.
95. **Smith, D. C.**, Materials used for construction and fixation of implants, in *Biology and Technology of Oral Prosthetic Implants,* Oral Science Reviews, Vol. 5, Melcher, A. H., and Zarb, G. A., Eds., Munksgaard, Copenhagen, 1974, 23.
96. **Bergman, B.**, The effects of prosthodontic materials on oral tissues, in *The Scientific Basis of Reconstructive Dentistry,* Oral Science Reviews, Vol. 10, Melcher, A. H. and Zarb, G. A., Eds., Munksgaard, Copenhagen 1974, 75.
97. **Sunami, Y. and Ishikawa, E.**, COP, a new alloy for surgical implants, *Acta Med. Okayama,* 31, 71, 1977.
98. **Homsy, C. A., Stanley, R. F., Anderson, M. S., and King, J. W.**, Reduction of tissue and bone adhesion to cobalt alloy fixation appliances, *J. Biomed. Mater. Res.*, 6, 451, 1972.
99. **Laing, P. G., Ferguson, A. B., Jr., and Hodge, E. S.**, Tissue reactions in rabbit muscle exposed to metallic implants, *J. Biomed. Mater. Res.*, 1, 135, 1967.
100. **Michel, R., Hofmann, J., and Zilkens, J.**, Influences of metal implants on trace element contents in human and mammalian tissue and organs, Paper IAEA-SM-227/10, in *Proc. Int. Symp. Nuclear Activation Techniques in the Life Sciences,* Vienna, May 22 to 26, 1978.
101. **Ferguson, A. B., Jr., Laing, P. G., and Hodge, E. S.**, The ionization of metal implants in living tissues, *J. Bone Jt. Surg. Am. Vol.*, 42, 77, 1960.
102. **Ferguson, A. B., Jr., Akahoshi, Y., Laing, P. G., and Hodge, E. S.**, Characteristics of trace ions released from embedded metal implants in the rabbit, *J. Bone Jt. Surg. Am. Vol.*, 44, 323, 1962.

103. **Wright, J. K. and Axon, H. J.,** Clinical and metallurgical observations on the corrosion of stainless steel screws used in orthopaedic surgery, *Nature (London),* 173, 1186, 1954.
104. **Hickman, J., Clarke, E. G., and Jennings, A. R.,** Structural changes in bone associated with metallic implants, *J. Bone Jt. Surg. Br. Vol.,* 40, 799, 1958.
105. **Gold, P.,** The electrical phenomena in bone: a review, *J. Periodontol.,* 38, 119, 1967.
106. **Williams, D. F.,** The properties and medical uses of materials, *Biomed. Eng.,* 6, 152, 1971.
107. **Hobkirk, J. A. and Rusinak, K.,** Metallic contamination of bone during drilling procedures, *J. Oral Surg.,* 36, 356, 1978.
108. **Brettle, J. and Hughes, A. N.,** A metallurgical examination of surgical implants which have failed in service, *Injury,* 2, 143, 1970.
109. **Cohen, J.,** Corrosion testing of orthopaedic implants, *J. Bone Jt. Surg. Am. Vol.,* 44, 307, 1962.
110. **Rae, T.,** A study on the effects of particulate metals of orthopaedic interest on murine macrophages in vitro, *J. Bone Jt. Surg. Br. Vol.,* 57, 444, 1975.
111. **Rae, T.,** Action of wear particles from total joint replacement prostheses on tissues, in *Biocompatibility of Implant Materials,* Williams, D. F., Ed., Sector Publishing, London, 1976, 55.
112. **Rae, T.,** The use of foetal rat knee joints in organ culture as a means of measuring the tolerance of tissues towards material used for orthopaedic implants, in *Organ Culture in Biomedical Research,* Balls, M. and Monnickendam, M. A., Eds., Cambridge University Press, Cambridge, 1976, 179.
113. **Rae, T.,** The haemolytic action of particulate metals (Cd, Cr, Co, Fe, Mo, Ni, Ta, Ti, Zn, Co-Cr alloy), *J. Pathol.,* 125, 81, 1978.
114. **Elves, M. W., Wilson, J. N., Scales, J. T., and Kemp, H. P. S.,** Incidence of metal sensitivity in patients with total joint replacements, *Br. Med. J.,* 4, 376, 1975.
115. **Deutman, R., Mulder, T. J., Brian, R., and Nater, J. P.,** Metal sensitivity before and after total hip arthroplasty, *J. Bone Jt. Surg., Am. Vol.,* 59, 862, 1977.
116. **Cramers, M. and Lucht, K.,** Metal sensitivity in patients treated for tibial fractures with plate of stainless steel, *Acta Orthop. Scand.,* 48, 245, 1977.
117. **Fisher, A. A.,** Safety of stainless steel in nickel sensitivity, *J. Am., Med. Assoc.,* 221, 1279, 1972.
118. **Brown, G. C., Lockshin, M. D., Salvati, E. A., and Bullough, P. G.,** Sensitivity to metal as a possible cause of sterile loosening after cobalt-chromium total hip-replacement arthroplasty, *J. Bone Jt. Surg. Am. Vol.,* 59, 164, 1977.
119. **Ovrutsky, G. D. and Ulyanov, A. D.,** Allergy to chromium using steel dental prosthesis, *Stomatologia (Moskva),* 55, 60, 1976.
120. **McDougall, J.,** Malignant tumour at the site of bone plating *J. Bone Jt. Surg. Br. Vol.,* 38, 709, 1956.
121. **Heath, J. C., Freeman, M. A. R., and Swanson, S. A. V.,** Carcinogenic properties of wear particles from prostheses made in cobalt chromium alloy, *Lancet,* 1, 564, 1971.

Chapter 10

BIOLOGICAL PROPERTIES OF MOLYBDENUM

D. F. Williams

TABLE OF CONTENTS

I. INTRODUCTION

Molybdenum is rarely used as a pure metal or even as a major alloying addition in any engineering situation, but does have widespread use as a minor, but nevertheless very important, addition in certain alloys. It is in this context that molybdenum is relevant to biocompatibility for it is a constituent of both the 316L austenitic stainless steel, at up to 4%, and the casting cobalt-chromium alloys, where it is usually present at 5 to 7%. Since it exists in solid solution in these alloys, and since even these nominally corrosion-resistant alloys corrode at a slow but finite rate in aqueous media, implant-derived molybdenum must inevitably find its way into the tissue.

The effects of molybdenum in the tissue have been largely ignored in discussions of the biocompatibility of these alloys, presumably because of its secondary importance to the major elements present in the alloys. The biological properties of molybdenum are extremely interesting, however. In spite of its general obscurity and low concentration in the earth's crust, it is an essential dietary element necessary for the function of some enzymes. Like all essential elements, it is toxic in large doses and there is a well-defined syndrome associated with its toxicity. Finally, molybdenum figures prominently in some of the more important antagonist reactions among metals in the body. It is worthwhile, therefore, reviewing the biochemical, physiological, and toxicological properties of this element in order to provide general background for the biocompatibility of molybdenum-containing alloys.

II. MOLYBDENUM IN ANIMAL AND HUMAN TISSUES AND FLUIDS

A small but finite level of molybdenum is found in virtually all animal tissues. According to Kienholz there is little variation between species[1] and the comprehensive study by Tipton and Cook showed that there was little accumulation in any organ with age.[2] The normal values in adult man given by Tipton and Cook were 0.14 ppm dry tissue in muscle, 0.14 ppm in brain, 0.15 ppm in lung, 0.20 ppm in spleen, 1.6 ppm in kidney, and 3.2 ppm in liver. The higher level in the liver is consistent with findings in various animals which typically have molybdenum liver levels of 2 to 4 ppm.[1-3] However, as noted later, molybdenum metabolism is closely related to the body burden of inorganic sulfate so that the liver level is not indicative of overall molybdenum status.

The level of molybdenum in blood and plasma has been in some doubt. Certainly in ruminants the level ranges from 1 μg Mo/100 mℓ to 6μg/100 mℓ, depending on the dietary sulfate level,[4] while these levels can be raised significantly if the molybdenum intake is increased. Several authors have reported values for molybdenum in human plasma or serum. Webster gave a value of 0.56 μg/100 mℓ[5] but de la Cruz[6] reported a level of 25.7 μg/100 mℓ. Most workers have found intermediate levels,[7-10] although even lower levels of 0.1 μg/100 mℓ and 0.058 μg/100 mℓ have been reported by Baert et al.[11] and Versieck et al.[12] It is likely that experimental error is responsible for much of this variation, with a large number of techniques, including emission spectrometry, X-ray spectrochemistry, and neutron activation analysis, being used.

III. MOLYBDENUM METABOLISM

A. Absorption

Hexavalent water-soluble compounds of molybdenum, such as sodium and ammonium molybdate, are readily absorbed from the intestinal tract.[13] In addition, even insoluble compounds such as the trioxide and calcium molybdate are absorbed if included in sufficiently large concentrations in the diet.[4] The exact amount absorbed is strongly dependent on copper and sulfate levels in the diet and in the tissues, as discussed below.

B. Excretion

As expected from the high rate of intestinal absorption, excretion is largely via the urine;[14] this is true, in fact, whether the molybdenum is administered orally or intravenously. There is also a significant amount of excretion in the bile.[15]

C. Effects of Copper and Sulfate on Molybdenum Metabolism

Much has been written on the relationship between molybdenum, copper, and sulfate in tissues, especially those of ruminant animals, following the early discoveries by Dick.[16] Basically, an increase in the sulfate content of the diet limits intestinal absorption and increases urinary molybdenum excretion, it being assumed that inorganic sulfate interferes with, and possibly prevents, transport of molybdenum across membranes. This would increase urinary molybdenum excretion, because the raised sulfate concentration in the ultrafiltrate of the kidney glomerulus blocks molybdenum reabsorption. Molybdenum appears to be able to interfere with copper metabolism; this subject has been recently reviewed by Underwood,[4] and is discussed further at a later point in this chapter on toxcity.

IV. BIOLOGICAL ACTIVITY OF MOLYBDENUM

A. Essentiality of Molybdenum

It was first shown in 1953 that molybdenum is an essential dietary trace element[17,18] when the flavoprotein enzyme xanthine oxidase was demonstrated to be a molybdenum-containing metalloenzyme that is dependent for its activity on the presence of the metal. It is difficult, however, to define minimum dietary requirements for the metal, and studies have shown that animals are able to grow normally, reproduce, and oxidize xanthine on diets containing extremely low molybdenum levels.[19]Most foods contain sufficiently high levels of molybdenum so that it is easy to satisfy these low requirements. The daily intake for adult humans has been variously estimated as 48 to 96 μg Mo[20] to 350 μg.[21]

B. Molybdenum-Containing Metalloenzymes

There are three principal molybdenum-containing metalloenzymes whose formation and activity depend on this metal: xanthine oxidase,[17,18] aldehyde oxidase,[22] and sulfite oxidase.[23] A review of the structure, function, and mechanisms of these metalloenzymes has recently been published by Spence.[24] Xanthine oxidase is involved in the formation of uric acid from purines. Aldehyde oxidase catalyzes the oxidation of aldehydes and sulfite oxidase catalyzes the last stage in the oxidation of sulfur-containing amino acids. In addition, molybdenum is contained in and is necessary for the action of the enzymes nitrogenase and nitrate reductase.

There has been much work performed recently on the mechanisms involved in the activity of these molybdenum-containing enzymes. As noted earlier by Rae, it is likely that the function of molybdenum in xanthine oxidase and aldehyde oxidase is to transfer electrons between substrate and iron. These reactions have been extensively studied by Bray[25,26] who observed the direct transfer of hydrogen atoms from substrate molecules to xanthine oxidase. In the reduced functional enzyme, Mo(V) interacts with two enzyme-bound protons. During the reaction, hydrogen from the substrate C-8 position, after transfer to the enzyme, appears strongly coupled to molybdenum.

Animal experiments have shown that diets deficient in molybdenum do result in lowered xanthine oxidase[18] which leads to a reduced ability to oxidase xanthine to uric acid and sulfite oxidase,[27] but generally the effects are not particularly serious.

C. Molybdenum and Dental Caries

According to Jenkins,[28] Adler first concluded that molybdenum was associated with an anticariogenic effect following epidemiological studies in Hungary.[29] Since then, this relationship has been investigated on many occasions,[28,30-32] the majority of which confirm this hypothesis. Although there is a reasonable amount of evidence on this subject, it is not all conclusive and some doubt remains as to the significance of molybdenum. Certainly the dentine and, to a greater extent, the enamel rapidly take up molybdenum administered to experimental animals[33] but there is no direct evidence to associate this with a lower enamel solubility.

V. TOXICITY OF MOLYBDENUM

Molybdenum is generally thought to be of low toxicity, but the subject is complicated by a significant species difference in animals and the molybdenum-copper-sulfate interactions already mentioned. The reader is referred to Underwood for an extensive review of molybdenum toxicology.[4] Cattle and rabbits appear to be particularly intolerant of high dietary molybdenum levels. For example, in rabbits the symptoms of molybdosis include growth retardation, Alopecia, dermatosis, severe anemia, and bony deformities.[34]

Frequently, raised molybdenum intakes are associated with a fall in plasma thyroxine, liver glucose 6-phosphate activity, and liver sulfide oxidase activity. This latter fact suggests that the raised molybdenum level causes an accumulation of sulfide in the tissues through a fall in sulfide oxidase activity and that this leads to the symptoms of anemia and diarrhea frequently observed.

There is very little evidence to suggest that molybdenum is toxic to humans in low doses. That cattle are extremely susceptible but not humans is probably due to the fact that the molybdenum appears to exert its effect in the rumen, where it is converted into soluble thiomolybdates. According to Browning,[35] industrial toxicity of molybdenum is virtually nonexistent, although the high activity of xanthine oxidase associated with raised molybdenum levels can give abnormally high serum uric acid concentrations, leading to a predisposition to gout.[4]

According to Luckey and Venugopal,[36] the toxicity of individual Group VI metals decreases from $Se > Te > Cr^{6+} > Mo > W > Cr^{3+}$, so that molybdenum does have a low order of toxicity. Without further explanation they cite symptoms of acute molybdenum toxicity as diarrhea, coma, and cardiac failure and indicate that high molybdenum levels inhibit the activities of ceruloplasmin, cytochrome oxidase, glutaminase, choline esterase, and sulfite oxidase. Excessive intake of molybdenum was noted to interfere with calcium and phosphorus metabolism and to induce osteoporosis and other tissue changes. The same authors state that molybdenum exhibits a mild teratogenic effect, but is not involved in carcinogenesis.

VI. SUMMARY

The evidence reviewed above suggests that molybdenum is an essential trace element where slight deficiences or excesses can, through the effect on certain enzymes, induce clinically observed symptoms. However, within limits, these effects are not very significant. There is little or no data available on the direct biocompatibility of molybdenum, but there is no reason to suppose that it has any effects of remarkable significance.

REFERENCES

1. **Kienholz, E.**, in *Transport and the Biological Effects of Molybdenum,* Chappell, W. R., Ed., University of Colorado Press, Denver, 1974, 148.
2. **Tipton, I. H. and Cook, M. J.**, Trace elements in human tissues. III. Adult subjects from the United States, *Health Phys.,* 9, 103, 1963.
3. **Gray, L. F. and Daniel, L. J.**, Effect of the copper status of the rat on the copper-molybdenum-sulfate interaction, *J. Nutr.,* 84, 31, 1964.
4. **Underwood, J.**, Molybdenum, in *Trace Elements in Human and Animal Nutrition,* 4th ed., Academic Press, New York, 1977.
5. **Webster, P. O.**, Trace elements in serum and urine from hypertensive patients before and during treatment with chlortholidone, *Acta Med. Scand.,* 194, 505, 1973.
6. **de la Cruz, B.**, in *Trace Elements in Relation to Cardiovascular Diseases,* International Atomic Energy Agency, Vienna 1973, 19.
7. **Mertz, D. P., Koschnick, R., Wilk, G., and Pfeilsticker, K.**, Untersuchungen uber den stoffwechsel von spurenelemenlen biem Menschen serumwerte von Co, Ni, Ag, Cd, Cr, Mo, Mn, *Z. Klin. Chem.,* 6, 171, 1968.
8. **Bala, Y. M. and Lifshits, V. M.**, Content of trace elements in the blood in leukemia and anemia. II. Molybdenum and chromium, *Fed. Proc. Transl. Suppl.,* 25, 370, 1966.
9. **Webb, J., Kirk, K. A., Jackson, D. H., Niedermeier, W., Turner, M. E., Rackley, C. E., and Russell, R. O.**, Analysis by pattern recognition technique of changes in serum level of 14 trace metals after acute myocardial infarction, *Exp. Mol. Pathol.,* 25, 322, 1976.
10. **Niedermeier, W. and Griggs, J. H.**, Trace metal composition of synovial fluid and blood serum of patients with rheumatoid arthritis, *J. Chronic Dis.,* 23, 527, 1971.
11. **Butt, E. M., Nusbaum, R. E., Gilmour, T. C., Didio, S. L., and Mariano, S.**, Trace metal levels in human serum and blood, *Arch. Environ. Health,* 8, 52, 1964.
12. **Baert, N., Cornelis, R., and Hoste, J.**, Molybdenum in human blood, *Clin. Chim. Acta,* 68, 355, 1976.
13. **Versiek, J., Hoste, J., Barbier, F., Vanballenberghe, L., de Rudder and Cornelis, R.**, Determination of molybdenum in human serum by neutron activation analysis, *Clin. Chim. Acta,* 87, 135, 1978.
14. **Comar, C. L.**, in *Copper Metabolism,* McEloy, W. P. and Glass, B., Eds., John Hopkins Press, Baltimore, 1950.
15. **Caujolle, F.**, L'elimination biliare du Molybdene, *Bull. Soc. Chem. Biol.,* 19, 827, 1937.
16. **Dick, A. T.**, Molybdenum and copper relationships in animal nutrition, in *Inorganic Nitrogen Metabolism,* McElroy, W. D. and Glass, B., Eds., John Hopkins Press, Baltimore, 1956.
17. **de Renzo, E. C., Kaleita, E., Heytler, P., Oleson, J. J., Hutchings, B. C., and Williams, J. H.**, Identification of the xanthine oxidase factor as molybdenum, *Arch. Biochem. Biophys.,* 45, 247, 1953.
18. **Richert, D. A. and Westerfeld, W. W.**, Isolation and identification of the xanthine oxidase factor as molybdenum, *J. Biol. Chem.,* 203, 915, 1953.
19. **Higgins, E. S., Richert, D. A., and Westerfeld, W. W.**, Molybdenum deficiency and tungstate inhibition studies, *J. Nutr.,* 59, 539, 1956.
20. **Robinson, M. F., McKenzie, J. M., Thomson, C. D., and Van Rij., A. C.**, Metabolic balance of zinc, copper, cadmium, iron, molybdenum and selenium in young New Zealand women, *Br. J. Nutr.,* 30, 195, 1973.
21. **Schroeder, H. A., Balassa, J. J., and Tipton, I. H.**, Essential trace metals in man: molybdenum, *J. Chronic Dis.,* 23, 481, 1970.
22. **Mahler, H. R., Mackler, B., Green, D. E., and Bock, R. M.**, Studies on metalloflavoproteins. III. Aldhelyde oxidase, *J. Biol. Chem.,* 23, 465, 1954.
23. **Cohen, H. J., Fridovich, I., and Rajogopalan, K. V.**, Hepatic sulfite oxidase: a functional role for molybdenum, *J. Biol. Chem.,* 246, 374, 1971.
24. **Spence, J. T.**, Reactions of molybdenum co-ordination compounds: models for biological systems, in *Metal Ions in Biological Systems,* Vol. 5, Seigel, H., Ed., Marcel Dekker, New York, 1976, chap. 6.
25. **Bray, R. C. and Knowles, P. F.**, Electron spin resonance in enzyme chemistry; the mechanism of action of xanthine oxidase, *Proc. R. Soc. London Ser. A,* 302, 351, 1968.
26. **Gutteridge, S., Tanner, S. J., and Bray, R. C.**, The molybdenum centre of native xanthine oxidase, *Biochem. J.,* 175, 869, 1978.
27. **Johnson, J. L., Rajagopalan, K. V., and Cohen, H. J.**, Molecular basis of the biological function of molybdenum, *J. Biol. Chem.,* 249, 859, 1974.

28. **Jenkins, G. N.,** Molybdenum and dental caries, *Br. Dent. J.,* 122, 435, 1967.
29. **Adler, P. and Straub, J.,** A water-borne caries protective agent other than fluorine, *Acta Med. Acad. Sci. Hung.,* 4, 221, 1953.
30. **Adler, P. and Porcsalmy, I.,** Recent experiments on the caries protective action of molybdenum added to drinking water, *Arch. Oral Biol.,* 4, 193, 1961.
31. **Malthur, R. S., Ludwig, T. G., and Healy, W. B.,** Effects of trace elements on dental caries in rats, *N. Z. Dent. J.,* 60, 291, 1964.
32. **Hadjimarkos, D. M.,** Effect of trace elements on dental caries, *Adv. Oral Biol.,* 3, 253, 1968.
33. **Bawden, J. W. and Hammarstrom, L. E.,** Autoradiography of ^{99}Mo in developing rat teeth and bone, *Scand. J. Dent. Res.,* 84, 168, 1976.
34. **Arrington, L. R. and Davis, G. K.,** Molybdenum toxicity in the rabbit, *J. Nutr.,* 51, 295, 1953.
35. **Browning, E.,** in *Toxicity of Industrial Metals,* 2nd ed., Butterworths, London, 1969, 246.
36. **Luckey, T. O. and Venugopal, B.,** in *Metal Toxicity in Mammals, Physiologic and Chemical Basis for Metal Toxicity,* Plenum Press, New York, 1977.

Chapter 11

BIOLOGICAL EFFECTS OF TITANIUM

D. F. Williams

TABLE OF CONTENTS

I. INTRODUCTION

From the biological point of view, titanium is a strange element. It is widely distributed in the earth's crust, but yet is found in only minute amounts in animal and plant tissues. There is no evidence to suggest it is an essential trace element. On the other hand, it appears to be extremely well tolerated by tissues and has a very low order of toxicity. It may, therefore, be described as a physiologically indifferent element. It has been used widely in implant surgery because of its excellent corrosion resistance and apparent biocompatibility,[1-4] as reviewed by this author elsewhere in this series.[50] Even here, however, the literature on the surgical uses of titanium is remarkable for its lack of objective assessments of the tissue response to the metal, possibly because of the way it is so readily and benignly accepted by tissues. In this chapter the known toxicology of titanium and its compounds is discussed and placed into the context of the use of titanium as a biomaterial.

II. OCCURRENCE, PROPERTIES, AND USES OF TITANIUM

A. Occurrence

Titanium is a relatively abundant element, occurring widely in the earth's crust in concentrations up to 10%. It occurs mainly as the ores ilmenite ($FeTiO_3$) and rutile (TiO_2). Extraction from these ores is a difficult process and involves conversion to titanium tetrachloride as an intermediate step followed by reduction.

B. Properties

Titanium has a relatively low specific gravity for a metal (4.5 g/cc) which, coupled with good mechanical properties, gives a very favorable strength per weight ratio. It is thermally very stable and may be regarded as one of the most corrosion resistant of metals.

C. Uses

Most industrial uses of titanium utilize the high strength: weight ratio which is particularly useful in the aircraft industry. The corrosion resistance has also led to its use in chemical plant and other situations involving aggressive media. Fortunately titanium carries some of its good properties, such as corrosion resistance, with it when alloyed and some of the titanium-transition metal alloys offer extremely good property combinations.

D. Biomedical Uses

Titanium and titanium alloys comprise one of the three main classes of metallic materials used in surgical prostheses. Initial experiments were performed in 1940. Numerous papers reported excellent results with commercially pure titanium in orthopedic fracture plates and oral surgery in the succeeding years.[6-8] The material was adopted for general surgical use in the U.K. in the mid 1960s and a few years later in the U.S.

While considerable success had been achieved with pure titanium, it was realized that this metal did not possess the same mechanical properties as either Vitallium® (the cobalt-chromium alloy then widely used in orthopedics and dentistry) or stainless steel. Since some titanium alloys had superior mechanical properties and possessed equivalent corrosion resistance, experiments with some of these alloys were performed. One alloy in particular, the Ti-6%Al-4%V alloy, was extensively investigated and gradually introduced into surgical use. Today this alloy and, to a lesser extent, the pure metal are used for many different types of surgical implant, not only in orthopedics but also in dental, maxillofacial, cardiovascular, and neurosurgery.

III. TITANIUM IN ANIMAL AND HUMAN TISSUES

Although there is normally a reasonable correlation between the abundance of any element in the earth's crust and the amount contained in animal tissues, the widespread natural distribution of titanium contrasts with a very low level in tissues. As noted by Browning,[9] early attempts to determine tissue levels of titanium, such as those performed by Lehmann and Herget,[10] failed to find any measurable amount, even after dosage with titanium dioxide. Maillard and Ettori,[11] using more sensitive methods, found the element in all organs and tissues in several animals at concentrations ranging from 1.5 to 11 μg/100 g fresh tissue. In man, the spleen and adrenals accumulated the most titanium (around 10 μg/100 g) while the liver (3 μg) and kidneys (1.5 μg) contained lower amounts.

Tipton and Cook[12] included titanium in their determinations of trace metal levels in adult humans, finding less than 0.05 μg/g in all organs except the lungs.

An extensive survey of titanium in animal and plant tissues was carried out by Schroeder's group and published in 1963.[13] A detailed analysis of results of lung concentrations in humans showed that while only 58% of neonates showed any demonstrable lung titanium content, all subjects over the age of 11 years did so, with concentrations rising from a mean of 145 ppm ashed tissue at 11 to 20 years to greater than 500 ppm above 51 years of age. No such accumulation was evident in the kidney, skin, or aorta, both the percentage occurrence (20 to 80%) and concentration (7 to 43 ppm) being fairly constant throughout life. Some differences in racial and geographical concentrations of titanium in liver, kidney, and lung were found, more titanium being generally found in specimens from the Middle East and India than elsewhere.

A comparison of titanium contents of animal tissues with human tissues showed a generally higher level in the former, although there were considerable variations. Roughly equivalent concentrations in human and animal milk were reported by Grebennikov.[14]

IV. METABOLISM OF TITANIUM

Data concerning the metabolism of titanium are extremely sparse. From the work of Schroeder[13] discussed above, it would seem that since there is an accumulation of titanium in human lung tissue during life, but no similar accumulation in other tissues, absorption through the gastrointestinal tract is very poor, while pulmonary absorption is more significant. Titanium is certainly present in some foods and therefore available to the gastrointestinal tract in normal subjects. While milk and cheese, lamb and beef, and some vegetables and fruits appeared to be devoid of titanium, greater than 1 μg/g were found in butter, pork, some fish, corn oil, and pepper.[13] It was calculated that a daily institutional diet contained 105.4 μg titanium.

Experiments in which mice were given drinking water containing 5 ppm titanium as the potassium oxalate did show, however, that increased titanium dietary intake results in accumulation in the organs of these animals.[13] The heart, lungs, spleen, and kidneys all showed considerably greater amounts, up to 32 μg/g, which were three to ten times the levels in control mice. In a further series of experiments using different mice,[13] similar values summarized in Table 1 were obtained, the heart again containing the greatest concentration of titanium.

A study of titanium metabolism in sheep has been reported by Miller et al.[16] who dosed two lambs orally and one intravenously with 3 μCi ^{44}Ti. Following intravenous administration, the ^{44}Ti disappeared from the blood plasma exponentially, 18.4% remaining in the plasma after 48 hr. The majority of the titanium had been transported

Table 1
TITANIUM CONCENTRATION IN VARIOUS ORGANS OF MICE AFTER EXPOSURE TO DRINKING WATER CONTAINING 5 PPM TITANIUM AND OF CONTROL MICE (VALUES IN μg/g WET WT)[15]

	Kidney	Liver	Heart	Lung	Spleen
Experimental mice					
Male	2.86	1.81	8.80	4.81	6.83
Female	2.89	2.05	4.10	1.66	3.70
Control mice					
Male	0.33	0.38	0.34	0.13	0.94
Female	0.55	0.67	1.08	0.66	1.10

to bone and cartilagenous tissue (24 to 8%) muscle (13.3%) and various internal organs, especially the kidney. Less than 10% was found in the feces and only 3% in the urine. Following oral administration, over 96% was recovered in the feces, much of the remainder at 48 hr still being contained in the digestive tract. None could be detected in the blood or urine.

In contrast to these studies which indicate very poor titanium absorption, Tipton et al.[16] have demonstrated considerable absorption with excretion via both urine and feces.[17] Huggins and Froehlich[18] studied the fate of intravenously injected titanium dioxide in female rats. The removal of TiO_2 from the blood again occurred at an exponential rate. Six hours after a single injection, the liver (4.13 mg/g) and the spleen (3.0) both contained elevated TiO_2 levels. The amount in the spleen increased considerably during the following year, but a remarkable concentration of 127 mg/g was recorded in the celiac lymph nodes which are found in the upper abdomen and which filter lymph from the liver.

V. BIOLOGICAL ACTIVITY OF TITANIUM AND TITANIUM COMPOUNDS

The experimental and clinical evidence concerning the biological activity of titanium and its compounds is neither voluminous nor coherent. The available facts are collated in this section.

A. Evidence for Essentiality

According to Schwarz,[19] there are 20 additional elements that have to be taken into serious consideration as potential trace elements, and of these, titanium is the most attractive because it is located at the end of the series of eight essential transition elements (V, Cr, Mn, Fe, Co, Ni, Cu, Zn). There is no conclusive evidence, however, of any essentiality, and Schroeder et al.,[13] although admitting it is not entirely impossible, argue against it, largely on the basis that all known essential metals in adults are present at birth in similar or higher concentrations, whereas half the newborn humans examined had no demonstrable titanium. They also argue that the sporadic nature of the distribution of titanium in older tissues indicates nonessentiality since it is difficult to believe that a metal to which man is exposed in food, water, and air, but which is not always found in his tissues, has an essential metabolic role. Underwood, as recently as 1977, states that no evidence has appeared that absorbed titanium performs any vital function in animals or that it is a dietary essential for any living organism.[20]

B. Effects of Titanium After Ingestion

Experiments in animals by Ereaux,[21] aimed at determining the safety of titanium compounds as skin preparations, showed that oral administration of titanium salts had no toxic effects but rather caused an improvement in health. Vernetti-Blina had previously reached the same conclusion that titanium oxide is inert following ingestion.[22]

Schroeder and colleagues[13,23] have studied the effects of several metals, including titanium, on the growth and survival of mice following oral administration. Both male and female rats given 5 ppm titanium in drinking water gained weight more rapidly, and when mature were heavier than the controls. A similar pattern was followed by mice on a chromium-supplemented diet, but not by those receiving lead, cadmium, and nickel. Mortality of male mice was variously affected by the majority of these metals but not by titanium. A comparison of the half-lives (time when 50% were dead) showed that the adverse effects decreased in the following order: nickel > lead > cadmium > titanium = control ≫ chromium, once again indicating the physiological indifference to this metal.

C. Effects After Inhalation

There have been some suggestions that exposure to titanium dust can lead to fibrosis of the lung, although this is by no means proven and it is likely that titanium has a minimal role in the pneumoconioses compared to other substances. Vernetti-Blina[22] could find no effect on respiratory function in workers exposed to titanium dust over a long period of time and similar conclusions were reached by Lundgren and Ohman.[24] While Moschinski et al. found some evidence of fibrosis of the lung in some workers exposed to the oxide dust, they concluded that titanium pneumoconiosis was improbable.[25] Husten also showed that slight fibrosis may result from repeated inhalation of very high concentrations of dust, but did not claim that the results were conclusive.[26] More recently, Uragoda and Pinto[27] undertook a radiological and clinical investigation of 130 workers at an ilmenite extraction plant in Ceylon. The workers were exposed to a number of minerals, principally including the titanium ores ilmenite and rutile. There was no significant difference in the incidence of radiological lesions of the chest between their workers and a control group drawn from the general population. The few respiratory symptoms found in the workers were attributed to the inhalation of dust per se rather than the specific effect of these titanium-bearing minerals. These results agree well with those of Christie et al.[28] and Reed.[29]

In contrast to this, Elo et al.[30] did find that industrially processed titanium dioxide behaves as a mild irritant in pulmonary interstitium. They reviewed some factory workers employed in the processing of titanium dioxide pigments and analyzed lung specimens. They found significantly higher titanium levels, deposits of titanium being associated with cell destruction and slight fibrosis. Titanium dioxide was found in the lymphatics, suggesting clearance via the lymphatic system and particles were found within lysosomes of alveolar macrophages. The authors stated that nothing was known of the relationship between the accumulation of titanium dioxide particles within phagocytic cells and other pathological findings such as fibrosis. Since their analyses showed that the concentration of several other metals was also slightly higher than in the control samples, they could not rule out the possibility that the clinicopathological findings may have been due to the toxic effects of these other substances rather than titanium.

The same group of workers has studied the pulmonary deposits of titanium dioxide further by light microscopy and EDAX techniques.[31] They found, in addition to titanium, the alveolar macrophages contained aluminium and silica, although morphologically the particulate matter was different. It was suggested that the fibrosis was due to this silica rather than the titanium dioxide. It was further noted that the titanium dioxide particles stayed within macrophages for a long period of time, suggesting they had very little adverse effect on these cells.

Gross et al. have presented evidence that fibrosis due to titanium compounds in the lung is quite different to that induced by asbestos and is only found where there are massive focal accumulations of particles.[32] They administered intratracheal injections of titanium phosphate fibers to rats and hamsters and found a slight, dose-related response at concentrations of 50 and 20 mg/kg in rats, but no effect at 2 mg/kg in rats or at any of the three levels in hamsters.

The lungs of the rats given 50 mg/kg titanium phosphate contained fiber bundles, the response to which included the formation of numerous nodular foci composed of mononucleated cells with abundant cytoplasm but poorly defined cell borders. The center of the foci were often formed by syncytical masses and giant cells. These foci occupied and obliterated the lumens of respiratory bronchioles and alveolar ducts and were associated with reticulin proliferation. Some of the intraluminal stroma was converted to collagen, but this remained as a loose structure in contrast to the dense scar that forms in response to asbestos. The response was less significant with the smaller doses. The lack of fibrogenic response in the hamster lungs indicates this material is nonfibrogenic in this species, in contrast to the significant collagen deposition associated with 10 mg/kg asbestos.

D. Effect of Titanium on Cells

Rae[33,34] has reported tissue culture studies of particulate metals including titanium and has found this metal to have minimal effect. Using mouse peritoneal macrophages, the activity of the enzyme lactate dehydrogenase was measured in the supernatant after exposure to particulate metals as a measure of the integrity of the cell membrane. While cobalt, nickel, and a cobalt-chromium alloy were all associated with raised LDH levels after as little as 3 hr, molybdenum, chromium, and titanium produced no increase in extracellular LDH above control values at any time. Similarly, the intracellular level of glucose-6-phosphate dehydrogenase (G6PD) was measured as an indication of nucleic acid synthesis and phagocytic ability. Again, while cobalt and nickel were associated with a decrease in G6PD activity over a period of 48 hr, titanium, along with molybdenum and chromium, had no demonstrable effect.

Rae[35] has also studied the hemolytic action of particulate metals, finding that nickel and cobalt produce a significant effect, while little or no hemolysis of erythrocytes is found with a variety of metals including titanium. It is interesting in this context to note the observation that leukocytes may contain a high level of titanium. This was first reported by Carroll and Tullis,[36] who noticed high levels of both titanium and zinc in the leukocytes of blood that had been subjected to dialysis. No such concentrations, which were estimated as between 10^6 and 10^8 atoms titanium per cell, were found in peripheral blood, but fresh bone marrow aspirates gave positive results. Rather variable results were obtained by McCue,[37] but differences in the titanium content of leukocytes from patients with Hodgkins disease and leukemia compared to healthy subjects were reported.

E. Response to Implanted Titanium

The benign response to implanted titanium has already been mentioned and detailed accounts of the tissue response are included elsewhere in this series.[50-52] It is of relevance to point out here, however, that virtually all observations point to an extremely good tissue tolerance to both pure titanium and its alloys. It is of some significance that the response is minimal even where large amounts of titanium-bearing deposits are observed in the tissue surrounding the implants.[38-40] Galante and Rostoker[41] and Brown and Mayor[42] are among the very few investigators who have suggested that there is a more significant response to titanium compared to other implantable alloys.

F. Effects of Titanium on the Skin

Titanium has little effect on the skin and the very fact that titanium compounds such as the sulfate and dioxide are used as topical preparations in the treatment of skin diseases is often used as proof of the lack of toxicity or irritancy by titanium. As far as the author is aware, there have been no reports of hypersensitivity reactions to titanium or titanium compounds.

G. The Role of Titanium in Caries Prevention

There has been some discussion in the literature concerning the prevention of dental caries by titanium compounds. This subject has been discussed by Bibby and Little[44] who noticed that the tetrafluorides of Group IV metals were reasonably effective in reducing the solubility of enamel. They indicated that the chemical literature suggests many types of reaction between the organic material of teeth and titanium are possible, especially those resulting in the formation of organic titanates. Regulati et al. showed that a typical application of titanium tetrafluoride significantly increased the rate of fluoride uptake by enamel and decreased the acid solubility, but this did not reduce the caries incidence.[45] The situation remains unresolved.

H. Miscellaneous Reactions and Observations

A further example of the biologically inert behavior of titanium has been provided by Zehnder and Wuhrmann[46] working with bacteria. In order to prepare media with a low oxidation-reduction potential for the culture of anerobic bacteria, it is necessary to add a reducing agent as well as removing all traces of oxygen. While sulfides, cysteine, and thioglycolic acid may be used, they tend to be toxic to microorganisms. Titanium seemed an attractive alternative, and titanium (III) citrate was used with various bacterial cultures. While faculative anerobic bacteria were unable to grow in the presence of Ti(III), growth started after there was complete oxidation of the Ti(III), and there was no inhibition by Ti(IV). Obligate anerobes all grew in the presence of Ti(III) citrate, and it was concluded that these titanium compounds were biologically inert.

Although there is very little data published on the biochemical reactions of titanium, Ti(III) has been shown to reduce folate. Kawai and Scrimageour[47] demonstrated that both folate and 7,8-dihydrofolate can be reduced by $TiCl_3$ above 30°C, the reaction proceeding by the formation of a complex between the metal and the pterin coenzyme.

Titanium is not thought to be a carcinogen, although some titanium compounds have induced tumors in experimental animals. Luckey and Venugopal[48] report that titanium dioxide suspended in trioctanoin induced fibrosarcomas when injected intramuscularly into rats, although no tumor developed after intratracheal or intraperitoneal injections of titanium phosphate in either rats or hamsters.[32] Luckey and Venugopal[48] also report experiments of Furst and Haro[49] who demonstrated a variety of malignant neoplasms at the site of injections in liver and spleen of titanocene, which is titanium complexed with dichlorodicylcopentadiene. It was not determined, however, whether the carcinogenicity was due to the organic part of the molecule or to the titanium.

VI. CONCLUSIONS

The evidence presented in this review suggests that titanium is essentially a physiologically indifferent metal. It appears neither to be an essential dietary element nor to have any significant toxic or irritant effects on tissues.

REFERENCES

1. **Williams, D. F.**, Titanium as a metal for implantation. I. Physical properties, *J. Med. Eng. Tech.*, 1, 195, 1977.
2. **Williams, D. F.**, Titanium as a metal for implantation. II. Biological properties and clinical applications, *J. Med. Eng. Tech.*, 1, 266, 1977.
3. **Williams, D. F.**, The properties of titanium and its uses in cardiovascular surgery, *J. Cardiovas. Tech.*, 20, 52, 1978.
4. **Williams, D. F.**, Titanium in oral and maxillofacial surgery, *Proc. Inst. Maxillo Facial Tech.*, in press.
5. **Bothe, R. T., Beaton, L. E., and Davenport, H. A.**, Reaction of bone to multiple metallic implants, *Surg. Gynecol. Obstet.*, 71, 598, 1940.
6. **Leventhal, G. S.**, Titanium, a metal for surgery, *J. Bone Jt. Surg. Am. Vol.*, 33, 473, 1951.
7. **Beder, O. E. and Eade, G.**, An investigation of tissue tolerance to titanium metal implants in dogs, *Surgery*, 39, 470, 1956.
8. **Beder, E. E. and Ploger, W.**, Intra-oral titanium implants, *Oral Surg. Oral Med. Oral Pathol.*, 12, 787, 1959.
9. **Browning, E.**, *Toxicity of Industrial Metals*, 2nd ed., Butterworths, London, 1969, 332.
10. **Lehmann, K. B. and Herget, L.**, Studien uber die hygienischen Eigenschaften des Titanoxyds und des Titanweiss, *Chem. Ztg.*, 82, 793, 1927.
11. **Maillard, L. C. and Ettori, J.**, Dosage de Titane de l'organisme, *C. R. Acad. Sci. Paris*, 202, 594, 1936.
12. **Tipton, I. H. and Cook, M. J.**, Trace elements in human tissue. II. Adult subjects from the United States, *Health Phys.*, 9, 103, 1963.
13. **Schroeder, H. A., Bolassa, J. J., and Tipton, I. H.**, Abnormal trace metals in man: titanium, *J. Chronic Dis.*, 16, 55, 1963.
14. **Grebennikov, E. P., Soroka, V. R., and Sabadash, E. V.**, Trace elements in human and animal milk, *Fed. Proc. 23 (Trans. suppl.)*, T 461, 1963.
15. **Schroeder, H. A., Bolassa, J. J., and Vinton, W. H.**, Chromium, lead cadmium, nickel and titanium in mice: effect on mortality, tumors and tissue levels, *J. Nutr.*, 83, 239, 1964.
16. **Miller, J. K., Madsen, F. C., and Hansard, S. L.**, Absorption, excretion and tissue deposition of titanium in sheep, *J. Dairy Sci.*, 59, 2008, 1976.
17. **Tipton, I. H., Stewart, P. L., and Martin, P. G.**, Trace elements in diets and excreta, *Health Phys.*, 12, 1683, 1966.
18. **Huggins, C. B. and Froehlich, J. P.**, High concentration of injected titanium dioxide in abdominal lymph nodes, *J. Exp. Med.*, 124, 1099, 1966.
19. **Schwarz, K.**, New essential trace elements, in *Trace Element Metabolism in Animals*, Vol. 2, Hoeskra, W. G., Suttie, J. W., Ganther, H. E., and Mertz, W., Eds., University Park Press, Baltimore, 1974, 355.
20. **Underwood, J. J.**, *Trace Elements in Human and Animal Nutrition*, 4th ed., Academic Press, New York, 1977, 452.
21. **Eneaux, C. P.**, Clinical observations on the use of titanium salts in the treatment of dermatitis, *Can. Med. Assoc. J.*, 73, 47, 1955.
22. **Vernetti-Blina, L.**, Richerche clinica e sperimentale sull Assido di Titanio, *Riforma Med.*, 47, 1516, 1928.
23. **Schroeder, H. A., Vinton, W. H., Bolassa, J. J.**, Effect of chromium, cadmium and other trace metal on the growth and survival of mice, *J. Nutr.*, 80, 39, 1963.
24. **Lundgren, K. D. and Ohman, H.**, Pneumokoniose in der Hartmetallindustrie, *Virchows. Arch. Pathol. Anat. Physiol.*, 325, 284, 1954.
25. **Moschinski, G., Jurisch, A., and Reind, W.**, Die Lungenversanderungen bei sinter hartmetall Arbeitern, *Arch. Gerverbepath. Gerverbehyg.*, 16, 697, 1959.

26. **Husten, K.,** Hartmetallfibrose de Lunge, *Arch. Gerverbepath. Gerverbehyg.,* 16, 721, 1959.
27. **Uragoda, C. G. and Pinto, M. R. M.,** An investigation into the health of workers in an ilmenite extracting plant, *Med. J. Aust.,* 22nd Jan, 167, 1972.
28. **Christie, H., MacKay, R. J., and Fisher, A. M.,** Pulmonary effects of inhalation of titanium dioxide in rats, *Am. Ind. Hyg. Assoc. J.,* 24, 42, 1963.
29. **Reed, C. E.,** A study of the effects on the lung of industrial exposure to zirconium dust, *Arch. Ind. Hyg.,* 13, 578, 1956.
30. **Elo, R., Maata, K., Uksila, E., and Arstila, A. V.,** Pulmonary deposits of titanium dioxide in man, *Arch. Pathol.,* 44, 417, 1972.
31. **Maata, K. and Arstila, A. V.,** Pulmonary deposits of titanium dioxide in cytologic and lung biopsy specimens, *Lab. Invest.,* 33, 342, 1975.
32. **Gross, P., Kociba, R. J., Sparschu, G. L., and Norris, J. M.,** The biologic response to titanium phosphate, *Arch. Pathol. Lab. Med.,* 101, 550, 1977.
33. **Rae, T.,** Action of wear particles from total joint replacement prostheses on tissues, in *Biocompatibility of Implant Materials,* Williams, D. F., Ed., Pitman Medical, London, 1976, 55.
34. **Rae, T.,** A study of the effects of particulate metals of orthopaedic interest on murine macrophages in vitro, *J. Bone J. Surg., Br. Vol.,* 57, 444, 1975.
35. **Rae, T.,** The Haemolytic action of particulate metals, *J. Pathol.,* 125, 81, 1978.
36. **Carroll, K. G. and Tullis, J. L.,** Observations of the presence of titanium and zinc in human leucoytes, *Nature (London),* 217, 1172, 1968.
37. **McCue, J. P.,** Titanium analysis of human blood, *Biochem. Med.,* 7, 282, 1973.
38. **Meachim, G. and Williams, D. F.,** Changes in non-osseous tissue adjacent to titanium implants, *J. Biomed. Mater. Res.,* 7, 555, 1973.
39. **Emneus, H., Stenram, V., and Baecklund, J.,** X-ray spectrographic investigations of the soft tissue around titanium and cobalt alloy implants, *Acta Orthop. Scand.,* 30, 226, 1960.
40. **Perren, S. M., Russenberger, M., Steinemann, S., Murller, M. E., and Allgower, M.,** A dynamic compression plate, *Acta Orthop. Scand.,* Suppl. 125, 31, 1969.
41. **Galante, J. and Rostoker, W.,** Corrosion-related failure in metallic implants, *Clin. Orthop. Relat. Res.,* 86, 237, 1972.
42. **Brown, S. A. and Mayor, M. B.,** The biocompatibility of materials for internal fixation of fractures, *J. Biomed. Mater. Res.,* 12, 67, 1978.
43. **Kato, L. and Gozsy, B.,** Stimulation of the cell-linked defence zone of the skin, *Can. Med. Assoc. J.,* 73, 31, 1955.
44. **Bibby, B. G. and Little, M. F.,** Fluorine and titanium and the organic material of dental enamel, *J. Dent. Res.,* 54, B.137, 1975.
45. **Regulati, B., Schait, A., Schmid, R., and Muhlemann, H. R.,** The effect of titanium, aluminium and fluoride on rat caries, *Helv. Ondontol. Acta,* 18, 92, 1974.
46. **Zehnder, A. J. B. and Wuhrmann, K.,** Titanium(III) citrate as a non-toxic oxidation-reduction buffering system for the culture of obligate anaerobes, *Science,* 194, 1165, 1976.
47. **Kawai, M. and Scrimgeour, K. G.,** Reduction of folate by Ti^{3+}, *Can. J. Biochem.,* 50, 1183, 1972.
48. **Luckey, T. D. and Venugopal, B.,** *Metal Toxicity in Mammals,* Vol. 1, Plenum Press, New York, 1977, 139.
49. **Furst, A. and Haro, R. T.,** A survey of metal carcinogenesis, *Prog. Exp. Tumor Res.,* 12, 102, 1969.
50. **Williams, D. F., Ed.,** Titanium and titanium alloys, in *Biocompatibility of Clinical Implant Materials,* Vol. 1, CRC Press, Boca Raton, Fla., in press.
51. **Meachim, G. and Pedley, R. B.,** The tissue response at implant sites, in *Fundamental Aspects of Biocompatibility,* Vol. 1, Williams, D. F., Ed., CRC Press, Boca Raton, Fla., in press.
52. **Solar, R. J.,** Materials for cardiac pace-maker encapsulation, in *Biocompatibility in Clinical Practice,* Vol. 2, Williams, D. F., Ed., CRC Press, Boca Raton, Fla., in press.

Chapter 12

METABOLISM AND TOXICITY IN VANADIUM

Michael G. Crews and L. L. Hopkins

TABLE OF CONTENTS

I. INTRODUCTION

Vanadium has been shown to be a nutritionally essential trace element for certain laboratory animals and as such is probably essential for the human as well. Evidence of essentiality as well as other data show that the body is capable of handling moderate amounts of ionic forms of this element without harm. As only small amounts of the ionic form of vanadium would probably enter the metabolic system of the body from metallic or plastic implant devices, it is doubtful that any harmful effects would be found systemically. However, once a high level of soluble vanadium is within the tissues, it is highly toxic.

II. IMPLANTATION AND INJECTION

Traces of vanadium are found in all steel alloys used for medical appliances, although titanium alloys, such as titanium, used as implant structural members, may contain as much as 4% vanadium.[1,2] The release of traces of vanadium from metallic implants is not thought to present a real problem in terms of the amounts of ion released into the system due to chronic exposure to physiological fluids, but data on this loss in modern alloys are lacking.[3] Polymer manufacturers utilize vanadium compounds as catalysts in the production of polymers derived from ethylene and propylene as well as ethylene-propylene-diene types of rubber.[2] The use of vanadium catalysts in the production of polymers and in the key coding (identification) of products may lead to the release of vanadium into the systemic circulation, either from implants or from i.v. delivery systems, but the amount of metal released is probably very low due to the nonwetting surface properties of most polymers.

III. ESSENTIALITY

Vanadium has been shown to be nutritionally essential for both the rat and the chick.[4] Chicks fed a diet low in vanadium (10 ppb) demonstrated reduced growth of wing and tail feathers when maintained on purified diets,[4,5] whereas rats fed a diet low in vanadium displayed reduced body growth.[6] Vanadium has yet to be shown to be essential for man, but its essentiality for laboratory animals,[7] indicated by its presence in certain tissues at low levels and some evidence of homeostatic control over vanadium at low levels, would appear to suggest essentiality for man as well.

IV. DIETARY LEVELS

Vanadium content of certain institutional diets is estimated to range from 12.4 to 30.1 μg/day.[8] Current research indicates levels of vanadium in water to range from 0.04 to 0.85 ppb[9] and in food from <1 to 93 ppb.[10]

V. ABSORPTION

Vanadium is poorly absorbed from the gastrointestinal tract with only 0.1 to 1.0% of oral dosages being absorbed in humans.[7] Absorption of vanadium fume or dust via the lungs appears to be both rapid and complete, with vanadium particles in the lungs of chronically exposed animals being no higher than lungs of singly exposed animals.[2] Although it has been estimated that for man up to 7.9% of total daily intake of vanadium is by the respiratory route, this should be considered with caution as there are little or no data on the pulmonary effects of lifetime exposures to airborne vanadium at current urban levels.[11]

VI. TRANSPORT

Vanadium first appears in the circulatory system unbound to serum protein,[4] but in rats 28% was found to be bound to the serum protein transferrin 4 hr after intravenous injection. Attempts to bind vanadium to human serum transferrin in vitro were unsuccessful until it was understood that vanadium, like iron, requires the presence of a suitable anion to successfully bind to the protein.[4,12] It now appears that either carbonate or bicarbonate ions serve as the anion in the binding process. Normal levels of vanadium in human blood have been found to range up to 2.0 μg/100 mℓ,[13] but most were under 1.0 μg/100 mℓ.

VII. LIPID METABOLISM

Vanadium at pharmacological levels has been shown to reduce cholesterol synthesis in both young men and animals.[7,14] The action of the element on the biosynthesis of cholesterol appears to be an inhibition of squalene synthetase which occurs in young men and animals but not in patients having hypercholesterolemia or ischemic heart disease. The variable effects of vanadium are probably due to the influence of the element on the mitochondrial acetoacetyl coenzyme A deacylase, whereby activation of the enzyme produces a depressed conversion of acetate to cholesterol in young animals but not in older animals. Plasma cholesterol turnover rate was higher in chicks fed toxic diets containing 100 ppm vanadium than in those fed the basal diet.[15] Although unclear,[7] vanadium deficiency in animals appears to alter blood cholesterol levels as well.

In animals vanadium appeared to decrease the synthesis of acetyl coenzyme A,[14] a precursor of fatty acids, but in man the effect was reversed with increased levels of serum fatty acids being observed. In the rat the incorporation of $^{32}PO_4^{=}$ into liver phospholipids was decreased with intraperitoneal injection of vanadium. Vanadium-deficient chicks displayed higher plasma triglycerides.[7] Male guinea pigs have been shown to respond to increasing pharmacological levels of dietary vanadium with increased serum triglycerides and phospholipid levels.[16] Clearly, vanadium affects lipid metabolism in man and animals, but it is unclear as to the relative effect of pharmacological and deficient levels on the various lipid components in the various metabolic pools.

VIII. IRON METABOLISM

Conflicting results were reported on the effects of vanadium administration on hemopoiesis.[14] In normal physiological states, vanadium administration did not seem to affect hemoglobin concentration of animals or humans, but in rats who were nutritionally anemic and fed supplementary vanadium and iron, the depressed hemoglobin levels were more quickly corrected than in rats fed iron without vanadium. Vanadium-deficient rats have been shown to have significantly increased packed cell volume,[6] blood iron, and bone iron. The lowering of hemoglobin levels in animals treated with vanadium pentoxide was reversed by the subsequent supplementation of L-ascorbic acid.[17]

IX. OTHER METABOLISM

Data suggest that vanadium may be involved in sulfur amino acid metabolism.[18] When rats were fed 25 to 1000 ppm of vanadium as the pentoxide,[14] the hair cystine content was found to be lowered below control values. Chronic exposure to vanadium in man also produced lower cystine and cysteine levels in their fingernails. Rat liver coenzyme A was reduced by sodium metavanadate either in the diet or by intraperitoneal injection, causing a reduction in the conversion of cystine to cysteine and decreasing the synthesis of coenzyme A. Diets containing less than 0.75% methionine have produced growth retardation in vanadium-deficient chicks.[5] Vanadium at high levels also affects various other components of metabolism, such as succinoxidase,[14] monoamine oxidase, and oxidative phosphorylation such as ATP synthesis, coenzyme Q, and thioctic acid.

X. INTERACTIONS

Vanadium has been shown to affect or be affected by ascorbic acid,[17,19-22] glucose, ethanol, and chromium in experimental animals. The administration of 200 mg/kg of L-ascorbic acid to rats has been shown to diminish the toxic effects (histological and enzyme alteration) of vanadium pentoxide administration but not the effect of reduced growth.[17] Chicks fed diets containing 0.2% ascorbic acid had a complete reversal of growth inhibition observed in their controls fed 20 ppm vanadium as NH_4VO_3.[22] In male rats, the intraperitoneal injection of high levels of ammonium metavanadate solution has been shown to produce a significant drop, as determined by radiorespirometry in the metabolism of ^{14}C-glucose to ^{14}C-carbon dioxide for short lengths of time (3 hr or less).[19] Fish maintained in water containing 1 μM of Na_2VO_4 for 5 days displayed markedly shortened times to produce acute ethanol intoxication.[20] Chicks supplemented with 500 ppm chromium as the acetate and 5 ppm vanadium (NH_4VO_3) displayed symptoms of vanadium toxicity, whereas chicks fed similar levels of vanadium (5 ppm) without chromium were not affected.[21] This suggests that high dietary chromium levels make relatively small amounts of vanadium toxic. Ascorbic acid has also been found to be a rapid and effective antidote for experimental vanadium poisoning in mice, rats, and dogs.[14]

XI. EXCRETION

High levels of vanadium absorbed by workers through the lungs have been shown to be excreted through the urine.[23] Limited research has reported normal human urinary vanadium levels to range from undetectable to 11.6 $\mu g/\ell$ of urine.[2,7] Urine levels of vanadium were monitored in a study where patients were fed 4.5 mg vanadium per day as oxytartarovanadate, and it was observed that excretion of vanadium increased with the increased dietary level.[7] In rat studies where ^{48}V was administered intravenously,[7] 46% of the isotope was excreted via the urine in the first 24 hr and an additional 9% appeared in the feces after 96 hr.

XII. PULMONARY

Toxic levels of airborne vanadium are sometimes encountered in the use of either vanadium or vanadium compounds and in by-product residues of vanadium bearing fossil fuels.[2,14,23,24] Exposure to vanadium-bearing dust or fume has produced both acute and chronic toxicity in exposed personnel. Exposures to vanadium-bearing aerosols have been documented. These varied from short-term unidentified high-level exposures that produced toxicity symptoms in ½-hr to 5-year exposures at levels of 0.84 mg/m^3 of air. Heavy exposure to vanadium-containing dust or fume has produced an array of symptoms which include conjunctivitus,[14,23] rhinitis, sore throat, persistent dry cough, wheezing, lassitude, depression, tongue discoloration (green), skin irritation, chest pain, bronchospasm, dizziness, ringing in ears, nausea, vomiting, deep bone pain, headache, blurred vision, and palpitations of the heart. Controlled exposure of men to vanadium oxide dust produced the pulmonary irritation and loose cough symptoms of vanadium toxicity even at levels as low as 0.1 mg vanadium as V_2O_5 per cubic meter of air for 8 hr. Exposure for 8 hr to moderate levels of vanadium (1 mg/m^3) produced other pulmonary symptoms of toxicity, but the onset of these symptoms was delayed for up to 24 hr. Toxicity of vanadium from either the cation or the anion form of the element has shown to increase with increasing oxidation state ($V^{+5} > V^{+4} > V^{+3}$).[14] Vanadium aerosols displayed greater toxicity symptoms depending on the degree of dispersion and solubility. The effects of particulate forms of vanadium were studied in vitro on alveolar macrophages, and those macrophages that were exposed to small concentrations of various chemical forms of vanadium were not as viable and possessed depressed phagocytic activity compared to controls.[11] Cytotoxicity was directly related to solubility in the order $V_2O_5 > V_2O_3 > VO_2$. These observations may help to explain the commonly observed secondary infections in pulmonary exposure to vanadium aerosols. Human lung tissue has been shown to reach high levels of vanadium in later life and in people living in the Middle East.[2] No specific chronic lesions in lungs of exposed men have been observed, and X-ray examination of subjects previously exposed to high levels of vanadium via the pulmonary route appeared uneventful and normal.[2,14]

XIII. BLOOD PRESSURE

Experimentation with dogs has demonstrated that IA vanadium (Na_3VO_4) acts as a vasoconstrictor and that increased coronary resistance occurs.[25] Rats given 5 mg vanadium (Na_3VO_4) per kilogram of body weight i.p. or i.v. exhibited increased urine output and sodium loss while displaying increased blood pressure.[26] Rats given higher levels of vanadium i.v. were shown to be hypotensive.

XIV. GASTROINTESTINAL TRACT AND OTHER ORGANS

In two separate studies human subjects were fed graded levels of vanadium (ammonium vanadyl tartrate and oxytartarovanadate) ranging from 4.5 to 18 mg/day with no toxic effects observed other than some cramps and diarrhea at the higher levels.[7] In the human studies conducted using 4.5 mg/day, the approximately 11 ppm vanadium in the diet represents a level of vanadium that would not be expected to have shown toxicity based on animal experimental data. Rats fed 3 to 4 mg vanadium (V_2O_5) per kilogram body weight displayed severe growth depression and pathological changes in the liver and kidneys.[17] High levels of vanadium have been reported to produce,[14] in various animal species, fatty dystrophy of the cells of the liver, kidney, and myocardium, perivascular edema in the myocardium, a lowered white cell count, and decreased respiration of liver and brain tissue. Various biochemical changes were also noted.

XV. DIFFERENTIAL DIAGNOSIS

Respiratory symptoms due to exposure to vanadium in the toxic range are not easily distinguishable from acute respiratory infections,[2,14,23] asthma, and other diseases. Medical and work histories should be obtained to determine whether the subject has been exposed to vanadium-bearing aerosols. Where appropriate laboratory facilities

Table 1[27]
TOXICITIES OF SOME VANADIUM-CONTAINING COMPOUNDS

vanadic acid, ammonium salt		
orl-rat	LD_{50}	160 mg/kg
scu-rat	LD_{50}	15 mg/kg
vanadic acid, monosodium salt		
orl-rat	LD_{Lo}	200 mg/kg
ipr-rat	LD_{Lo}	10 mg/kg
vanadic acid, trisodium salt		
scu-rat	LD_{Lo}	50 mg/kg
scu-rbt	LD_{Lo}	9280 μg/kg
vanadium pentoxide		
ihl-hmn	TC_{Lo}	100 μg/m³
orl-mus	LD_{50}	23 mg/kg
scu-rbt	LD_{Lo}	20 mg/kg
vanadium tetrachloride		
orl-rat	LD_{50}	160 mg/kg
vanadium tribromide		
scu-rbt	LD_{Lo}	20 mg/kg
vanadium trichloride		
orl-rat	LD_{50}	350 mg/kg
orl-mus	LD_{50}	23 mg/kg
vanadium trioxide		
orl-mus	LD_{50}	130 mg/kg

Note: hmn = human, ihl = inhalation, ipr = intraperitoneal, kg = kilogram, LC_{Lo} = lowest published lethal concentration, LD_{50} = lethal dose 50% kill, LD_{Lo} = lowest published lethal dose, m^3 = cubic meter, mg = milligram, mus = mouse, orl = oral, rat = rat, rbt = rabbit, scu = subcutaneous, TC_{Lo} = lowest published toxic concentration.

are available, vanadium concentrations in blood and/or urine should be determined. Chronic low-grade exposure to toxic levels of vanadium may be detected using the decrease in cystine content of fingernails.

XVI. TOLERANCE AND TOXICITY LEVELS

Presently, the only limits to industrial exposure to vanadium involve the pulmonary route. Current limit values are 0.5 mg/m³ of air for dust and 0.05 mg/m³ for fumes as the pentoxide.[2,14] Table 1 gives some of the relative toxicities of the various vanadium compounds. Additional compounds are listed in the Registry of Toxic Effects of Chemical Substances.[27] Clearly there are mechanisms to protect the body from vanadium toxicity. Toxicity has been found to be low by mouth, intermediate by respiratory tract, but high when given parentally.[14] In Table 1 it will be noted that the toxicity for vanadic acid, both the ammonium and monosodium salts, for the rat is over ten times as great by the parental route as compared to that orally. Although unknown for man, it is known that the levels necessary to produce toxicity vary according to species.[14] For example, In Table 1 the LD_{50} of vanadium trichloride orally for the rat is 350 mg/kg vs. 23 for the mouse. Likewise, the lowest published lethal dose of vanadic acid (trisodium salt) given subcutaneously in the rat is 50 mg/kg vs. 9280 μg/kg for the rabbit. The rat and mouse are relatively resistant to vanadium compounds, while the rabbit is especially sensitive.[14] Both the cation and anion are toxic, and increasing valence leads to increasing toxicity. Solubility of vanadium compounds is directly related to cytotoxicity.[11] Although adequate data are not available, it could be postulated that because of the low levels of vanadium present in medical devices and the low solubility at physiological pH, there is little possibility of systemic toxicity. No evidence was found indicating that vanadium or its compounds was a cause of mutagenicity.[2] Tissue concentrations of vanadium in the 1-ppm range can induce serious derangement of basic metabolic processes that might not be manifested clinically or be felt by the subject.[2,14]

XVII. REVIEWS

For further information on the metabolism and toxicity of vanadium, the References 7, 14, 28, and 29 are suggested.

REFERENCES

1. **Lemons, J. E., Niemann, K. M. W., and Weiss, A. B.,** Biocompatibility studies on surgical-grade titanium-, cobalt-, and iron-base alloys, *J. Biomed. Mater. Res., Symp.,* 7, 549, 1976.
2. **Zenz, C.,** Vanadium and its compounds, in *Occupational Medicine, Principles and Practical Applications,* Zenz, C., Ed., Year Book Medical Publishers, Chicago, 1975, 702.
3. **Cahoon, J. R.,** On the corrosion products of orthopedic implants, *J. Biomed. Mater. Res.,* 7, 375, 1973.
4. **Hopkins, L. L., Jr.,** Essentiality and function of vanadium, in *Proc. 2nd Int. Symp. Trace Element Metabolism in Animals,* Hoekstra, W. G., Suttie, J. W., Ganther, H. E., and Mertz, W., Eds., University Park Press, Baltimore, 1974, 397.
5. **Nielsen, F. H., Myron, D. R., and Uthus, E. O.,** Newer trace elements — vanadium (V) and arsenic (As) deficiency signs and possible metabolic roles, in *Proc. 3rd Int. Symp. Trace Element Metabolism in Man and Animals,* Kirchgessner, M., Ed., Technische Universität München, Freising-Weihen Stephan, Germany, 1978, 244.
6. **Strasia, C. A.,** Vanadium: Essentiality and Toxicity in the Laboratory Rat, Ph.D. thesis, Purdue University, Lafayette, 1971.
7. **Underwood, E. J.,** *Trace Elements in Human and Animal Nutrition,* 4th ed., Academic Press, New York, 1977, 388.
8. **Myron, D. R., Zimmerman, T. J., Shuler, T. R., Klevay, L. M., Lee, D. E., and Nielsen, F. H.,** Intake of nickel and vanadium by humans, *Am. J. Clin. Nutr.,* 31, 527, 1978.
9. **Söremark, R.,** Vanadium in some biological specimens, *J. Nutr.,* 92, 183, 1967.
10. **Myron, D. R., Givand, S. H., and Nielsen, F. H.,** Vanadium content of selected foods as determined by flameless atomic absorption spectroscopy, *J. Agric. Food Chem.,* 25, 297, 1977.

11. **Waters, M. D., Gardner, D. E., and Coffin, D. L.**, Cytotoxic effects of vanadium on rabbit alveolar macrophages, *Toxicol. Appl. Pharmacol.*, 28, 253, 1974.
12. **Campbell, R. F. and Chasteen, N. D.**, An anion binding study of vanadyl(IV) human serotransferrin, *J. Biol. Chem.*, 252, 5996, 1977.
13. **Allaway, W. H., Kubota, J., Losee, F., and Roth, M.**, Selenium, molybdenum and vanadium in human blood, *Arch. Environ. Health*, 16, 342, 1968.
14. **Zenz, C., Hopkins, L. L., and Byerrum, R. U.**, Biological effects of vanadium, in *Medical and Biological Effects of Environmental Pollutants — Vanadium*, Byerrum, R. U., Ed., National Academy of Sciences, Washington, D.C., 1974, 46.
15. **Hafez, Y. and Kratzer, F. H.**, The effect of dietary vanadium on fatty acid and cholesterol synthesis and turnover in the chick, *J. Nutr.*, 106, 249, 1976.
16. **Bruffy, G. R. and Dowdy, R. P.**, Effect of dietary vanadium on cholesterol metabolism in guinea pigs, *Fed. Proc. Abstr.*, 38, 450, 1979.
17. **Chakraborty, D., Bhattacharyya, A., Majumdar, K., and Chatterjee, G. C.**, Effects of chronic vanadium pentoxide administration on L-ascorbic acid metabolism in rats: influence of L-ascorbic acid supplementation, *Int. J. Vitam. Nutr. Res.*, 47, 81, 1977.
18. **Neilsen, F. H. and Myron, D. R.**, Evidence which indicates a role for vanadium in labile methyl metabolism in chicks, *Fed. Proc. Abstr.*, 35, 683, 1976.
19. **Meeks, M. J., Landolt, R. R., Kessler, W. V., and Born, G. S.**, Effect of vanadium on metabolism of glucose in the rat, *J. Pharm. Sci.*, 60, 482, 1971.
20. **Grisolia, S. and Guerri, C.**, Interaction of lithium and vanadium, *Fed. Proc. Abstr.*, 38, 378, 1979.
21. **Hunt, C. and Nielsen, F. H.**, The interaction between vanadium and chromium in the chick, *Fed. Proc. Abstr.*, 38, 449, 1979.
22. **Hill, C. H.**, Studies on the ameliorating effect of ascorbic acid on mineral toxicities in the chick, *J. Nutr.*, 109, 84, 1979.
23. **Proctor, N. H. and Hughes, J. P.**, *Chemical Hazards of the Workplace*, J. B. Lippincott, Philadelphia, 1978, 503.
24. **Key, M. M., Henschel, A. F., Butler, J., Ligo, R. N., and Tabershaw, I. R., Eds.**, *Occupational Diseases, A Guide to Their Recognition*, U.S. Government Printing Office, Washington, D.C., 1977, 405.
25. **Steffen, R. P., Inciarte, D. J., Swindall, B. T., Johnston, J., and Haddy, F. J.**, The effect of vanadate on the dog heart, *Fed. Proc. Abstr.*, 38, 1440, 1979.
26. **Bonventre, J. V., Roman, R. J., and Lechene, C.**, Effect of vanadate on renal sodium and water excretion, *Fed. Proc. Abstr.*, 38, 1042, 1979.
27. **Christensen, H. E. and Fairchild, E. J., Eds.**, *Registry of Toxic Effects of Chemical Substances*, U.S. Government Printing Office, Washington, D.C., 1976, 1190.
28. **Hopkins, L. L., Jr.**, Nutrient deficiencies in animals: vanadium, in *CRC Handbook Series in Nutrition and Food*, Sect. E, Vol. II, Rechcigl, M., Jr., Ed., CRC Press, West Palm Beach, 1978, 357.
29. **Hopkins, L. L., Jr.**, Vanadium, in *Geochemistry and the Environment*, Vol. 2, Mertz, W., Ed., National Academy of Sciences, Washington, D.C., 1977, 93.

Chapter 13

BIOLOGICAL PROPERTIES OF ALUMINUM

C. Allen Bradley

TABLE OF CONTENTS

I. INTRODUCTION

Two major reviews have been written on aluminum in the environment and its effect on human health. The first was in 1957 by Campbell et al.[1] and the second by Sorenson et al.[2] in 1974. In the brief time since the last review, major additions to the literature concerning the potentially serious toxicity of aluminum have been made. Thus the renewed interest in the element aluminum.

The conclusions of the 1957 review represented the popularly accepted opinion concerning the safety of aluminum at that time:

1. The quantities of aluminum which are required within the animal organism to induce harmful effects are large compared with those found in the body tissues under the severest conditions of ordinary or occupational exposure to aluminum and its compounds, or compared with the quantities of certain other metals known to be dangerous.
2. Despite the abundance of aluminum compounds in the earth's crust and in the natural environment, and despite the artificial increase in the environment of modern man, the absorption of aluminum into the bodies of animals and man occurs to only a slight extent.
3. The quantities of aluminum and its organic compounds which are absorbed into the body are independent, for all practical purposes, of the quantities in the environment — specifically, in the food and in the atmosphere. Concerning the use of aluminum, Kehoe stated: "There is no reason for concern, on the part of the public or of the producer and distributor of aluminum products, about hazards to human health from well established and extensive, current use of such products. Nor need there be concern over the more extended uses which would seem to be in the offing." The toxicologist of today would hardly be as expansive in his statement of the safety of aluminum.

II. OCCURRENCE OF ALUMINUM IN THE ENVIRONMENT AND TISSUES

Aluminum is an ubiquitous element which is found in the soil, in vegetation, and in all vertebrates. In animals, generally, it is present in minute amounts, less than 1g in the entire body of man.

In view of the abundance of aluminum in the earth's crust (8.13%), its presence in all soils, though in greatly varying amounts, can be taken for granted.[3] The literature reflects a great deal of continuing research on the forms in which aluminum occurs in soil and on its reactions and interactions with other elements. Webber has pointed out that while the soil contains up to 10% aluminum, the plants grown in it contain only a microgram per gram of plant tissue.[4]

In spite of mounting concern over air pollution, particularly since the mid-1960s, relatively few investigations have included aluminum in atmospheric surveys. Most investigations now include aluminum, which may be explained in part by the development of newer analytical methods, particularly neutron activation, which are capable of determining nanogram quantities of pollutants in the air. The determination of pollutants in atmospheric precipitations is a more recent interest. Examples of aluminum in the atmosphere ranges from 0.005 $\mu g/m^3$ in Hawaii to 0.57 $\mu g/m^3$ in Ann Arbor, Michigan.[5,6] One should expect industrialized areas to have higher aluminum concentrations. Automotive exhausts contribute greatly to the atmospheric aluminum level. Oelschlager reported the content of 26 trace elements in automobile emission; the aluminum content was 35 mg/kg of dust-form exhaust.[7]

In summarizing the sources and chemistry of aluminum in natural waters, Hem observes that this, the most abundant metallic element, is highly resistant to removal from minerals by solution during weathering and remains behind persistently in the process of rock decomposition to form the clay mineral in soils and the greater part of shales and similar hydrolyzate sediments.[8] The concentrations of aluminum in most natural waters are negligible. Only where the pH is maintained below five can aluminum be present in large amounts (7100 $\mu g/m\ell$). This condition may occur from industrial waters, mine drainage, or natural phenomena such as in acid spring waters or volcanic regions.

As for aluminum in food, exhaustive tests conducted since 1890, when it became available through the Hall-Heroult processes in adequate quantities, have shown that negligible amounts of aluminum are dissolved by food. Also, its effect on the food is negligible, for most of its salts are colorless and, in the concentrations in which they occur, tasteless (in contrast to those of iron or copper) and have not been demonstrated to affect the nutritive qualities of food.[9]

A. Use of Aluminum in Packaging, Containers and Cooking Vessels

Another source of aluminum in the environment is the ever-increasing use in packaging, containers, and cooking vessels. The suitability of aluminum in the processing and packaging of foods and pharmaceuticals has been under investigation since around 1890. Aluminum has been used for canning of food extensively in Europe without ill effect. In the U.S. the use of aluminum is increasing every year. The aluminum beverage cans and ends (particularly the easy-open lids) have been the single most important factors of this growth, although flexible packaging, semirigid food containers, consumer foil, and particularly convenience packaging have contributed considerably. It has been shown that food and drugs stored in aluminum for prolonged periods are seldom adversely affected, even though the aluminum content of the ingredients often increases.[10,11]

Aluminum is used extensively by the pharmaceutical and cosmetic industries. It has been shown that aluminum tubes coated on the inside with a synthetic lacquer and used for various hydrophilic ointments remained intact after 1½ years storage at room temperature, while tin tubes were corroded.[12] Aluminum aerosol cans for drugs and cosmetics have been shown to have favorable characteristics and few limitations, being good in appearance, light weight, and free from external rusting.

1. Home Food Preparation and Storage

Never before has there been such an array of cookware available to the homemaker as there is today, particularly that made of metals. It should be pointed out that the chief requisite of a cooking utensil is its ability to transfer heat quickly and evenly from the source to the food. This is best met by metals. In regard to aluminum, its excellent conductivity of heat is given as the reason for its popularity. It is now generally agreed that aluminum is safe. "Many scientific studies have determined that the amount of aluminum ingested as a result of preparing food in aluminum cookware — even when soda is used in the cooking — is so small as to be of no significance in comparison with the amount of aluminum consumed from other sources."[13,14]

In relation to nutrient loss due to food being prepared or processed in aluminum containers, it has been concluded that the processing and packaging of foods in aluminum does not contribute to the destruction of nutrients over and above that caused by the handling of the food in the home or in the food industry. Indeed, aluminum has been found suitable for the synthesis and storage of ascorbic acid, the most labile of the vitamins, to which it is very resistant.[9,15] Other vitamins and yeast are also processed and handled in aluminum for the same reason.

2. Use of Aluminum for Water Storage

The use of metallic aluminum for the storage of water was the subject of probably the first inquiry in 1891 into the safety of aluminum in water and food processed or stored in aluminum.[16] Not much more can be added to the earlier review except that aluminum has a very good resistance to most waters and even to some considered very reactive, such as those of higher sulfur content.

3. Use of Aluminum for Treatment of Drinking Water

Aluminum compounds and certain aluminum-bearing minerals represent the major coagulants commonly used in the treatment of drinking water. Aluminum sulfate is the principal coagulant. Bentonite is also used. Surprisingly, in modern purification practices, use of aluminum compounds as coagulants results, generally, in the presence of less rather than more aluminum in the drinking water than in raw water.[17]

4. Aluminum Compounds in Food

The Food and Drug Administration has long allowed aluminum-containing baking powder to be used in food. "Aluminum compounds have a number of uses as direct food ingredients, as for example, in aluminum baking powder and in the use of alum in pickles to keep them firm. These uses are generally recognized as safe by scientists qualified to evaluate the safety of food additives."[13]

5. Aluminum in Human Food and Maximum Allowable Concentrations

Because of the abundance of aluminum, its presence in all foods, although in small amounts, may be taken for granted. Until 1971, aluminum had been listed among the nonessential elements.[18] Some are now considering aluminum for inclusion among the growing number of essential elements.[19] An exhaustive listing of aluminum concentrations in foods and beverages has been compiled by Sorenson et al.[2]

In the U.S., the FDA has established no tolerance for the content of aluminum in finished food and beverage products other than those shown for food additives, all of which are by prior sanction and/or generally regarded as safe (GRAS).

In France, the Society of Expert Chemists of France proposed a MAC of 20 mg/kg of food products as received in markets and up to 100 mg/kg for food products processed or stored in aluminum.[20]

B. Aluminum Concentrations of Human Tissues

It should be borne in mind that the concentration of aluminum found in tissues reflects the geochemical environment of the individuals and of locally raised food and meat products. This has been shown by various investigators.[21]

Although the biological role of aluminum has not been clearly established, it now seems likely that it is an essential metal. Many investigators have found it in measurable concentrations, together with other trace metals in all tissues, which appear to change with changes in biological activity by the tissues. A brief review of aluminum metabolism in various tissues and fluids is discussed.

1. Adrenal

Only a limited amount of work has been done concerning changes in adrenal activity and aluminum concentration. Soroka,[22] using dogs, analyzed the content of several metals including aluminum in the adrenal following excitation of the central nervous system (CNS) with caffeine and depression with ether anesthesia. The results showed that both CNS excitation and depression resulted in a decrease of about 50 to 65% of the normal aluminum concentration of the adrenal glands within 1 hr after administration of the drugs, with a gradual return to normal during the second hour.

2. Blood

Since 1957, there have been many publications dealing with the changes in the aluminum content in human blood. In 1958, Silvestri reported that in human serum,[23] aluminum was a predominant normal metallic constituent. Sabadash demonstrated that the aluminum concentration in blood of normal individuals changed in a circadian fashion.[24] It was high at 9 a.m. and low at 6 p.m. However, young males in the state of inactivity were observed to have no change in serum aluminum content.[25] Diurnal variation in blood has been observed to be the same in dogs as it is in man.[24,26-28] The results of these experiments suggested that the aluminum concentration in blood is regulated by the autonomic nervous system. At rest, the parasympathetic nervous system predominates, resulting in an increased accumulation of aluminum in the brain and a decrease in blood and cerebrospinal fluid (CSF). While in the awake state or under adrenal stimulation, the sympathetic system reverses these changes. In contrast, adrenalin and cortisone were reported to lower the serum aluminum content in rabbits.[29]

Other observations, with regard to the metabolism of aluminum which suggest its essentiality, were those of Leonov and Milosha that aluminum could be found in the human fetus.[30,31] Blood aluminum and other trace metals increased between 4.5 and 6 months of age. At 8 months of intrauterine life, the elements showed a second period of maximum concentration and dropped sharply following delivery. However, blood of premature infants contained more aluminum than those delivered at term.[32] At delivery, it was shown that newborn blood contained a higher level of aluminum than did the mothers.[33] The content in the blood of female newborns was somewhat higher than that of males.

Blood aluminum has also been observed to increase in pregnant women throughout normal pregnancy.[34-36] These changes have also been demonstrated in gravid dogs and cows.[37,38] However, in early toxicosis blood aluminum was markedly increased while late toxicosis was associated with a severe decrease in mothers and infants.[39,40] Similar changes were observed in placental blood in toxicosis and other disease states associated with pregnancy.[41]

Lifshits was able to distinguish leukemia from leukemoid reactions on the basis of change in whole blood aluminum concentration.[42] In the period of exacerbation of chronic leukemia and in the developed phase of acute leukemia, the concentration of aluminum was elevated. The most marked changes in concentration were related to the increases associated with the formed elements of the blood which did not occur in leukemoid reactions. In Hodgkins disease,[32] serum aluminum was elevated. The concentration of aluminum in blood and tissue of patients with a variety of neoplasms has been demonstrated to be different from control subjects.[43,44] Some neoplastic tissues had elevated aluminum levels while others had depressed levels. This suggested that aluminum did not cause neoplastic growth. The effect of therapy on blood aluminum levels in these diseases is unknown. Studies of the effects of alkylating drugs on the aluminum concentration in the blood of rabbits demonstrated that the content in the plasma and erythrocytes was markedly elevated with both mannoumustine, a cytotoxic cyclic nitrogen mustard, and triethylenethiophosphoramide, another cyclic alkylating agent.[45]

Various brain and nerve-related disorders have been studied with regard to changes in trace metal content of the blood. Organic nerve damage, viral neuroinfections, pain syndromes of different origin, and syringomyelia have been demonstrated to be associated with a fall in blood aluminum, while blood aluminum values increase in encephalitis.[46-49] With the first three of these, there was a concomitant increase of aluminum in the cerebrospinal fluid (CSF). Brain tissues from abscesses and following cerebral vascular accidents were also observed to contain less aluminum than normal.[50] However, brain and spinal cord trauma resulted in an increase of aluminum in both blood and CSF. An elevation in blood aluminum level has also been reported in senility.[51]

Some aspects of cardiovascular disease have also been studied with regard to changes in blood aluminum levels. With myocardial infarction and coronary atherosclerosis, aluminum levels were markedly elevated,[52,53] but decreased in angina pectoris.[54] Blood aluminum levels in rabbits with experimental atherosclerosis could not be correlated with the disease state.[55]

Blood aluminum levels have also been studied in gastrointestinal disorders. Over half of the patients with colonic stases had elevated blood aluminum levels.[56] However, 23% of the more severe cases had decreased aluminum levels as did patients with adenocarcinoma of the colon.[43] The work of Soroka and Sabadash with dogs suggested that the reflex nature of the intestinal tract may depend in part upon trace element metabolism.[57] They demonstrated that an increase in aluminum concentration in the blood occurred following stimulation of the large colon and the magnitude of the change depended upon the duration of the stimulus. Gastric ulcers,[58] gastrogenic iron deficiency, and pernicious anemias were also associated with a general lowering of blood aluminum. However, in cases of stomach carcinoma the blood aluminum contents were elevated.[59]

Adults with acute kidney disease and children with acute and chronic nephritis all had markedly increased blood aluminum levels.[60,61] With acute disease, the male children had higher levels than the females, but this was reversed in the chronic disease state. Interestingly, in chronic renal failure, $Al(OH)_3$ is the agent of choice for the prophylaxis of hyperphosphatemia.[62,63]

Patients with infectious hepatitis,[64] cirrhosis of the liver, malignant liver neoplasia, and chronic hepatocholangitis also had elevated blood aluminum levels.[32] However, in cholelithic disease and in cancer of the biliary tract, a decrease in blood aluminum concentration has been reported.[65]

Pulmonary diseases as studied by several authors have also been distinguished on the basis of blood aluminum level changes. The aluminum level in blood or serum of individuals with berylliosis,[66] massive pulmonary infarction,[52] pneumonia, and diffuse fibrosis was elevated.[67] However, patients with bronchogenic carcinoma had depressed levels.

Studies regarding disease associated with pancreatic disorders have demonstrated that in patients with diabetes mellitus, the whole blood content of aluminum was less than in normal individuals.[68,69] The reduction was dependent in all cases on the severity and duration of the disease.

Patients with chronic liver and bile duct involvement showed increased plasma bilirubin and aluminum concentration in whole blood and plasma.[70] However, the concentration of aluminum in the formed elements was below normal. These data suggested that the elevated plasma aluminum concentration in cases of diabetes with liver involvement was due to the lack of excretory activity of the liver and bile ducts. Change in the concentration of aluminum in the blood of diabetics was to a certain degree influenced by its migration from cellular content into plasma.[69] The observed decrease in blood aluminum without liver involvement was attributed to a loss of cellular aluminum in the urine and bile. Improvement in the condition of the patients with treatment was accompanied by a normalization of the aluminum content in the blood.

In patients with cystic fibrosis, the sweat, urine, duodenal fluid, teeth, finger- and toenails as well as hair, in addition to plasma and serum, contained higher concentrations of aluminum than normal.[71]

Changes in aluminum content of the blood were studied in numerous skin diseases: lichen planus, lupus erythematosis, cutaneous tuberculosis, eczema,[72,73] infectious eczema,[74] bullous pemphigoid, benign familial pemphigus,[75] psoriasis,[76] and lepromatous leprosy.[77] Generally, aluminum was below normal, except in the case of psoriasis, where it was higher than normal. In most cases there was either partial or complete restoration to normal levels after therapy. In the case of lepromatous leprosy patients, a direct correlation was found between the decrease in blood and increase in urine aluminum concentrations with increasing severity of disease.[78] This led the authors to suggest that these decreases were associated with the migration of aluminum and other metals from the blood into the infection foci for participation in proliferative processes.

3. Bone

A relatively small number of studies have been undertaken concerning the aluminum content in bone and the role it plays with regard to bone metabolism. Bone regeneration in fractured frog and pigeon femurs as well as rabbit autografts has been studied with regard to changes in bone aluminum concentration.[79-82] Aluminum accumulated during early regeneration in the cortical callus of the fracture zones and autografts, and in some instances increased up to seven times the control concentration before returning to normal levels at the end of the repair period. It was also reported that the aluminum concentration decreased in the opposite uninjured bone depending upon the degree of damage to the fracture zone. The extremities beyond the fracture and the uninjured portions of the bone participated actively in these metabolic process of bone repair.

Subsequently, it was demonstrated that the content of aluminum in an immobilized rabbit femur increased considerably during the first days of immobilization and fluctuated in the remaining time up to the 360th day.[83] During this period, the contents of aluminum in the humerus and femur of nonimmobilized limbs fluctuated in the same manner, but the fluctuation was less pronounced in the humerus. The fact that these changes occurred in bones of limbs that had not been immobilized suggested that immobilization of one limb resulted in mobilization of aluminum in bone tissue of the entire skeleton.

The increase in bone aluminum level associated with vitamin deficiency-induced rickets in the rat has been reversed by supplementation with vitamin D.[84,85]

Synovial fluid from rheumatoid arthritis patients has been observed to contain more aluminum than normal while the serum content decreased.[86-89] However, in infectious nonspecific polyarthritis, the serum level increased twofold.

4. Cardiovascular Tissue

Very little work has been done concerning aluminum and its role in the metabolism of cardiac tissue. Sauer et al.[90] reported a negative correlation between the aluminum content in drinking water and the number of deaths due to human myocardial and coronary heart disease. Biochemical studies have demonstrated that aluminum has a specific activating effect on the myocardial succinate oxidase system.[91] Activation was only observed in nicotinamide adenine dinucleotide (NAD) requiring succinate systems containing cytochrome c.[92]

In an analysis of the ash from the aorta, calcified plaques, and sclerotic aorta without calcification obtained from elderly individuals, Berki found the concentration of aluminum to be much greater than in the aortic tissue from younger individuals.[93] Popov observed that the ignition residue and aluminum concentration increased in the aorta with age,[94] but that after age 60 the aluminum content in this tissue declined abruptly. The hearts of guinea pigs with pericarditis were also shown to have elevated aluminum levels.[95]

5. Eye

The concentration of aluminum in the eye has been found to decrease according to phylogenetic order, the lowest being obtained for humans and the highest for fish.[96] Aluminum in the optic nerve of men and women, obtained following accidental death, was less than in blood and decreased with age.[97,98] However, aluminum in the lens tissue from rats has been observed to increase with age.[99]

6. Kidney

The metabolism of aluminum in the kidney in health and in disease states has been studied in detail. A direct correlation has been found for aluminum in the geochemical environment and in urinary calculi obtained from humans, but no correlation was found for X-ray transparency of the stones.[100-102]

Significant increases in kidney aluminum concentrations in two types of human kidney diseases, lipid nephrosis and tuberculoma, have been reported.[103] An increase in kidney aluminum may also be a reflection of an increase of aluminum in the blood and urine of patients with kidney diseases. Patients with bladder tumors have been shown to have elevated urine aluminum levels.[104]

A study of rabbits and humans from the same environment suggested that the normal kidney contents of aluminum were comparable.[105] Aluminum in kidney tissue of normal dogs has been demonstrated to be distributed between filterable forms which were regulated autonomically.[106,107]

An increase of urinary excretion of aluminum has been effected with kidney denervation,[108] drugs which increase the cardiac output,[109] diuretics,[110-112] as well as ligation of the biliary duct,[113] the usual predominant excretory route. Changes in urinary excretion of aluminum following administration of various chelates have also been studied in the normal animal.[114]

7. Liver

The livers from patients with iron storage associated with Bantu siderosis, hemochromatosis, and sickle cell anemia as well as in alcoholics with cirrhosis and early hemochromatosis have been shown to contain less aluminum than normal.[115-117] This observation could be reproduced in rabbits and rats with the administration of iron.

Studies in young and adult cattle have demonstrated that liver aluminum content was related to the geographic location of their pastures.[118] Diurnal variation in liver content has been observed in rats, which appeared to be regulated autonomically.[119] Similar studies in dogs have also suggested autonomic control of storage of aluminum in the liver and release into the blood as well as its excretion via the bile duct.[120,121] Biochemical studies have failed to show an effect of aluminum on rat liver mitochondria and microsomes as well as guinea pig liver mitochondria.[122,123]

8. Lung

The concentrations of aluminum normally found in lung tissue are the highest found in all adult tissues studies and are, in part, the result of inhalation of aluminum-containing dusts which may accumulate with age.[124,125] In addition to this variation, sex and geochemical environment are important modifiers of normal lung aluminum content.

9. Male Genital Tissue

Aluminum has been found in the prostate, seminal fluid, testicle, and seminal vesicle of normal males. Marked elevations of aluminum have been observed in hypertrophied and neoplastic prostate glands.[51,126]

10. Milk

The normal aluminum content of human milk has been compared to the content found in various metabolic and disease states. Aluminum increases during gestation and in the lactation period. Individuals with hypogalactosis initially had high levels which fell to levels observed in nonpregnant women or lower.[32,127,128] Milk from mothers with premature deliveries also contained more aluminum than the milk from mothers who delivered at term.[32]

11. Muscles and Tendons

A small amount of work has been done concerning aluminum metabolism in these two tissues. Rabbit leg immobilization which caused pronounced muscle and tendon atrophy was also associated with a decrease of aluminum in both tissues.[129] Studies of the filterable and nonfilterable forms of aluminum in skeletal muscles of dogs, anesthetized with ether and stimulated with caffeine, demonstrated an increase in the filterable bound form in the excited state and an increase in the nonfilterable form in the depressed state.[130]

12. Nerve Tissue

In addition to adult brain data provided by Tipton and others,[12,131-133] measurable concentrations of aluminum in brain tissue of the human embryo have been reported, which gradually increased during gestation and reached the highest level in the newborn.[30] These data offer evidence that aluminum may be essential with regard to brain tissue development and activity. This conclusion was supported by the observations in dogs that the diurnal variations in brain and CSF concentrations of aluminum were regulated by the autonomic nervous system.[26,134] Aluminum was released from the brain in excitation and accumulated with depression of CNS activity. Studies have demonstrated that the decrease in free and bound forms of brain aluminum in excitation was associated with a marked depression of the free form.[135] Conversely, in narcosis, both free and bound forms increased in brain tissue, but the increase in the free form was greater. Other physiological studies suggest that aluminum may be associated with nerve conduction and transmission.[136,137] Studies of both acetylcholinesterase and cholinesterase have demonstrated an activation effect of aluminum.[138]

Studies of brain neoplastic tissue have demonstrated an increase in aluminum content.[50]

Crapper et al. have reported that very high levels of aluminum, 9 to 11 μg/g (dry weight),[139] were found in brain tissue from patients who had Alzheimer's disease. Their age-matched controls had levels of 0.23 to 0.40 μg/g (dry weight). This observation, along with an earlier observation that senility was associated with high blood ash aluminum levels, 5450 μg/g, may be of etiological significance.[51]

13. Pancreas

The aluminum content of pancreatic juice collected from healthy individuals has been studied with regard to amylase, trypsin, and lipase activities.[140] Patients with increased pancreatic activity associated with chronic pancreatitis and cholangiohepatitis had higher levels of aluminum.

In vitro studies of dialysis-inactivated crystalline trypsin demonstrated a 70% activation with the addition of aluminum.[141] However, addition of $Al_2(SO_4)_3$ to human, pig, rat, and dog amylases was inhibitory. This excess aluminum-induced inhibition could be prevented by the addition of albumin.[142]

14. Skin

High aluminum concentrations have been found in the skin of the human embryo. This interesting find may be additional evidence which supports the view that aluminum is an essential metal. Its concentration, although somewhat less, was relatively constant during maturity and old age.[143]

Caution must be observed in interpreting the results of post mortem skin aluminum determinations, since it has been reported that the aluminum content of the skin increases up to 18 to 24 hr after death.[144]

Fingernail clippings from 17 normal male and female donors, with ages ranging from 1 month to 55 years, have been analyzed for aluminum with spark source mass spectrometry.[145] In this survey of trace elements, the aluminum values found in fingernails were very high compared to the values found in hair.

15. Spleen

The concentration of aluminum in the blood flowing from the spleen has also been demonstrated to be regulated by the autonomic nervous system.[146] Stimulation of the CNS in dogs caused a decrease in spleen aluminum with an increase of this metal in venous blood, while narcosis brought about a decrease in blood with an increase in the spleen.

16. Tooth

Aluminum metabolism associated with the development, growth, and prevention of dental cavities has received considerable attention. It is generally accepted that aluminum is a normal constituent of deciduous and permanent teeth which fluctuates in the various anatomical portions of the tooth with age.[147,148] Discolored areas of sound teeth have been observed to contain markedly elevated concentrations of both aluminum and fluoride. Human dental calculus has also been shown to contain high concentrations of aluminum which have been shown to inhibit alkaline phosphatase and alkaline phosphomonoesterase from human dental plaque.[149]

There is a basic recognition of geographic variation and caries incidence.[150] Studies of geochemical differences and caries incidence have suggested that elevated aluminum contents of the soil, drinking water, and vegetables were associated with low caries incidence.[151,152] Food or water supplemented with aluminum and fluoride were cariostatic when given to rats during the period of enamel mineralization.[153]

17. Uterus

Uterine washings from rats, rabbits, and sheep obtained at estrus and in the midluteal phase contained significantly different concentrations of aluminum.[154] In rats and rabbits, estrus phase washings contained a higher concentration of aluminum; however, sheep had much higher concentrations of aluminum in the uterine washings obtained during the luteal phase.

III. METABOLISM AND PHYSIOLOGY

The intake of aluminum is chiefly by mouth, from foods and beverages, but also by the lungs, and from the atmospheric dust content. It is present in the natural diet, in amounts varying from very low in animal products to relatively high in plants.

The total amount of aluminum in the normal diet per day has been estimated to be between 10 and 100 mg. The question whether foods cooked in aluminum or to which aluminum has been added (for example, in baking powder) has been discussed. Studies on the absorption of aluminum from the gastrointestinal tract have been made in both normal patients and patients with chronic renal failure. It has been shown that the aluminum ion is absorbed in normal man, but the absorption is greater in chronic renal failure. Absorption of aluminum from the gastrointestinal tract in normal man has little effect on plasma phosphorous levels, but in renal failure an associated fall in plasma phosphorous levels has been found, the reason for which is unknown.[155] There have been few studies of renal excretion of aluminum. It has been shown, however, that in normal man aluminum is excreted in the urine and that it is handled by overall tubular reabsorption.[156] The earlier studies that state no aluminum was excreted by way of the kidney were possibly due to poor and difficult methods of analysis.

IV. TOXICITY

Man has had every opportunity to be exposed to aluminum due to its ubiquitous nature. It is found in the soil, vegetation, and in most animal cells. As pointed out, the widespread use of aluminum in cooking utensils and cans has added to the possibility of ingestion of the aluminum ion by man. After the initial fears of toxicity, little has been written of it as a potentially hazardous substance until the past few years. Inhibitions of glycolysis and phosphorylation have been suggested as the most significant toxic reactions of aluminum-containing compounds.[157] Enzymatic conversion of isocitric acid to α-ketoglutarate in the presence of nicotinamide adenine dinucleotide phosphate (NADP) was inhibited in a dose response manner by aluminum.[158] On the other hand, the rate of enzymatic decarboxylation of pyruvic acid could be increased with aluminum.[159] A study of the aluminum inhibition of yeast and guinea pig hexokinase has pointed out that Mg-dependent phosphorylating enzyme was most easily inhibited when Mg was present in suboptimal levels.[160] A competing complexation reaction between ATP and aluminum was suggested as the mechanism which was observable with rate-limiting concentrations of Mg. Possibly all phosphate-transferring systems involving ATP and Mg may be biological targets for excess aluminum. In other in vitro studies, the aluminosilicate clay, bentonite, has been demonstrated to complex with free purines and pyrimidines obtained as breakdown products of ribonucleic acid (RNA), deoxyribonucleic acid (DNA), ATP, ADP, and AMP.[161,162] This clay has also been shown to complex with and inhibit nucleoside phosphototransferases and RNase.[162,163] In the rat, both with and without renal failure, aluminum salts administered orally or parenterally cause the development of experimental porphyria.[164] Berlyne has confirmed the fluorescence of the rat brain, testes, and bone after aluminum salt administration. The experimental porphyria in rats caused by aluminum administration gives a unique porphyrin pattern. It was investigated using the aluminum isotope ^{28}Al. When the isotope was injected intracardiacally, 42% found its way into the liver. It was found by Berlyne that aluminum was specifically taken up by the nuclei of the hepatocytes and that the DNA of the nucleus had a specific affinity for aluminum.[156] This may be the fundamental basis of the production of the defective enzymes in the liver which are a causal factor in experimental rat porphyria. In man, there is no evidence of porphyria developing as a result of aluminum ingestion.

The toxicities of bentonite, kaolin, $Al_2(SO_4)_3$ and $Al(OH)_3$ have been studied with rabbits, chicken, sheep, cows, and hogs with regard to their use as feed additives.[165,166] In general, beneficial effects, such as growth stimulation without adverse effects on Ca and P metabolism, were observed with low levels, 1 to 2% of the diet, of these aluminum compounds. Higher levels in the feed decreased growth and caused muscle weakness and death with marked disturbances of Ca and P metabolism. Similar studies with large amounts of $Al(OH)_3$ and Al_2O_3 in chickens and rats demonstrated that these compounds were able to affect adversely Ca and P metabolism and to cause impaired growth with the production of rachitic bone changes.[167,168] Associated with these changes was an increase in bone aluminum levels.

The use of $Al(OH)_3$ in the prophylaxis of hyperphosphatemia has been questioned because of toxicity reported by Berlyne et al.[169,170] The toxicity seen in uremic rats (5/6 nephrectomy) given subcutaneous (s.c.) injections or drinking water containing 10 g $AlCl_3$ per liter, drinking water containing 10 or 20 g $Al_2(SO_4)_3$ per liter, or 1.98 g of $Al_2(SO_4)_3$ per kilogram per day given i.p., were periorbital bleeding, lethargy, anorexia, and death. An attempt to reproduce these or toxic manifestations in rats with chronic renal failure (7/8 nephrectomy) demonstrated that $Al(OH)_3$ administration leads to increased deposition of aluminum in bone, impaired growth, and rachitic bone changes in normal rats, according to Thurston et al.[171] This effect could be corrected by phosphate supplements. In contrast, no impairment of growth was produced by $Al(OH)_3$ in hyperphosphatemic uremic rats, and no other previously reported pathological abnormalities could be demonstrated. It was pointed out that patients with chronic renal failure treated with $Al(OH)_3$ for prolonged periods had bone aluminum levels similar to those in uremic rats. The conclusion was that, although aluminum was deposited in the bone in chronic renal failure, $Al(OH)_3$ was nontoxic if hypophosphatemia was avoided.

Death in rats which had been reported following i.p. injection of $Al(OH)_3$ was suggested to have been the result of peritonitis by both Baily and Verberckmoes.[63,167] This was consistent with the earlier findings of Kay and Thornton following the injection of $Al(OH)_3$ in mice.[172] The anorexia reported earlier was consistent with the observation that uremic rats given $AlCl_3$ in their drinking water did not drink the water and died of severe dehydration.[167] Lethargy would be expected with peritonitis or dehydration. The periorbital bleeding could not be independently reproduced by Thurston et al.,[171,173] and they suggested it was the result of a possible local infection. Failure to confirm the originally reported toxicity suggested that elimination of $Al(OH)_3$ from therapeutic use in the treatment of renal disease was unwarranted, according to Bailey et al.,[62] Sherrard,[174] and Thurston et al.[171,173]

Regular ingestion of 5.5 mg of Al per kilogram of body weight in food and drink by humans has been suggested to cause no adverse effects because the aluminum was present in the less toxic colloidial form.[175] This is inconsistent with the observation that the daily intake of more than 3 mg of Al per kilogram of body weight has been found to increase fecal elimination of phosphate with a concomitant decrease in urinary excretion. Maximum drinking water levels of the soluble $AlCl_3$ and $Al(NO_3)_3$ were recommended on the basis of animal experiments to be 4 mg and 0.1 mg/ℓ, respectively, in the latter case to avoid an effect on the palatability of water. In studies of $Al(NO_3)_3$, the effect of this form of nitrate was determined by measuring reduced working capacity and duration of narcotic reflex inhibition in rabbits, with daily oral doses of 0.1, 1, and 10 mg/kg for 6 months. The working capacity of the rabbits and their oxygen uptake was significantly affected only by the 10-mg/kg dose. Rabbits treated with $Al(NO_3)_3$ at 0.1, 1, and 10 mg/kg did not exhibit any narcotic effect.

The adjuvant activities of alum (potassium aluminum sulfate) and $Al(OH)_3$ have been investigated in guinea pigs and rabbits.[176] Following s.c. injection of antigenic proteins mixed with the aluminum compounds, there was an increase in antibody production. Separate s.c. injections of antigen and aluminum compounds at separate injection sites of $Al(OH)_3$ alone failed to produce the same increase in antibody synthesis. Subcutaneous injections of an aluminum anode material, used for biogalvanic elements as pacemaker energy sources, and Al_2O_3 also failed to cause an immune response.[177]

It has been noted that the aluminum contents of the spleen and bone marrow increase during immunization without the metal adjuvants, suggesting a possible role for aluminum in the immune response.[176] This observation was consistent with gradual increase of aluminum in blood plasma of rats following s.c. injection of various tissue homogenates. These homoantigens caused a gradual increase in aluminum which reached its highest level 3 to 6 days after transplantation.[178]

It has been shown that the blood Al levels in 30 to 36 workers in a factory exposed to Al_2O_3 dust were two to three times the level of the controls.[205] Serum intestinal alkaline phosphatase, acid phosphatase, as well as ATP levels were significantly reduced. A number of consistent observations has been made with regard to the in vitro effects of aluminum-containing compounds in blood or serum. An antithrombogenic effect has been observed with alum and kaolin.[179] In addition, other aluminosilicates have been shown to be either fibrinolytic or hemolytic.[180,181]

Aluminum-induced changes in the physical properties of components or chemicals similar to those found in blood have also been studied. The viscosity of transfusion gelatin has been shown to increase with the addition of large amounts of Al_2O_3 and $AlCl_3$.[182] Similarly, a pectinic mucopolysaccharide was studied with regard to coagulation following the addition of ionic aluminum.[183]

V. CLINICAL EXPERIENCE

There appears to be appreciable toxicity of aluminum in spite of all the protestations that it is harmless.

Neuroscientists became interested in aluminum in 1942 when Kopeloff found that in monkeys after direct application of aluminum to the cortex, there was produced a chronic state of convulsive reactivity with recurrent seizures that simulate epilepsy in man.[184-186] In 1965 Klatzo reported that aluminum induced neurofibrillary degeneration in rabbits.[187-189]

DeBoni examined the possibility that systemic aluminum may cross an apparently intact blood brain barrier and may induce an encephalopathy with neurofibrillary degeneration (NFD).[190] The route of access of aluminum to brain parenchyma is unknown, but the tissue compartment which concentrates aluminum is the nuclear chromatin in rabbits.[191]

Presenile and senile dementia is characterized by the presence of neurons with neurofibrillary changes, thus a possible link to Alzheimer's disease, since it was observed that aluminum is also a cause of neurofibrillary degeneration changes.[192]

Crapper observed that some brain regions in Alzheimer's disease contain elevated aluminum concentrations,[193] and suggests that concentrations greater than 4 μg/g dry weight occur in regions of the cerebral cortex likely to exhibit neurofibrillary degeneration.[194]

By far, the most serious toxicity to appear is the dialysis encephalopathy syndrome. During the last 10 years or so, aluminum-containing phosphate-binding gels have been widely used a method of controlling serum phosphorus levels in uremic patients on dialysis. The orally administered aluminum is assumed to be excreted as an insoluble aluminum phosphate in the feces, with little, if any, of this complex actually absorbed by the patient. In 1972, Clarkson performed balance studies on a group of uremic patients receiving aluminum-containing phosphate-binding gels and found that these patients experienced a positive aluminum balance of 100 to 568 mg/day during the study period.[195] Berlyne reported elevated serum aluminum levels in three of six patients on dialysis receiving oral aluminum salts, whereas Clarkson found that serum aluminum levels were not consistently affected by aluminum administration in the uremic patients in their study.[196] Parsons described elevated bone aluminum levels in a number of patients on chronic hemodialysis who received aluminum-containing antacids.[197] Using a flameless atomic absorption technic, Alfrey found that the mean muscle aluminum was 14.8 ppm and the trabecular-bone aluminum was 98.5 ppm in patients on dialysis and maintained on phosphate binding-alumium gels, as compared with 1.2 and 2.4 ppm in controls.[198] Brain gray-matter aluminum values in a group of uremic patients on dialysis who died of a neurologic syndrome of unknown cause were 22 ppm as compared with 6.5 ppm in a group of uremic patients on dialysis who died of other causes and 2.2 ppm in control subjects. This fact lead Alfrey to conclude that since gray-matter aluminum was higher in all patients with the dialysis-associated encephalopathy syndrome than in any of the control subjects or other uremic patients on dialysis, that this syndrome may be due to aluminum intoxication.

Mayor has shown that there is a significant correlation of serum parathyroid hormone and aluminum retention in humans and suggests that both Alzheimer's disease and dialysis encephalopathy may be explained in part by the parathyroid hormone acting or by orally ingested aluminum.[199]

Although techniques and results vary, all groups agree that there is aluminum retention in patients receiving aluminum hydroxide while on chronic dialysis. Over the last few years there have been reports of a relentlessly progressive form of dementia in chronic-dialysis patients. This disorder is characterized by speech disturbance, dyspraxia, tremor, psychosis, and personality changes, resulting in death in 6 to 7 months.[200-202]

It appears to occur with great frequency in certain clinics. In one area, it was the most common cause of mortality, accounting for the deaths of 7 out of 28 patients maintained for more than 3 years on dialysis.[198] A substantial increase in the aluminum content of the gray matter of patients who died of this disorder was compared to those dying of other causes. It seemed possible at this point that aluminum is a potential toxin. It was pointed out in *Lancet* in 1976 that many centers used aluminum hydroxide and that progressive dementia was not a serious problem and called for a study of the apparent discrepancy of results.[203]

It was not until 1978 that a partial answer to some of the discrepancies began to appear. Dunea reports that a possible source of aluminum intoxication is the water used for preparing the dialysis fluid.[204] The outbreak of dialysis dementia may have been related to an excess aluminum load because of a change in the city's method of water treatment. It was reported that only some of the patients exposed became demented and that others that use deionizers continue to observe this syndrome. Elliott and his group found much the same syndrome in Scotland.[205] Their results provide further evidence to suggest aluminum intoxication is the cause of the dialysis encephalopathy syndrome. The 13 patients with encephalopathy received home dialysis in three areas of Scotland with a high aluminum content in the water supply. None of the 40 patients who dialysed in Glasgow, where the aluminum content of the water supply is negligible, developed encephalopathy. Aluminum is added to most water supplies in the west of Scotland, with the exception of Glasgow, and it is the undoubted source of the high aluminum content in the water of the suspect areas. The Elliott group observed an increased incidence of severe renal osteomalacic disease in their encephalopathic group. Four of the thirteen (31%) patients developed symptoms and radiological appearances typical of osteomalacia, as compared with an incidence of about 10% in the rest of the home dialysis populations. This confirms the findings of the other studies in which a more definite relation between bone disease and encephalopathy is reported.[206,207]

It was observed that hemoglobin values in encephalopathic patients fell during the year before neurological systems developed. There is preliminary evidence that aluminum may be toxic to enzymes concerned in hemin biosynthesis. Elliott's findings suggest that aluminum retention occurs in all patients with renal impairment. The ingestion of aluminum hydroxide may be a contributory factor, but the major source is the high aluminum content of the water supply from which the dialysis fluid is prepared and thus the probable cause of dialysis encephalopathy.

VI. SUMMARY

In view of the abundance of aluminum in the earth's crust, its presence in all soils, though in greatly varying amounts, can be taken for granted. Due to this abundance of aluminum and its versatility, its concentration is ever increasing in man's environment due to modern day methods of packaging, distribution, and utilization of such essentials as food, cosmetics, and pharmaceuticals. There was little hesitancy to use aluminum, for it appeared to be relatively nontoxic. The same can probably be said of today's use with one or two exceptions. When aluminum is consumed or used in dialysis procedures in large amounts for prolonged periods, signs and symptoms of encephalopathy may appear. There is believed to be a causal relationship here, and if there is, then indeed, aluminum is potentially toxic.

ACKNOWLEDGMENT

The author is deeply indebted to Dr. John Sorenson, Associate Professor, Department of Biopharmaceutical Sciences, University of Arkansas, College of Pharmacy, for his data base used in certain sections of this review.

REFERENCES

1. **Campbell, I. R. et al.,** Aluminum in the environment of man. A review of its hygienic status, *Arch. Ind. Health,* 15, 359, 1957.
2. **Sorenson, R. J. et al.,** Aluminum in the environment and human health, *Environ. Health Proscept.,* 8, 3, 1974.
3. **Emmons, W. H., et al.,** *Geology: Principles and Processes,* 5th ed. McGraw-Hill, New York, 1960, 60.
4. **Webber, M. D.,** Aluminum in plant tissue grown in soil, *Can. J. Soil Sci.,* 51, 471, 1971.
5. **Hoffman, G. L., Duce, R. A., and Zoller, W. H.,** Vanadium, copper, and aluminum in the lower atmosphere between California and Hawaii, *Sci. Technol.,* 3, 1207, 1969.
6. **Rahn, K. A.,** Sources of Trace Elements in Aerosols. Approach to Clean Air, U.S. Atomic Energy Commission COO, 1705, 9, U.S. National Technical Information Service, 1971.
7. **Oelschlager, W.,** Composition of dustlike vehicle emissions and their effect on farm livestock, *Staub-Reinhalt. Luft (in English),* 32(6), 9, 1972.

8. **Hem, J. D.**, Study and interpretation of the chemical characteristics of natural water, *U.S. Geol. Surv. Water Supply Pap.*, 1473, 1, 1959.
9. **Juniere, P. and Sigwalt, M.**, *Aluminum. Its Application in the Chemical and Food Industries,* W.C. Barnes, Transl., Chemical Publishing, New York, 1964.
10. **Hall, J. E. and Mulvaney, T. E.**, Aluminum ends for beer cans, *Brew. Dig.*, 38(11), 48, 1963.
11. **Ullmann, F.**, Metallbestimmungen in Dosenbier, *Schweiz. Brau. Rundsch.*, 76(6), 104, 1965.
12. **Steiger, K. E. and Lehmann, H.**, Korrosion von Zinntuben durch das Fullgut, *Pharm. Acta Helv.*, 39, 622, 1964.
13. U.S. Food and Drug Administration, Safety of Cooking Utensils, FDA Fact Sheet, U.S. Food and Drug Administration, Washington, D.C., July 1971.
14. U.S. Food and Drug Administration, Teflon and aluminum cooking utensils, Nutrition sense & nonsense, *FDA Consumer,* 6(7), 16, 1972.
15. Aluminum Association, *Aluminum with Food and Chemicals,* Compatibility Data on Aluminum in the Food and Chemical Process Industries, 2nd ed., Aluminum Association, New York, 1969.
16. **Plagge and Lebbin,** Canteens and cooking vessels of aluminum, 1893, in *Mellon Institute of Industrial Research.* A Select, Annotated Bibliography on the Hygienic Aspects of Aluminum and Aluminum Utensils, Bibliographic Ser. Bull. No. 3, 1933, 24.
17. **Barnett, P. R., Skougstad, M. W., and Miller, K. J.**, Chemical characterization of a public water supply, *J. Am. Water Works Assoc.*, 61, 61, 1969.
18. **Schroeder, H. A. and Nason, A. P.**, Trace-element analysis in clinical chemistry, *Clin. Chem.*, 17, 461, 1971.
19. **Frieden, E.**, The chemical elements of life, *Sci. Am.*, 227(1), 52, 1972.
20. **Jaulmes, P. and Hamelle, G.**, Presence et taux des oligo-elements dans les aliments et les boissons de l'homme, *Ann. Nutr. Aliment,* 25B, 133, 1971.
21. **Tipton, I. H. et al.**, Trace elements in human tissue. III. Subjects from Africa, the Near and Far East and Europe, *Health Phys.*, 11, 403, 1965.
22. **Soroka, V. R.**, Participation of adrenal glands in the metabolism of some trace elements during medicinal excitation and inhibition of the central nervous system, *Probl. Endokrinol. Gormonoter.*, 10(3), 76, 1964; *Chem. Abstr.*, 61, 13783, 1964.
23. **Silvestri, U.**, Richerche spettrografiche sulla composizione in elementi in tracce del siero di sangue, *Boll. Soc. Ital. Biol. Sper.*, 34, 1745, 1958.
24. **Sabadash, E. V.**, Daily rhythm of trace element content in the blood humans, *Nauk. Zap. Stanislav. Derzh. Med. Inst.*, 3, 67, 1959; *Chem. Abstr.*, 59, 7949, 1963.
25. **Ivanov, I. I., Korovkin, B. F., and Mikhaleva, N. P.**, Biochemical serum indexes during prolonged hypodynamia, *Probl. Kosm. Biol.*, 13, 99, 1969; *Chem. Abstr.*, 73, 23200, 1970.
26. **Voinar, A. O., Soroka, V. R., and Sabadash, E. V.**, Neiro-gumoral'naya regulyatsiya obmena mikroelementov v organizme. (neuohumoral regulation of trace element metabolism in the organism.), *Mikroelem. Nerv. Sist. Baku,* 1963, 6, 1966.
27. **Soroka, V. R.**, Trace-element metabolism in the brain as judged by sinusostomy in stimulated central nervous activity, *Ukr. Biokim. Zh.*, 31, 435, 1959; *Chem. Abstr.*, 54, 14403, 1960.
28. **Soroka, V. R.**, Soderzhanie nekotorykh mikroelementov v spinnomozgovoi zhidkosti pri vozbuzhdenii i tormozhenii tsentral'noi nervnoi sistemy. (Effect of stimulation and inhibition of the central nervous system on the concentration of some trace elements in cerebrospinal fluid.), *Tr. Stalinabad. Gos. Med. Inst.*, 19, 187, 1961.
29. **Rasin, M. S.**, Effect of adrenaline and cortisone on the trace element content of blood serum, *Ukr. Biokim. Zh.*, 36(4), 506, 1964; *Chem. Abstr.*, 61, 14996, 1964.
30. **Leonov, V. A.**, The amount of microelements in blood and various internal organs of man, *Vestsi Akad. Navuk B. SSR Ser. Biyal. Navuk*, 1, 151, 1956; *Chem. Abstr.*, 51, 5251, 1957.
31. **Mikosha, A.**, Trace elements in human embryos, *Nauk. Zap. Stanislav. Derzh. Med. Inst.*, 3, 85, 1959; *Chem. Abstr.*, 59, 7969, 1963.
32. **Sviridov, N. K.**, Alyuminii v patologii cheloveka. (Aluminum in human pathology.), *Lab. Delo,* 12, 699, 1966.
33. **Grebennikov, E. P.**, Soderzhanie medi, kremniya, alyuminiya, titana i margantsa v krovi novorozhdennykh. (Contents of copper, silicon, aluminum, titanium and manganese in the blood of newborns.), *Pediatriya (Moscow),* 38(10), 29, 1960.
34. **Bibileishvili, Z. V.**, Change in trace element concentration and activity of some enzymes in blood of females, *Soobshch. Akad. Nauk Gruz. SSR,* 39(3), 58, 1965; *Chem. Abstr.*, 64, 7131, 1966.
35. **Grebennikov, E. P.**, Trace elements in women during pregnancy and pregnancy complicated by toxicosis, *Pediatr. Akush. Ginekol.*, 1, 36, 1964; *Chem. Abstr.*, 62, 12305, 1965.
36. **Laptieva, E. D.**, Trace elements in blood of donors and parturient women with physiological blood loss during deliver, *Zdravookhr. Tadzh.*, 3, 578, 1967.

37. **Soroka, V. R. and Grebennikov, E. P.**, The effect of pregnancy on the silicon, aluminum, and titanium content of the liver and hepatic blood of dogs, *Byull. Eksp. Biol. Med.*, 52(8), 62, 1961; *Chem. Abstr.*, 56, 9280, 1962.
38. **Kirchgessner, M.**, Der Mengen und Spurenelementgehalt von Rinderblut, *Z. Tierernaehr. Futtermittelkd.*, 12, 156, 1957.
39. **Grebennikov, E. P.**, Copper, manganese, silicon, aluminum and titanium in the blood of pregnant women suffering from early toxicosis, *Mikroelem. Selsk. Khoz. Med.*, p. 587, 1963; *Chem. Abstr.*, 62, 10989, 1965.
40. **Grebennikov, E. P.**, Content of some trace elements and proteins in the blood of women suffering from vomiting of pregnancy, *Sov. Med.*, 27(2), 134, 1964; *Chem. Abstr.*, 61, 3552, 1964.
41. **Mischel, W.**, Chemical composition of the human placenta, *Zentralbl. Gynaekol.*, 78, 1089, 1966; *Chem. Abstr.*, 51, 18185, 1957.
42. **Lifshits, V. M.**, Soderzhanie nekotorykh mikroelementov v tsel'noi krovi u bol'nykh leikozami i u bol'nykh s leikemoidnymi reaktsiyami. (The contents of some trace elements in whole blood in patients with leukemia and with leukemoid reactions.), *Nek. Vop. Kardiol. Mikroelementy*, p. 85, 1964.
43. **Mulay, I. L., et al.**, Trace-metal analysis of cancerous and noncancerous human tissues, *J. Nat. Cancer Inst.*, 47(1), 1, 1971.
44. **Nuryagdyev, S. K.**, Soderzhanie mikroele mentov u bol'nykh rakom do i posle lechneiya. (Content of trace elements in patients with cancer before and after treatment), *Vop. Onkol.*, 17(5), 7, 1971.
45. **Timakin, N. P., Malyutina, G. N., and Goldberg, E. E.**, Effect of Degranol and thio-TEPA on the content of strontium, barium, aluminum and magnesium in the plasma and formed elements of the blood, *Biol. Deistvie Tsitostaticheskikh Prep. Mater. Konf. Tsent. Nauch. Issled, Lab. Tomsk Med. Inst.*, 1967, 131, 1968; *Chem. Abstr.*, 73, 118869, 1970.
46. **Del'va, V. A.**, Some trace elements in the fluids and blood in health and disease, *Tr. Stalinsk. Med. Inst.*, 9, 109, 1957; *Chem. Abstr.*, 54, 17657, 1960.
47. **Del'va, V. A.**, Changes in the concentration of aluminum of the blood and cerebrospinal fluid in some pathologic processes, *Tr. Donetsk. Gos. Med. Inst.*, 19, 193, 1961; *Chem. Abstr.*, 58, 4918, 1963.
48. **Mil'ko, V. I. and Kirienko, M. G.**, Trace elements in the blood of patients with syringomyelia treated with radioactive iodine, *Vrach. Delo*, 2, 883, 1967; *Chem. Abstr.*, 67, 20554, 1967.
49. **Chukavina, A. I. and Osintseva, V. S.**, Content of microelements (copper, aluminum, iron) in the blood, serum and urine in tick-borne encephalitis with a biphasic course, *Zh. Nevropatol. Psikhiat. im. S.S. Korsakova*, 68, 1175, 1968; *Biol. Abstr.*, 50, 71387, 1969.
50. **Del'va, V. A.**, Chemistry and pathochemistry of trace elements in the human brain, *Mikroelem. Selsk. Khoz. Med.*, p. 580, 1963; *Chem. Abstr.*, 62, 15160, 1965.
51. **Saakashvili, T. G.**, Concentrations of manganese, silicon, aluminum, copper, and zinc in the blood of patients with prostate-gland cancer, *Soobshch. Akad. Nauk Gruz. SSR*, 42(2), 361; *Chem. Abstr.*, 65, 11104, 1966.
52. **Hegde, B., Griffith, G. C., and Butt, E. M.**, Tissue and serum manganese levels in evaluation of heart muscle damage. A comparison with serum glutamicoxalacetic transaminase, *Proc. Soc. Exp. Biol. Med.*, 107, 734, 1961.
53. **Arleevskii, I. P.**, K dinamike urovnya nekotorykh bioelementov v krovi pri koronarnom ateroskleroze. (The dynamics of the levels of some bioelements in blood in the presence of coronary atherosclerosis), *Kardiologiya*, 7, 1967.
54. **Oblath, R. W. and Griffith, G. C.**, Treatment of angina pectoris with a new monoamine oxidase inhibitor, pivalybenzhydrazine, *Am. J. Cardiol.*, 6, 1132, 1960.
55. **Arleevskii, I. P.**, Electrolyte balance in the blood in experimental atherosclerosis, *Byull. Eksp. Biol. Med.*, 62(11), 58, 1966; *Chem. Abstr.*, 66, 36056, 1967.
56. **Accart, R. A. and Mauverney, R. Y.**, Blood electrolytes (whole blood) in right colonic stases, *J. Med. Lyon*, 41, 867, 1960; *Chem. Abstr.*, 60, 4597, 1964.
57. **Soroka, V. R. and Sabadash, E. V.**, Effect of stimulation of rectal receptors on the contents of certain trace elements in the blood, *Mater. Vses. Simp. Mikroelem. Nerv. Sist.*, 41, 1966; *Chem. Abstr.*, 67, 1391, 1967.
58. **Pristupa, C. B.**, Sdvigi v soderzhanii mikroelementov pri yazvennoi bolezni zheludka. (Shifts in trace element contents in patients with gastric ulcers), *Zdravookhr. Beloruss.*, 4, 28, 1962.
59. **Lifshits, V. M.**, Trace elements in the blood of anemic patients, *Nek. Vop. Kardiol., Mikroelementy*, p.88, 1964; *Chem. Abstr.*, 65, 11093, 1966.
60. **Sernyak, P. S., et al.**, Vliyanie ekstrakorporal'nogo gemodializa na soderzhanie nekotorykh mikroelementov v krovi bol'nogo s ostroi pochechnoi nedostatochnost'yu. (The effect of extracorporeal hymodialysis on the levels of some microelements in the blood of a patient with acute kidney insufficiency), *Klin. Khir.*, p.6, 1967.

61. **Usov, I. N.,** Med. alyuminn l kremnii v krovi pri nefropatiyakh u detei. (Copper, aluminum and silicon in the blood of children suffering from kidney disease), *Zdravookhr. Beloruss.*, 9, 35, 1963.
62. **Bailey, R. R., et al.,** The effect of aluminum hydroxide on calcium, phosphorus and aluminum balances and the plasma parathyroid hormone in patients with chronic renal failure, *Clin. Sci.*, 41, 5P, 1971.
63. **Verberckmoes, R.,** Aluminum toxicity in rats, *Lancet*, 1, 750, 1972.
64. **Boudarev, L. S.,** Metabolism of trace elements in Botkin's disease (infectious hepatitis), *Klin. Med. (Moscow)*, 40(2), 66, 1962; *Chem. Abstr.*, 58, 6079, 1963.
65. **Morita, J.,** Relation between hair color and the metalic elements in domestic animals, *Tottori Nogakkaiho*, 12, 112, 1960; *Chem. Abstr.*, 60, 4552, 1964.
66. **Schepers, G. W. H.,** The mineral content of the lung in chronic berylliosis, *Dis. Chest*, 42, 600, 1961.
67. **Truknikov, G. V. and Fomin, A. A.,** Blood plasma iron, copper, manganese, aluminum, nickel and chromium content in inflammatory diseases of the lungs and bronchial asthma, *Sov. Med.*, 31, 137, 1968; *Biol. Abstr.*, 50, 74417, 1969.
68. **Kosenko, L. G.,** Content of some trace elements in the blood of patients suffering from diabetes mellitus, *Klin. Med. (Moscow)*, 43(4), 113, 1964; *Fed. Proc. (Transl. Suppl.)*, 24, 237, 1965.
69. **Kosenko, L. G.,** Distribution of some trace elements among the formed elements and plasma in patients with diabetes mellitus, *Vrach. Delo*, 1, 60, 1965; *Chem. Abstr.*, 62, 12292, 1965.
70. **Kosenko, L. G.,** Dynamics of the concentration of various trace elements in the blood of patients with diabetes mellitus and its dependence on the functional condition of the liver, *Ter. Arkh.*, 36(11), 37, 1964; *Chem. Abstr.*, 62, 10975, 1965.
71. **Kopito, L. and Shwachman, H.,** Spectroscopic analysis of tissues from patients with cystic fibrosis and controls, *Nature (London)*, 202, 501, 1964.
72. **Zatolokin, F. D.,** Examination of trace elements in blood in skin diseases, *Mikroelem. Sel. Khoz. Med., Ukr. Nauch-Issled. Inst. Fiziol. Rast., Akad. Nauk, Ukr. SSR, Mater. Vses. Soveshch.*, 4th, Kiev 1962, 610, 1963; *Chem. Abstr.*, 63, 10461, 1965.
73. **Zatolokin, F. D.,** Trace element content in the blood of eczematous patients, *Mikroelem. Estestv. Radioaktiv. Pochv.*, p.210, 1962; *Chem. Abstr.*, 59, 10626, 1963.
74. **Zakharov, I. Ya., Soroka, V. R., and Sabadash, E. V.,** Dynamics of some trace element levels in the blood of patients with microbial eczema and neurodermatitis circumscripta, *Vestn. Dermatoal. Venerol.*, 38(2), 9, 1964; *Chem. Abstr.*, 61, 3545, 1964.
75. **Torsuev, N. A., et al.,** Content of certain trace elements in the blood of patients with Lever's bullous pemphigoid, Sneddon-Wilkinson subcorneal postulosis and Halley-Halley benign familial pemphigus, *Vestn. Dermatol. Venerol.*, 44, 35, 1970; *Biol. Abstr.*, 52, 14385, 1971.
76. **Bokshtein, O. E.,** Vliyanie kortikosteroidnoi terapii na soderzhanie alyuminiya veritrotsitakh i syvorotke krovi bol'nykh psoriazom. (Effect of corticosteroid therapy on the concentration of aluminum in the erythrocytes and serum of psoriasis, *Vestn. Dermatol. Venerol.*, 34, 12, 1970.
77. **Zatolokin, F. D.,** Excretion of manganese and other trace elements in the urine in patients with lepromatous leprosy in the regressive stage, *Vop. Teor. Prakt. Med., Donetsk.*, p.217, 1966; *Chem. Abstr.*, 66, 103482, 1967.
78. **Zztolokin, F. D. et al.,** Contents of certain trace elements in the blood of patients with lepromatous leprosy, *Vopr. Leprol. Dermatol.*, 3, 74, 1965; *Chem. Abstr.*, 65, 7779, 1966.
79. **Belous, A. M.,** Quantitative content of trace elements in the regenerate during healing of the fractured femur in frogs, *Byull. Eksp. Biol. Med.*, 52(12), 96, 1961; *Chem. Abstr.*, 56, 14755, 1962.
80. **Belous, A. M.,** Content of some trace elements in healing of femur fracture, *Ukr. Biokhim. Zh.*, 23, 856, 1961; *Chem. Abstr.*, 56, 10756, 1962.
81. **Skoblin, A. P.,** Content of trace elements in various bones in autoplasty, *Tr. Nauch. 18th Konf. Inst. Ortopedii Travmatol. Ego Nauch-Oporn. Punktov Sovmestno s Khar'kovsk. Nauch. Obshchestvom Ortopedov-Travamatologov*, 18, 204, 1962; *Chem. Abstr.*, 60, 9755, 1964.
82. **Skoblin, A. P. and Belous, A. M.,** The concentration of copper, aluminum, titanium, and vanadium in the blood experimental animals during the healing of fractures, *Aktual. Vop. Klin. Lechen. Ortopedo-Travmatol. Bol'nykh*, 1965, 266; *Chem. Abstr.*, 65, 4346, 1966.
83. **Gordienko, V. M. and Belous, A. M.,** Content of certain trace elements in bone tissue during immobilization, *Ukr. Biokim. Zh.*, 35(3), 428, 1963; *Chem. Abstr.*, 59, 10577, 1963.
84. **Belan, M. G.,** Contents of certain trace elements in the tubular bones in rickets, *Mater. Nauch. Konf. Ukr. Inst. Ortopedii Travmatol. Ego Nauch-Oporn. Punktov, Sovmestno s Khar'kovsk. Nauch. Obshchestvom Ortopedov-Travmatologov*, 19, 117, 1962; *Chem. Abstr.*, 60, 5940, 1964.
85. **Kortus, J., Dibak, O., and Kotuliak, V.,** Effect of various nutritional factors and fluorine ions on aluminum retention in bony tissue of rats, *Cesk. Gastroentrol. Vyz.*, 17(4), 202, 1963; *Biol Abstr.*, 44, 18280, 1963.
86. **Niedermeier, W., Creitz, E. E., and Holley, H. L.,** Trace metal composition of synovial fluid from patients with rheumatoid arthritis, *Arthritis Rheum.*, 5, 439, 1962.

87. **Niedermeier, W., Griggs, J. H., and Johnson, R. S.,** Emission spectrometric determination of trace elements in biological fluids, *Appl. Spectrosc.*, 25, 53, 1971.
88. **Niedermeier, W. and Griggs, J. H.,** Trace metal composition of synovial fluid and blood serum of patients with rheumatoid arthritis, *J. Chronic Dis.*, 23, 527, 1971.
89. **Malozhen, B. G.,** Trace element concentrations in the blood rheumatic patients, *Mikroelem. Estestv. Radioaktiv. Pochv.*, p. 211, 1962; *Chem. Abstr.*, 59, 10626, 1963.
90. **Sauer, H. I., Parke, D. W., and Neill, M. L.,** Associations between drinking water death rates, in *Trace Substances in Environmental Health, Vol. 4,* Hemphill, D. D., Ed., University of Missouri, Columbia, 1971, 318.
91. **Rapoport, S.,** Uber die Wirkung von Al and Ca auf das Succinatoxydase-System des Herzmuskels, *Z. Physiol. Chem.*, 302, 156, 1955.
92. **Rapoport, S. and Nieradt, C.,** Zur Lokalisierung des Al-Effektes in der Atmungskette, *Z. Physiol. Chem.*, 302, 161, 1955.
93. **Berki, E., et al.,** Inorganic content of intact (healthy) and of atherosclerotic aortas, *Orv. Hetil.*, 106(5), 201, 1965; *Chem. Abstr.*, 62, 13642, 1965.
94. **Popov, K., Rusanov, E., and Balevska, P.,** Trace elements of the human aorta depending on age, hemodynamics and some diseases, *Int. Congr. Gerontol. Proc.*, 2, 393, 1966; *Chem. Abstr*,, 71, 20269, 1969.
95. **Stepan, J., Vortel, V., and Fridrich, E.,** Aluminum in guinea pig organs in normal and pathological states, *Cas. Lek. Cesk.*, 97, 214, 1958; *Chem. Abstr.*, 52, 20609, 1958.
96. **Shlopak, I. V.,** Trace elements in ocular tissue in physiological and pathological conditions, *Mikroelem. Sel. Khoz. Med., Ukr. Nauch-Issled. Inst. Fiziol Rast., Akad, Nauk Ukr. SSR, Mater. Vses. Soveshch.*, 1962, 632, 1963; *Chem. Abstr.*, 63, 10401, 1965.
97. **Semenov, G. I.,** Age-dependent dynamics of the manganese, silicon, aluminum, titanium, and copper lever in the optic nerve of humans, *Mater. Vses. Simp. Mikroelem. Nerv. Sist.*, 1963, 84, 1966; *Chem. Abstr.*, 67, 114819, 1967.
98. **Semenov, G. I.,** Age-varying levels of some trace elements in the optic nerve of man, *Mikroelem. Biosfere Ikh Primen. Sel. Khoz. Med. Sib. Dal'nego Vostoka, Dokl. Sib. Konf.*, 5557, 1967; *Chem. Abstr.*, 70, 757512, 1969.
99. **Swanson, A. A., Jeter, J., and Tucker, P.,** Inorganic cations in the crystalline lens from young and aged rats, *J. Gerontol.*, 23, 502, 1968.
100. **Maksudov, N. and Talipov, S. T.,** Preliminary examination of kidney stone composition, *Uzb. Khim. Zh.*, 6(4), 88, 1962; *Chem. Abstr.*, 58, 3758, 1963.
101. **Ioshin, O. I.,** Effect of the mineral composition of drinking water on that of urinary calculi, *Hyg. Sanit. (USSR)*, 36(1), 442, 1971.
102. **Tsintsadze, V. S.,** Trace element composition of kidney and ureter stones, *Soobshch. Akad. Nauk Gruz. SSR*, 27(1), 33, 1961; *Chem. Abstr.*, 56, 6558, 1962.
103. **Stepan, J.,** Vyskyt hliniku, vanadia, zirkonia, wolframu a rtuti v ledvinach za normalnich a patologickych stavu, *Cesk. Farm.*, 15(1), No. 1, 43, 1966.
104. **Kvirikadze, N. A.,** Content of trace elements in the urine of patients with bladder tumors, *Soobshch. Akad. Nauk. Gruz. SSR*, 45(1), 241, 1967; *Chem. Abstr.*, 66, 93172, 1967.
105. **Ferguson, A. B., Jr., et al.,** Trace metal ion concentration in the liver, kidney, spleen, and lung of normal rabbits, *J. Bone Jt. Surg. Am. Vol.*, 44, 317, 1962.
106. **Soroka, V. R.,** Effect of excitation and inhibition of the central nervous system of drugs on trace element metabolism in the kidneys, *Ukr. Biokhim. Zh.*, 35, 528, 1963; *Chem. Abstr.*, 59, 15698, 1963.
107. **Soroka, V. R.,** State of some trace elements in the kidneys with changes in the function of the central nervous system, *Ukr. Biokhim. Zh.*, 37, 274, 1965; *Chem. Abstr.*, 63, 6092, 1965.
108. **Soroka, V. W. and Semenov, G. I.,** Metabolism of trace elements Mn, Si, Al, Ti, and Cu in denervated kidneys, *Mikroelem. Sel. Khoz. Med., Akad. Nauk Ukr. SSR, Respub. Mezhvedom.*, P. 216, 1966; *Chem. Abstr.*, 66, 9237, 1967.
109. **Soroka, V. R.,** Effects of strophanthin and adoniside on renal elimination of some trace elements, *Farmakol. Toksikol. (Moscow)*, 26(3), 289, 1963; *Chem. Abstr.*, 60, 1017, 1964.
110. **Soroka, V. R.,** Effect of diuretic drugs on the metabolism of some trace elements in the kidneys, *Vop. Med. Khim.*, 6, 420, 1960; *Chem. Abstr.*, 55, 8651, 1961.
111. **Soroka, V. R.,** Some trace element metabolism in normal dog kidneys as indicated by angiostomy results, *Ukr. Biokim. Zh.*, 33, 570, 1961; *Chem. Abstr.*, 56, 1881, 1962.
112. **Soroka, V. R.,** Effect of urea loading on kidney trace element metabolism, *Ukr. Biokhim. Zh.*, 34, 834, 1962; *Chem. Abstr.*, 58, 8296, 1963.
113. **Soroka, V. R.,** The effect of bile duct ligation on the exchange of trace elements in the dog, *Arkh. Patol.*, 27(9), 58, 1965; *Chem. Abstr.*, 64, 2564, 1966.

114. **Sutton, D. A. and Marasas, L. W.,** Urinary excretion of aluminum after administration in chelated forms, *Experientia,* 15, 476, 1959.
115. **Butt, E. M., et al.,** Trace metal patterns in disease states. I. Hemochromatosis and refractory anemia, *Am. J. Clin. Pathol.,* 26, 225, 1956.
116. **Butt, E. M. et al.,** Trace metal patterns in disease states: Hemochromatosis, Bantu siderosis, and iron storage in Laennec's cirrhosis and alcoholism, in *Metal-Binding in Medicine,* Proc. Symp., Seven, M. J. and Johnson, L. A., Eds., J. B. Lippincott, Philadelphia, 1960, 43.
117. **Butt, E. M., et al.,** Trace metal patterns in disease states. III. Hepatic and pancreatic cirrhosis in alcoholic patients, with and without storage of iron, *Am. J. Clin. Pathol.,* 42, 437, 1964.
118. **Mozgovaya, E. N. and Arnautov, N. V.,** The trace element contents of the liver and pancreas of cattle, *Izv. Sib. Otd. Akad. Nauk SSSR,* 2, 104, 1960; *Chem. Abstr.,* 55, 1845, 1961.
119. **Sabadash, E. V.,** Daily rhythm of trace element content in liver, *Mikroelem. Sel. Khoz. Med., Ukr. Nauch-Issled. Inst. Fiziol, Rast., Akad. Nauk Urk. SSR, Mater. Vses. Soveshch.,* 4th Kiev 1962, 598, 1963; *Chem. Abstr.,* 64, 2518, 1966.
120. **Soroka, V. R.,** Metabolism of certain trace elements in the liver of normal dogs from angiostomy data, *Mikroelem. Sel. Khoz. Med.,* p. 600, 1963; *Chem. Abstr.,* 61, 15148, 1964.
121. **Soroka, V. R.,** Effect of biliary obstruction on the content of some trace elements in the blood and urine, *Patol. Fiziol. Eksp. Ter.,* 9(1), 63, 1965; *Chem. Abstr.,* 63, 1060, 1965.
122. **Baessler, K. H. and Unbehaun, V.,** Effect of aluminum on metabolic processes, *Arzneim. Forsch.,* 12, 124, 1962; *Chem. Abstr.,* 56, 16089, 1962.
123. **Biggs, M. H.,** Effect of calcium and aluminum ions on succinate oxidation by liver mitochondria, *N.Z. J. Sci.,* 6, 14, 1963; *Chem. Abstr.,* 59, 2030, 1963.
124. **Perry, H. M., Jr., et al.,** Variability in the metal content of human organs, *J. Lab. Clin. Med.,* 60, 245, 1962.
125. **Tipton, I. H., Johns, J. C., and Boyd, M.,** The variation with age of elemental concentrations in human tissue, in *Proc. 1st Int. Congr. Radiation Protection,* Pergamon Press, New York, 1968, 759.
126. **Saakashvili, T. G.,** The content of some microelements in hypertrophied prostate gland, *Soobshch. Akad. Nauk Gruz. SSR,* 31(2), 303, 1963; *Chem. Abstr.,* 60, 14970, 1964.
127. **Dinevich, L. S.,** Trace-element content of mother's milk, *Sb. Tr. Mold. Nauch-Issled. Inst. Epidemiol., Mikrobiol. Gig.,* 2, 187, 1956; *Chem. Abstr.,* 52, 16539, 1958.
128. **Dinevich, L. S.,** Aluminum in human milk, *Vop. Pitan.,* 19(2), 76, 1960; *Chem. Abstr.,* 57, 9049, 1962.
129. **Soroka, V. R.,** Effect of limb immobilization on the content of some trace elements in muscles and tendons, *Ukr. Biokhim. Zh.,* 33, 739, 1961; *Chem. Abstr.,* 56, 7895, 1962.
130. **Soroka, V. R.,** Forms of some trace elements in the skeletal muscles, *Byull. Eksp. Biol. Med.,* 57(5), 49, 1964; *Chem. Abstr.,* 62, 5639, 1965.
131. **Tipton, I. H. and Cook, M. J.,** Trace elements in human tissue. II. Adult subjects from the United States, *Health Phys.,* 9, 103, 1963.
132. **Tabakman, M. B.,** The physiological distribution of microelements in humans, *Sub-Med. Ekspertiza,* 10(4), 24, 1967; *Chem. Abstr.,* 68, 57868, 1968.
133. **Bertha, H., et al.,** Regional cation distribution in human brain, *Monatsh,* 93, 118, 1962; *Chem. Abstr.,* 56, 15996, 1962.
134. **Soroka, V. R.,** The fate of trace elements in the brain, muscles and liver, as shown by the results of sinusostomy and angrostomy during stimulation and inhibition of the central nervous system, *Bull. Exp. Biol. Med. (USSR),* 47, 196, 1959.
135. **Soroka, V. R.,** K voprosu o formakh prebyvaniya nekotorykh mikroelementov v mozgu pri razlichnykh funktsional'nykh sostoyaniayakh tsentral'noi nervnoi sistemy. (Forms of residence of certain trace elements in the brain in different functional states of the central nervous system), *Mater, Vses. Simp. Mikroelem. Nerv. Sist.,* p.125, 1966.
136. **Blaustein, M. P. and Goldman, E. E.,** The action of certain polyvalent cation on the voltage-clamped lobster axon, *J. Gen. Physiol.,* 51, 279, 1968.
137. **Magazanik, L. G. and Vyskocil, F.,** Dependence of acetylcholine desensitization on the membrane potential of frog muscle fibre and on the ionic changes in the medium, *J. Physiol.,* 210, 507, 1970.
138. **Patocka, J.,** The influence of Al^{+++} on cholinesterase and acetylcholinesterase activity, *Acta Biol. Med. Ger.,* 26, 845, 1971.
139. **Crapper, D. R., Krishnan, S. S., and Dalton, A. J.,** Brain aluminum distribution in Alzheimer's Disease and experimental neurofibrillary degeneration, *Science,* 180, 511, 1973.
140. **Nechaev, E. N. and Soroka, V. R.,** Correlation between trace elements and enzymic activity of pancreatic juice, *Lab. Delo,* 6, 334, 1967; *Chem. Abstr.,* 67, 71909, 1967.
141. **Langenbeck, W., Augustin, M., and Schafer, C.,** Metallic ions activating trypsin, *Z. Physiol. Chem.,* 324, 54, 1961; *Chem. Abstr.,* 55, 24868, 1961.

142. **McGeachin, R. L., Pavord, W. M., and Pavord, W. C.,** Inhibition of various mammalian amylases by beryllium and aluminum, *Biochem. Pharmacol.,* 11, 493, 1962.
143. **Nasel'skii, N. B.,** Effect of age on the trace-element composition of human skin, *Nauk. Zap. Stanislav. Med. Inst.,* 3, 90, 1959; *Chem. Abstr.,* 59, 14372, 1963.
144. **Kononenko, V. I.,** Quantitative changes of the dermal major and trace elements in the areas of cadaveric lividity at various times after death, *Sud-Med. Ekspertiza,* 12(4), 15, 1969; *Chem. Abstr.,* 72, 108972, 1970.
145. **Harrison, W. W. and Clemena, G. G.,** Survey analysis of trace elements in human fingernails by spark source mass spectrometry, *Clin. Chem. Acta,* 36, 485, 1972.
146. **Soroka, V. R. and Krivonos, A. Ya.,** Participation of the spleen in trace element metabolism in relation to central nervous system function, *Vop. Biol. Med. Khim., Mater. Nauch, Biokhim. Konf.,* 1, 58, 1968; *Chem. Abstr.,* 72, 2062, 1970.
147. **Losee, F., Cutress, T. W., and Brown, R.,** Trace elements in human dental enamel, in *Abstr. Univ. Mo. 7th Ann. Conf. on Trace Substances in Environmental Health,* University of Missouri, Columbia, 1973, 4.
148. **Yaroshenko, A. N.,** Trace element levels in hard dental tissues in the population of Arkhangelsk region, *Mater. Nausch. Sess. Arkhangel'sk. Gos. Med. Inst.,* 28, 1967, 351; *Chem. Abstr.,* 72, 29381, 1970.
149. **Retief, D. H., et al.,** Quantitative analysis of Mg, Na, Cl, Al, and Ca in human dental calculus by neutron activation analysis and high resolution gamma-spectrometry, *J. Dent. Res.,* 51, 807, 1972.
150. **Losee, F. L. and Ludwig, T. G.,** Trace elements and caries, *J. Dent. Res.,* 49, 1229, 1970.
151. **Vrbic, V., Logar, A., and Podobnik, B.,** Trace elements and dental caries, *Zobozdravst. Vestn.,* 25(5), 152, 1970; *Excerpta Med.,* (Sect. 17) 18, 2413, 1972.
152. **Ludwig, T. G., Adkins, B. L., and Losee, F. L.,** Relationship of concentrations of eleven elements in public water supplies to caries prevalence in American schoolchildren, *Aust. Dent. J.,* 15, 126, 1970.
153. **Briner, A., Cerasa, G., and Barros, L.,** Influence of certain chemical elements on the susceptibility of teeth to caries, *Odont. Chile,* 18(97), 36, 1970; *Oral Res. Abstr.,* 7, 1293, 1972.
154. **Heap, R. B.,** Some chemical constituents of uterine washings: A method of analysis with results from various species, *J. Endocrinol.,* 24, 367, 1962.
155. **Cam, J. M. et al.,** The effect of aluminum hydroxide orally on calcium, phosphorous and aluminum metabolism in normal subjects, *Clin. Sci. Molec. Med.,* 51, 407, 1976.
156. **Berlyne, G. M. and Rubin, J. E,** Aluminum ion: metabolism and toxicity, *J. Hum. Nutr.,* 31, 439, 1977.
157. **Bessarabova, R. V.,** Toxicology of aluminum compounds, *Vop. Gig. Tr., Prof. Patol. Toksikol. Prom. Sverklovsk. Oblasti, Sverdlovsk.,* 1955, 142; *Chem. Abstr.,* 54, 7894, 1960.
158. **Kratochvil, B., Boyer, S. L., and Hicks, G. P.,** Effects of metals on the activation and inhibition of isocitric dehydrogenase, *Anal. Chem.,* 39, 45, 1967.
159. **Langenbeck, W. and Schellenberger, A.,** Uber die Decarboxylierung der Brenztraubensaure durch Aluminiumionen, *Arch. Biochem. Biophys.,* 69, 22, 1957.
160. **Harrison, W. H., Codd, E., and Gray, R. M.,** Aluminum inhibition of hexokinase, *Lancet,* 2, 227, 1972.
161. **Shaw, J. G.,** Binding of purines and pyrimidines by bentonite, and inhibitor of nucleases, *Can. J. Biochem.,* 43, 829, 1965.
162. **Soroka, V. R.,** Bonding of metals with nucleic acids, *Mater. Respub. 1st Konf. Probl. Mikroelem. Med. Zhivotnovoid.,* 1968, 73; *Chem. Abstr.,* 74, 19413, 1971.
163. **Jacoli, G. G.,** Mechanism of inhibition of ribonuclease by bentonite, *Can. J. Biochem.,* 46, 1237, 1968.
164. **Berlyne, G. M., et al.,** Hyperaluminanemia from aluminum resins in renal failure, *Lancet,* 1, 564, 1972.
165. **Bolkvadze, P. D., et al.,** Effect of askan (bentonite of Askania) gel and its simultaneous use with protein hydrolyzate on the growth and development of rabbits, *Mater. Nauch. Konf., Gruz. Zootech-Vet. Ucheb-Issled. Inst.,* 1967, 133; *Chem. Abstr.,* 70, 26793, 1969.
166. **Day, E. J., Bushong, R. D., Jr., and Dilworth, B. C.,** Silicates in broiler diets, *Poult. Sci.,* 49, 198, 1970.
167. **Bailey, R. R.,** Aluminum toxicity in rats and man, *Lancet,* 2, 276, 1972.
168. **Thompson, A., Hansard, S. L., and Bell, M. C.,** The influence of aluminum and zinc upon the absorption and retention of calcium and phosphorus in lambs, *J. Anim. Sci.,* 18, 187, 1959.
169. **Berlyne, G. M. et al.,** Aluminum toxicity in rats, *Lancet,* 1, 564, 1972.
170. **Berlyne, G. M. et al.,** Aluminum Toxicity, *Lancet,* 1, 1070, 1972.
171. **Thurston, H., Gilmore, G. R., and Swales, J. D.,** Aluminum retention and toxicity in chronic renal failure, *Lancet,* 1, 881, 1972.

172. **Kay, S. and Thornton, J. L.,** The intraperitoneal injection of aluminum hydroxide in mice, *Arch. Pathol.*, 60, 651, 1955.
173. **Thurston, H. and Swales, J. D.,** Aluminum toxicity, *Lancet*, 1, 1241, 1972.
174. **Sherrard, D. J.,** Aluminum toxicity in rats and man, *Lancet*, 2, 276, 1972.
175. **Evenshtein, Z. M.,** Concerning the permissible content of aluminum in adults' food, *Hyg. Sanit. (USSR)*, 36(1), 299, 1971.
176. **Kotter, L., et al.,** Increasing the formation of antibodies against heated protein with the help of metallic compounds, *Zentralbl. Veterinaermed. Reihe B*, 13, 613, 1966; *Chem. Abstr.*, 66, 45035, 1967.
177. **Schaldach, M.,** Long-term use of biogalvanic elements as energy sources for cardiac pacemakers, *Thoraxchir. Vask. Chir.*, 18, 437, 1970; *Chem. Abstr.*, 74, 61482, 1971.
178. **Dunaev, V. G.,** Effect of homotransplantation on the metabolism of some trace elements in the blood, *Tr. Gor'k. Gos. Med. Inst.*, 26, 84, 1968; *Chem. Abstr.*, 72, 11057, 1970.
179. **Vulpis, N. and Giorgino, R.,** Some effects of metal cations on blood coagulation, *Thromb. Diath. Haemorrh.*, 8, 121, 1962.
180. **Funahara, Y., Nakamura, S., and Okamoto, S.,** Factors controlling coupling in contact fibrinolysis, *Kobe J. Med. Sci.*, 14(4), 281, 1968; *Chem. Abstr.*
181. **Zajusz. K., Paradowski, Z., and Dudziak, Z.,** Hemolytic activity of silica and other dusts, *Med. Pracy*, 19(1), 26, 1968; *Chem. Abstr.*, 69, 1468, 1968.
182. **Salahuddin,** Metal ion-binding capacity of gelatin and transfusion gelatin, *Recent Advan. Mineral Tannages. Pap. Symp.*, 1964, 165; *Chem. Abstr.*, 64, 16158, 1966.
183. **Kovalenko, S. L. and Kurilenko, O. D.,** Modern concepts of pectin substances, *Izv. Vyssh. Ucheb. Zaved., Pshch. Tekhnol.*, 5, 28, 1963; *Chem. Abstr.*, 65, 9626, 1966.
184. **Kopeloff, L. M., et al.,** Recurrent convulsive seizures in animals produced by immunologic and chemical means, *Am. J. Psychiatry*, 98, 881, 1942.
185. **Kopeloff, N., et al.,** The experimental production of epilepsy in animals, in *Epilepsy*, Hoch, P. H. and Knight, R. P., Eds., Grune and Stratton, New York, 1947, 163.
186. **Pacella, B. L., et al.,** Experimental production of focal epilepsy, *Arch. Neurol. Psychiatry*, 52, 183, 1944.
187. **Kaltzo, I. et al.,** Experimental production of neurofibrillary degeneration, *J. Neuropathol. Exp. Neurol.*, 24, 187, 1965.
188. **Terry, R. D. and Pena, C.,** Experimental production of neurofibrillary degeneration, *J. Neuropathol. Exp. Neurol.*, 24, 200, 1965.
189. **Wisniewski, H. et al.,** Experimental production of neurofibrillary degeneration, *J. Neuropathol. Exp. Neurol.*, 24, 139, 1965.
190. **DeBoni, U., Otvos, A., Scott, J., and Crapper, D.,** Neurofibrillary degeneration induced by systemic aluminum, *Acta Neuropathol.*, 35, 285, 1976.
191. **Deboni, U., Scott, J., and Crapper, D.,** Intracellular aluminum binding: A histochemical study, *Histochemie*, 40, 31, 1974.
192. **Wisnuwski, H. M. et al.,** Neurotoxicity of aluminum, in *Neurotoxicology*, Roizin, H. and Shiraki, R., Eds., Raven Press, New York, 1977, 313.
193. **Crapper, E. et al.,** Brain aluminum distribution in Alzheimer's disease and experimental neurofibrillary degeneration, *Science*, 180, 511, 1973.
194. **Crapper, D. R. et al.,** Aluminum, neurofibrillary degeneration and Alzheimer's disease, *Brain*, 99, 67, 1976.
195. **Clarkson, E. M. et al.,** The effect of aluminum hydroxide on calcium, phosphorus, and aluminum balances, the serum parathyroid hormone, concentration and the aluminum content of bone in patients with chronic renal failure, *Clin. Sci.*, 43, 519, 1972.
196. **Berlyne, G. M. et al.,** Hyperaluminemia from aluminum resins in renal failure, *Lancet*, 2, 494, 1970.
197. **Parsons, V. et al.,** Aluminum in bone from patients with renal failure, *Br. Med. J.*, 4, 273, 1971.
198. **Alfrey, A. C. et al.,** The dialysis encephalopathy syndrome, *N. Engl. J. Med.*, 294, 184, 1976.
199. **Mayor, G. H.,** Aluminum absorption and distribution: effect of parathyroid hormone, *Science*, 197, 1187, 1977.
200. **Rosenbek, J. C. et al.,** Speech and language findings in a chronic hemodialysis patient: a case report, *J. Speech Hear. Disord.*, 2, 245, 1974.
201. **Chokroverty, S. et al.,** Progressive dialytic encephalopathy, *J. Neurol., Neurosurg. Psychiatry*, 39, 411, 1976.
202. **Burks, J. S. et al.,** A fatal encephalopathy in chronic hemodialysis patients, *Lancet*, 1, 764, 1976.
203. Editoral Comment, *Lancet*, 1, 349, 1976.
204. **Dunea, G. et al.,** Role of aluminum in dialysis dementia, *Ann. Intern. Med.*, 88, 502, 1978.
205. **Elliott, H. L. et al.,** Aluminum toxicity during regular hemodialysis, *Br. Med. J.*, 1, 1101, 1978.

206. **Piesides, A. M. et al.**, Hemodialysis encephalopathy: Possible role of phosphate depletion, *Lancet*, 1, 1234, 1976.
207. **Elliott, H. L. et al.**, Aluminum toxicity syndrome, *Lancet*, 1, 1203, 1978.

Chapter 14

COPPER

Stephen P. Halloran

TABLE OF CONTENTS

I. INTRODUCTION

Copper was one of the first metals known to man, and instruments made from it have been found in Egyptian tombs dating back to the fifth millennium B.C. Its old name cyprium and later cuprum derive from the fact that it was mined on the island of Cyprus.

Since 1830 copper has been recognized as a normal consituent of blood; however, its medicinal use by the Egyptians pre-dates this by over 2000 years. During the time of Hippocrates, the Greeks found it of value in the treatment of pulmonary disease, and in the western world, the element was used by Paracelsus in the treatment of mental lues and lung diseases. An Indian custom of drinking water left standing overnight in copper pots is considered effective in the treatment of rheumatoid arthritis, a practice perhaps related to the western tradition of wearing copper bracelets.

Evidence that anemia responds to the "blue ash from liver tissue" was documented in 1875 and was later confirmed and extended by Hart et al.[1] who in 1928 found that the hemoglobin of anemic rats responded more acutely to a supplement of iron and copper than to iron alone. Recent research has established copper as an essential element in promoting hemopoiesis,[2-4] although its mode of action must still be considered speculative.

It was in 1945 that Glazebrook[5] recognized that hepatolenticular degeneration, an inherited disease first characterized by Wilson[6] in 1912, presented with large deposits of copper in the liver and brain. Today, disturbed copper homeostasis is recognized as a common feature of a plethora of diverse conditions and as a cardinal feature of two rare inborn errors of metabolism, those of Wilson's disease and a syndrome characterized by pili torti (kinky hair) and described by Menkes in 1962.[7]

The estimation of copper in blood has been invoked as a useful corollary in pregnancy testing, as an indication of fetal distress[8,9] of diagnostic potential in both Hodgkins' disease[10] and tumors of the digestive tract,[11] in the differential diagnosis of cerebral infarction and subarachnoid hemorrhage,[12] and of prognostic value in alcohol cirrhosis[13] and myocardial infarction.[14]

As a form of treatment, copper has found an accepted place as a supplement to parenteral fluids and cattle foods[15] and as a replacement therapy in patients with Menkes' kinky hair syndrome.[16] A recent paper by some Russian workers has claimed startling success using copper sulfate for treatment of secondary amenorrhoea — by electrophoresis![17]

II. ANALYTICAL DETERMINATION

A variety of techniques are available for the quantitation of copper in biological specimens. These techniques differ in their limits of sensitivity and in the specimen preparation required. The choice of technique will be influenced by the biological specimen used, be it serum, urine, cerebral spinal fluid, synovial fluid, liver tissue, hair, or nail, and the particular copper fraction required, be it free copper, amino acid bound copper, or that present in ceruloplasmin.

Spectrophotometric methods, which determined the optical density of a copper complex, were the methods of choice prior to the 1960s. The popular complexing agents include bathocuproine (2,9-dimethyl-4,7-diphenyl-1,10-phenanthroline) and its water-soluble sulfonated derivative, diphenylcarbazone, oxalyldihydrazide, zinc dibenzyl dithiocarbamate diphenylcarbazide in benzene, and 1,5-diphenylcarbohydrazide. The latter is the most sensitive and forms a 1:2 dye complex of high extinction coefficient.[18]

Colorimetry has been adapted for continuous flow equipment, and analysis of two elements simultaneously has been described by combining two dyes.[19] The chief limitations of spectrophotometry in trace metal analysis are those of sensitivity and, to a lesser extent, selectivity, the first of these making it unsuitable for the determination of the low concentrations of copper encountered in tissues, serum subfractions, and urine.

The development of analytical atomic absorption spectroscopy (AAS) by Walshe[20] in 1955 enabled a reduction in the limits of sensitivity for the determination of copper by several orders of magnitude. More recent refinements of his technique have enabled determinations into the nanogram range. Copper is ideally suited for AAS since its resonance line at 324.7 nm is well removed from other line interference and is detected in a region of the photomultiplier spectral window least subject to noise.[18] Preparation procedures vary greatly. Diluted serum can be aspirated directly into the flame, although the presence of the serum proteins gives rise to a high continuum radiation and alters the viscosity and thus the rate of aspiration. The factor may be compensated for by preparation of standard solutions in a similar matrix. Such simple preparations have been found suitable for combining the atomic absorption spectrophotomer and continuous flow equipment, although such methods have not been widely adopted.

In an alternative procedure to circumvent protein interference, the proteins are degraded in acids such as trichloracetic acid, which release copper into the supernatant and make it available for aspiration into the flame. A further reduction in interference is achieved by chelating copper to ammonium pyrrolidine dithiocarbamate and extracting the complex into an organic solvent. Protein-free copper can be estimated by this procedure if none is released from protein by acid degradation. The use of an organic solvent may increase the flame temperature and the change in viscosity alter the rate of aspiration; the standard should therefore be prepared in a similar solvent.

Determination of copper in tissues requires either acid extraction, when nitric acid is commonly used, or dry ashing. In both of these procedures, care is required to avoid contamination and loss of copper.

Reduced limits of detection are achieved by circumventing the inefficient aspiration into the nebulizer burner system, a process which loses 80% of that aspirated. Ward et al.,[21] described such a procedure in which the whole of the preparation is placed into a modified delves cup (a nickel crucible with a hole in the middle) and atomization produced by direct heating under a flame.

The temperature present in a flame (about 1200°C) is not ideal for atomization of metals of low volatility such as copper, but electrothermal atomization achieves temperatures in excess of 3000°C under more controlled conditions. L'vov[22] first introduced this technique in 1960 using a graphite tube furnace and now a variety of supports are in use including rods, wires, and cups. The great sensitivity of electrothermal atomization is countered to some extent by matrix interference from the support or container and from the sample, particularly if simple sample preparations are used. Poor reproducibility of this method has, however, been considerably improved by automatic sample dispensing and improved temperature regulation.

Evenson and Warren[23] have described a method for copper estimation using this technique with a claimed interbatch and extrabatch precision of 1 and 2%. Electrothermal atomization is also ideal for tissue copper determinations where extensive preparation would otherwise be required to enable aspiration into a flame and where low limits of detection are important. Evenson and Anderson have described such a method for liver copper quantitation.[24]

Delves[25] used an ingenious technique for the determination of copper present in the serum $\alpha 2$ globulin fraction, by which serum proteins were separated by cellulose electrophoresis and the $\alpha 2$ band cut out and its copper contents determined by an electrothermal technique. Similar protein fractionation and subsequent determinations have been described using starch block electrophoresis.

Anodic stripping, polarography, molecular absorption spectrophotometry, X-ray fluorescence, neutron activation, radioisotope dilution, and spark source mass spectrometry are further techniques used in copper determination.

Neutron activation is a popular tool employed for the determination of materials such as liver biopsies, hair, and nails. It uses the reaction $^{63}Cu(n,\gamma)^{64}Cu$ and determines as little as 10^{-9} g with a precision of 10%[18] Its most serious disadvantages are the long time required to allow decay of ^{24}Na and the necessary facility of an atomic pile.

Atomic absorption is restricted to a single element analysis; atomic emission, however, enables multiple analysis. Very high temperatures are required for emission and can be achieved by DC arc and spark sources; they are not however, an easy tool for routine use. An inductively coupled plasma ionization has enabled very high temperatures to be achieved in a controlled environment with precise determinations of several elements simultaneously. This technique described by Kiniseley et al.[26] promises to become popular for routine analysis, and already several commercial instruments are available. The facility for multiple element analysis has decided advantages in the analysis of biological fluids, where an element profile will render more complete information of the patient's nutritional state. This is a highly desirable situation for the ever-increasing number of patients receiving total parenteral nutrition and those receiving dialysis treatment.

The impressive development of analytical methodology during the past decade, with the increased sensitivity it has brought, has not made the analyst's task any simpler since greater precision and accuracy are demanded in the nanogram and subnanogram range. The collection and preservation of specimens has become of greater importance. Chrome and brass syringe fittings, glass collecting vials, cardboard stoppers, and water obtained from the ubiquitous copper pipe are all sources of contamination.

Stainless-steel syringe needles are to be recommended and borosilicate glass containers are superior to the more common soft soda glass variety for storing dilute copper solutions. Siliconizing glass surfaces is another recommended procedure to reduce leaching of copper from glass walls. Many plastics contain copper as accelerators, plasticisers, pigments, or fillers and ideally, high purity plastics such as polytetrafluoroethylene and polymethylmethacrylate should be used.[27]

Bacterial growth may alter trace element concentration by incorporating airborne ions.[27] Bacteriostats, such as streptomycin and thymol-fluoride, are effective, yet are potential sources of contamination. The author has found that freezing at −20°C is satisfactory for storage over a 2-year period. Thiers[28] expressed his thoughts forcefully: "Unless the complete history of the sample is known with certainty, the analyst is well advised not to spend his time analysing it".

III. DIET

Clinical symptoms associated with dietary copper deficiencies have never been described in man; indeed Scheinberg[29] maintains that the average western diet provides a substantial excess. Kwashiorkor and marasmus are known to reduce blood and liver copper concentrations, an effect which is remedied by copper supplement and aggravated by high calorie diets.[30] In these severe hypoproteinaemic states, the deficiency of copper may reflect defective amino acid transport of the metal across the mucosa, rather than a dietary deficiency of copper.

In contrast to adults, infants have been repeatedly shown to be susceptable to dietary copper deficiency.[31] In the neonate, fed entirely on milk for an extended period of time, hypochromic anemia develops, which will respond to copper supplements.[32] Milk has the low copper concentration of approximately 1.5×10^{-6} M and so the infant is almost entirely dependent on its liver reserve, which fortunately is reported to be about seven times greater than that of the adult (in contrast to its plasma concentration of one third adult levels).[33] Premature infants are particularly at risk since liver concentrations have not reached the normal adult level. This has prompted the American Academy of Pediatrics to consider the addition of copper to formulas for low birth weight infants.[34] Copper-deficient states in adults do occur following 3 to 17 weeks of total parenteral nutrition,[3] although the subnormal ceruloplasmin, copper, and hemoglobin levels present respond rapidly to copper supplement. In anorexia nervosa, a nutritional deficiency state par excellence, only a slight, yet significantly reduced plasma copper level has been reported, which further indicates the adult indisposition to induced dietary copper deficiency.[35]

Cattle, sheep, and lambs are also susceptible to copper deficiency as was recognized in 1931 in cattle grazing on copper-deficient pastures in Holland[36] and Florida.[37] This deficiency presents symptoms including anemia, vascular abnormalities, and depigmentation of hair and wool. These pasture deficiencies are commonly remedied by applying copper-containing fertilizers and using salt licks.[15] Copper calcium edate injections have also been used satisfactorily in cattle and ewes during mid-pregnancy, when the demand for copper is at its greatest. Copper supplements are not without their dangers. Several cases of chronic copper poisoning of cattle are reported in the literature, such as that reported in 1978 in Victoria,[38] where supplements of 10 to 20 g of copper sulfate daily in a dairy herd produced anorexia, liver damage, jaundice, dehydration, anemia intestinal haemorrhage, and finally, death.

Copper toxicity is far more common in multigastric than monogastric mammals, and in man reports of copper toxicity are uncommon. They are reported to occur following ingestion of citric juices left standing for several hours in copper vessels, and in patients receiving hemodialysis using cuprophane membranes.[29]

Symptoms associated with ingestion of copper above 15 mg/day include nausea, vomiting, diarrhea, and crampy abdominal pain.

Chuttani's study[39] of suicides in India by ingestion of gram quantities of copper sulfate shows mucosal ulceration and extensive liver and kidney damage. The deleterious effects of excess copper on human metabolism may be explicable in terms of its inhibitory action on microsomal membrane ATPase[40] and a variety of glycolytic enzymes.[41]

The toxicity brought about by drinking water from copper pipes indicates it to be an important nutritional source. However, the richest sources of dietary copper are in crustaceans, shellfish, lamb, and beef liver, followed by nuts, dried legumes, dried vines, some dried stoned fruits, and cocoa.[42] Studies among women in New Zealand showed that a liver-containing diet increased copper content fivefold to an average intake of 7.6 mg/day.[43] The United States National Research Council: Food and Nutrition Board recommend an intake of 2 mg of copper per day for an adult,[44] and the WHO's expert committee recommended 30 μg of copper per kilogram per day as being adequate for an adult male.[45]

The average western diet contains 1 to 3 mg/day whereas in India, with a diet rich in rice and wheat, 4.5 to 5.8 mg/day would be an average intake. The neonatal requirement for copper is considerably greater than that of the adult and the WHO has recommended 80 μg of copper per kilogram per day as a minimum.

Average dietary copper content gives little information about the nutritional value of the diet when it is known that zinc and molybdenum are important antagonists to copper absorption at the mucosa and when the sulfate, phytate, and fiber content of the diet contribute towards reduced bioavailability.

IV. INTESTINAL ABSORPTION

The quantity of copper available for absorption from the diet is dependent upon the foods' intrinsic content, its bioavailability, the presence of antagonists to the absorption mechanism, and the rate of passage of food.

The form copper takes in food is not well understood; it is suggested that the formation of macromolecular copper complexes contributes to its diminished utilization, and in this respect raw meat and the phytic acid content of soy bean have been most studied. Copper sulfide formation also renders copper unsuitable for facilitated transport across the mucosa. Chapman and Bell[46] have tested the uptake by beef cattle of ^{64}Cu from several inorganic compounds and found the following sequence of decreasing uptake:

$$CuCo_3 > Cu(NO_3)_2 > CuCl_2 > Cu_2O > CuO(powder) > CuO\ needles > Cu\ wire.$$

Ascorbic acid is known to depress Cu absorption and has been used to achieve deficiency states in chicks. Antagonism to absorption from the cations silver, zinc, cadmium, and mercury is likely to be mediated through competition with copper for binding sites on the intestinal metalloprotein, metallothionein.[47] This intracellular sulfhydryl-rich protein was first isolated from equine kidney and has now been isolated from other species and tissues including human kidney.[48] It has a molecular weight of 10,000 daltons, contains 30 to 35% cysteine, exists in at least three forms, and binds metals to the extent of 6 to 11% of its molecular weight by forming mercaptide bonds to the cysteinyl side chains. The binding of metallothionein to copper is less avid than that to zinc and cadmium; indeed these metals are able to displace copper from its binding to metallothionein.

The site of maximal absorption of copper varies slightly with species; in man the rapid appearance of orally administered ^{64}Cu indicates absorption by the stomach and upper small intestine.[49,50] Two modes of copper absorption from the gastrointestinal tract have been indicated.[51]

1. An energy-dependent mechanism of limited capacity responsible for the transport of copper-L-amino acid complexes across the mucosa. This mechanism is unlikely to be restricted solely to cellular membranes of this tissue.
2. A protein-mediated absorption, in which the passive diffusion of copper ions into the mucosa is followed by competition with cadmium and zinc for chelation to two proteins, a low molecular weight protein of the same estimated size and perhaps composition as metallothionein, and a higher molecular weight protein likely to be superoxide dismutase. Copper, which is later dissociated from the metallothionein-like protein, diffuses directly into the plasma, or, after amino acid complexing, is transported to the serosal side.

The metallothionein-like protein provides the greatest binding capacity. It may have a function both in absorption of copper from the gut by binding to it and providing a store for subsequent release into the plasma, and secondarily as a mucosal block becoming saturated at high concentrations and protecting against excessive absorption.

Menkes[7] described a copper deficiency syndrome which has now been shown to occur due to reduced intestinal absorption. Investigation of the cytosol copper in these patients shows it present in high concentrations. It has been postulated that it may be due (1) to a deficiency of an enzyme catlyzing the cleavage of copper, (2) to a high concentration of a binding protein (metallothionein or superoxide dismutase) or (3) to an increased capacity of this molecule to bind copper.[52] Recent work has demonstrated abnormal cadmium retention and metallothionein induction in skin fibroblasts from patients with Menkes' kinky hair syndrome.[52,53] This further implicates a role for metallothionein, both in the pathogenesis of Menkes' syndrome and in the normal transport of copper across the mucosa.[53]

V. PLASMA COPPER

Studies with ^{64}Cu indicate that when absorbed, it is bound to albumin and transported through the portal vessels to the liver. The albumin complex has been extensively studied and the binding sites fully characterized.[54] The first preferential binding site has a dissociation constant of 6.61×10^{-17} with copper II and consists of an amino nitrogen, the imidazole nitrogen from histidine in position three, and two peptide-bound nitrogen radicals.[55] The less avid binding sites in dog albumin are due to omission of a histidine residue at the three position, which explains the susceptibility of this species to copper toxicity.[56]

Ionic copper is in equilibrium with its albumin complex, other peptide complexes, and amino acid complexes. These fractions have been separated from plasma by ultrafiltration and thin layer chromatography. From them a tertiary coordinated complex of copper II, albumin, and histidine has been isolated. Amino acid bound copper, although quantitatively small, is probably vital in the transport of copper to the tissues; there is also considerable evidence of its importance in the facilitated transport of copper across cell membranes in human liver and erythrocytes.[94]

The major component (90 to 95%) of human serum copper is a dark blue protein, first isolated and characterized in 1948 by Holmberg and Laurell,[57] which they called ceruloplasmin. This labile α_2 glycoprotein with a 7% carbohydrate and 0.32% copper content is synthesized in the hepatocytes of the liver and is a member of the acute reacting proteins. The labile nature of ceruloplasmin makes estimation of its molecular weight difficult.

It was first reported to have a molecular weight of 150,000 as determined by sedimentation and diffusion.[57] Crystallographic data obtained by Magdoff-Fairchild et al.[58] indicate a value of 132,000 ± 4,100. This latter value has recently been confirmed by Ryden[59] using sedimentation equilibrium studies. Amino acid and carbohydrate composition suggest a molecular weight near to 130,000. Certainly its molecular weight is lower than that originally reported and is probably between 125,000 and 134,000.

Ceruloplasmin has been considered to contain eight atoms of copper per molecule.[60] However, recent work suggests six or seven atoms may be more correct. The paramagnetic properties of Cu^{2+} have lent themselves to extensive studies of ceruloplasmin using electron paramagnetic resonance (EPR). Such studies have revealed three types of copper present in the molecule, a paramagnetic type I (blue), type II (nonblue), and an EPR nondetectable diamagnetic copper component accounting for 56% of the total.[61]

Marriott and Perkins suggested the presence of two groups of four atoms, four cupric and four cuprous.[62] This is difficult to reconcile with the evidence of Carrico et al.[63] that all of the copper ions in ceruloplasmin can accept electrons. The conflicting data are extensively reviewed by Fee.[60]

Owen[64] has demonstrated that injected radioactive copper accumulates in the tissues only after it has emerged from the liver in ceruloplasmin. Hsieh and Frieden[65] demonstrated an increased activity of the copper-dependent enzyme cytochrome c oxidase in heart, spleen, lungs, and liver following injection of ceruloplasmin. Other workers, however, have shown that in vivo exchange of ceruloplasmin and nonceruloplasmin copper does not occur.[66] It is concluded that exchange of copper between ceruloplasmin and extrahepatic tissue probably involves a degradation of the protein, either on the cell membranes or intracellularly. The presence of a small quantity of apo-ceruloplasmin in plasma has been confirmed and estimated by radioimmunoassay techniques.[67] Its origin is obscure and may indicate degradation or abnormal synthesis of holo-ceruloplasmin.

Several investigations have indicated that ceruloplasmin (EC 1.16.3.1) is a multifunctional enzyme. It possesses oxidase activity towards ascorbate[68] and the aromatic diamine, ϱ-phenylenediamine. The latter has proved a valuable means of determining the activity of ceruloplasmin, which in most cases correlates well with its quantitation by immunological techniques, although a deviation has been reported in rheumatoid arthritis patients treated with penicillamine. Similar oxidase activity of ceruloplasmin towards adrenalin, noradrenalin, 5-hydroxytryptamine, and melatonin has suggested that it may have a function in controlling the plasma levels of certain amines. This hypothesis is supported by evidence of increased ceruloplasmin concentrations following stress and exercise.[69]

Recent experiments have indicated a role for ceruloplasmin in normal hemopoiesis.[70] The iron required by the bone marrow for chelation to protoporphorin is derived from ferritin, which is present in the cells of the intestine, liver, and reticuloendothelial system. It is mobilized by the oxidation of ferrous iron to ferric, facilitating its binding to plasma transferrin, which is the sole supplier of iron to the marrow cells. Nonenzymic oxidation of ferrous iron has been demonstrated to be inadequate in maintaining normal hemoglobin production.[71] Ferroxidase activity of ceruloplasmin was first observed in vitro by Curzon and O'Reilly.[2] Evidence for a physiological role in swine in promoting the transfer of iron from storage cells to transferrin has since been obtained by Ragen et al.[72] and Roeser et al.,[70] although such a role has yet to be demonstrated in man.

The enzymic activity of ceruloplasmin has led to the adoption of its pseudonym, ferroxidase I, by some workers.[71] A macromolecular copper complex which possesses similar ferroxidase activity to ceruloplasmin has recently been isolated from plasma and is designated ferroxidase II.[74] In contrast to ceruloplasmin's established oxidase activity, antioxidant properties have also been demonstrated in vitro by Al-Timimi and Dormandy[75] using a brain homogenate system developed by Stocks et al.[76] The physiological significance of its antioxidant activity remains to be elucidated.

Tables 1 to 4 list copper concentrations in man which have been reported in the literature. The female normally shows a higher plasma concentration than the male, although this was not confirmed by Yunice[85] in his study. The difference is considerably increased for women receiving estrogen-containing oral contraceptives. An average concentration among 20 women of 30.4 ± 6.8 μmol/ℓ was observed by Scudder[78] and values of 47.2 ± 11.0 μmol/ℓ,[86] 34.0 μmol/ℓ,[87] and 34.8 ± 2.2 μmol/ℓ[88] were reported in three other studies.

A similar increase in copper and ceruloplasmin occurs during pregnancy and Nielson[89] observed an increase from the third month of pregnancy to 42.5 μmol/ℓ at term among 31 pregnant women. Hambridge et al.[88] have found copper levels of 25.6 ± 1.0 μmol/ℓ at 16 weeks and 30.2 ± 0.85 μmol/ℓ at 38 weeks. These elevated concentrations return to normal during the first few weeks following parturition.

In the neonate, copper and ceruloplasmin are found in concentrations of less than one third normal adult levels and then rise gradually during the first weeks of life to within the adult range at 3 months and still higher at 8 months. Delves et al.[90] observed a gradual reduction in plasma copper throughout childhood and Yunice et al.[85] found that a significant increase occurs throughout adult life, although ceruloplasmin did not display a similar significant increase.

Table 1
SERUM COPPER CONCENTRATIONS IN MAN

Normal Subjects

Sex	Mean ± S.D. μmol/ℓ	No. of subjects	Year	Ref.
M	17.9 ± 1.9		1944	89
F	20.0 ± 2.6		1944	89
M	16.5 ± 2.5	40	1953	83
F	18.3 ± 2.5	23	1953	83
M	17.3 ± 2.5	120	1960	77
F	18.9 ± 2.8	85	1960	77
F	19.0 ± 1.8	23	1966	124
M	17.4		1966	87
F	20.1		1966	87
F	19.2 ± 3.2	20	1968	86
F	22.3	15	1969	125
M	18.7 ± 2.8	100	1970	80
F	19.9 ± 3.3	100	1970	80
F	20.3	91	1971	81
M	17.5 ± 2.7	33	1972	82
F	20.5 ± 3.6	33	1972	82
M	17.8 ± 3.0	180	1974	85
F	17.8 ± 3.1	44	1974	85
M	14.9		1974	88
F	17.5		1974	88
M/F	16.8 ± 3.8	46 (25♂, 21♀)	1975	14
M	16.4 ± 2.2	40	1977	78
F	19.9 ± 3.9	60	1977	78
F	19.4 ± 0.61		1978	151
M/F	12.4 ± 2.0	35	1978	157
M/F	18.1 ± 2.5	12 (8♂, 4♀)	1978	156
M/F	16.2	17	1978	161
M (years)				
3—11	95% range 13.38 — 28	564	1978	84
12—19	10.08 — 23	482	1978	84
20—74	10.89—25.5	1961	1978	84
F				
3—11	13.54—27	537	1978	84
12—19	8.17—29	436	1978	84
20—74	10.01—34	2142	1978	84

Table 2
COPPER LEVELS IN SUBJECTS RECEIVING ESTROGEN-CONTAINING CONTRACEPTIVE THERAPY

Mean ± S.D. (μmol/ℓ)	No. of subjects	Year	Ref.
40.7	25	1966	124
35.1		1966	87
48.7 ± 11.4	10	1968	86
37.9	15	1969	125
36.0 ± 2.2		1974	88
30.4 ± 6.9	20	1977	78

Table 3
COPPER CONCENTRATION IN PREGNANT SUBJECTS

Stage of pregnancy	Mean ± S.D. (μmol/ℓ)	No. of subjects	Year	Ref.
Term	43.7 ± 8.0	31	1944	89
	36.3	175	1969	125
16 weeks	25.6 ± 1.0		1974	88
38 weeks	30.3 ± 1.0	20	1974	88

Table 4
COPPER LEVELS MISCELLANEOUS DISEASE STATES

Disease	Sex	Mean ± S.D. μmol/ℓ	No. of subjects	Year	Ref.
Wilson's disease	M/F	9.9 ± 0.6	36	1960	77
	M	9.4 ± 0.8	21	1960	77
	F	10.6 ± 0.8	15	1960	77
Rheumatoid arthritis	M/F	25.9 ± 5.9	52	1978	79
Osteoarthritis	M/F	19.8 ± 5.1	32	1978	79
Myocardial infarction	M/F	19.4 ± 3.9	16	1975	14
Anorexia nervosa	M/F	13.5 ± 2.8	24	1978	151
Ca. of dig. organs	M/F	19.5 ± 0.5	35	1978	157

In addition to these age-related copper variations, Liftschitz and Henken[91] have established a circadian rhythm for both copper and ceruloplasmin in ten healthy subjects. Concentrations for the constituents rose above the mean at 10:00 a.m. and 2:00 p.m., were at the mean at 6:00 p.m. and 10:00 p.m., and below the mean at 2:00 a.m. and 6:00 a.m. A similar, but less marked circadian rhythm was observed for urinary copper excretion.

Plasma levels are influenced by many pathological conditions and some of these will be described in the final section of this chapter.

VI. ERYTHROCYTE COPPER

The mean erythrocyte concentration of copper is 15.4 μmol/ℓ, a concentration which tends to remain constant despite fluctuations in total plasma or hepatic copper.

$$2\ O_2^- + 2H^+ \xrightarrow{S.D.} O_2 + H_2O_2$$

FIGURE 1. Dismutation of superoxide anions by superoxide dismutase (S.D.).

Erythrocyte copper can be classified into two distinct fractions; copper which can be dialysed is designated the labile pool and that which is protein-bound and thus not able to be dialysed is called the stable pool. The latter, which comprises 60% of the erythrocyte copper, has been most investigated and contains the first cuproprotein isolated, a blue-green protein, isolated from ox erythrocyte by Mann and Keilin[92] in 1938. This protein had a molecular weight of 35,000 and contained 0.38% copper. Several similar proteins were subsequently isolated from human erythrocytes, brain, and liver and were designated erythrocuprein, cerebrocuprein, and hepatocuprein, respectively. In 1969, work by Carrico and Deutsch demonstrated that these proteins were identical and suggested that the proteins be designated cytocuprein;[93] this proved inappropriate with the discovery that the protein contained two atoms of zinc as well as the two atoms of copper per molecule. Nomenclature problems were simplified in 1969 with the characterization of a common enzymatic property,[94] that of catalyzing the dismutation of superoxide anions. On the basis of this function, the group of isoenzymes has been redesignated superoxide dismutases (EC 1.15.1.1), their reaction being shown in Figure 1.

Superoxide dismutases may play a vital role in protecting the cell from the damaging effects of superoxide radicals,[95] particularly in their effect on cytochrome c and other enzyme systems. The protective nature of this enzyme has been exploited in the chemotherapy of acute inflammatory conditions in animals and is undergoing clinical trials in man.[96] These conditions are associated with endogenously generated superoxide anions.[97] Knowledge of the enzymic properties of the superoxide dismutase is of short duration, and it is not surprising that no specific pathology of copper deficiency is as yet attributed to it.

Severe deficiency of dietary copper in rats has been shown to reduce erythrocyte superoxide dismutase activity by 40% in 45 days and reduce the plasma ferroxidase activity of ceruloplasmin by 15% in 15 days.[98] In swine made copper deficient, heme synthesis correlates with the activity of cytochrome oxidase, and Williams et al.[99] suggest that an intact electron transport system is necessary to reduce Fe III for heme synthesis. It has recently been reported that the microcytic hypochromic anemia induced by lead toxicity in man may itself impair copper metabolism, which will further contribute to reduced synthesis of heme.[100]

The labile copper pool has not been fully characterized. Evidence suggests that it consists of copper complexed to amino acids, and it is in this form that facilitated diffusion of copper into erythrocytes is thought to occur.[101]

VII. TISSUE COPPER

The total body content of copper in the human adult is about 100 to 150 mg. The highest concentrations are found in the liver, which itself contains 10% of the total body's content. High extrahepatic tissue concentrations are found in the brain, heart, and kidney and lower concentrations are found in the pancreas, lungs, intestine, spleen, and aorta.[102] Muscle and bone, although of low copper concentrations, because of their large mass contain approximately 50% of the total body copper.

A. Hepatic Tissue Copper

The liver plays a key role in copper homeostasis. Radioactive tracer studies have indicated that it possesses at least three distinct functions: excretion of copper in the bile, temporary storage in the hepatocytes, and incorporation of copper into ceruloplasmin and its release into the plasma.[103] It has been repeatedly demonstrated that biliary excretion is the major pathway for the loss of body copper. The examination of bile secretions has indicated the presence of two binding species,[105] a low molecular weight component, predominantly in hepatic bile, and a macromolecular component, predominantly of gall bladder origin.

The low molecular weight component is composed of amino acids and peptides bound to copper and is likely to be one of the forms in which the metal is transported across the bile caniculi. Macromolecular copper species in bile are probably composed of nonspecific combinations of copper with bile proteins, as well as catabolized products of ceruloplasmin released from hepatic lysosomes.[106]

The enterohepatic circulation of copper is of minimal importance and is dependent on the degree of protein-binding, since the macromolecular copper complexes are not available for reabsorption. Work has now shown that bile flow influences the rate of copper excretion, a mechanism which among other factors is influenced by adrenal steroid secretions. The recent development of a noninvasive intestinal perfusion technique for the study of bile flow[107] may catalyze further work in this field.

The distribution of copper in the hepatocyte has been extensively studied.[108] Ten per cent is found in the microsomal fraction from which ceruloplasmin and other copper proteins are derived; twenty per cent lies in the nuclear fraction, binding to polynucleotides, 20% is present in the fraction composed of the mitochondria and lysosome, and the 50% remaining is present in the cytosol, a little of which is bound to enzymes, but the majority as three forms of metallothionein similar to the intestinal protein already described.[109]

Whanger and Weswig[110] have presented evidence that the small amount of copper in the microsomal fraction is that utilized in ceruloplasmin synthesis and in doing so is subject to antagonism by silver, cadmium, molybdenum, and zinc. A larger quantity is found bound to nucleic acids and several basic proteins. Evidence suggests that it may have a role in bridging polynucleotide strands.[111]

Electron probe analysis and electron microscopic examination have demonstrated, in copper overloaded livers, that the metal is sequestered within pericanicular, acid phosphatase-rich lysosomes prior to biliary excretion.[112] Porter has described a protein of molecular weight 5000 to 10,000 (neonatal hepatic mitochondrocuprein), which is found predominantly in neonatal liver.[113] He suggests that it functions as a copper reservoir in the synthesis of cytochrome oxidase during neonatal development, although its presence in Wilson's disease suggests that it may have a protective role in copper overload. Hepatic copper concentrations increase during intrauterine life, reaching a maximum at or before birth and thereafter decreasing to adult levels within 2 years of life. The subcellular distributions of copper during this period suggests that, having saturated cytosol binding sites, copper becomes bound to lysosomes and incorporated into proteins of the mitochondria and nucleus, a process which is reversed as the concentration of hepatic copper returns to adult levels.

Very considerable evidence indicates that copper has a stimulating action on the synthesis of ceruloplasmin. Neifakh et al.[114] have shown that copper salts can induce biosynthesis of ceruloplasmin in both human and monkey liver slices. In vivo experiments with rats[115] and studies in man, following copper treatment of children with Menkes' kinky hair syndrome and in copper poisoning, all show increases of ceruloplasmin following administration of copper. Pathological conditions associated with decreased biliary excretion and a subsequent increase of hepatic copper concentrations such as biliary atresia and biliary cirrhosis are also accompanied by raised serum ceruloplasmin levels. The high concentration of hepatic copper in the neonate is initially associated with low ceruloplasmin levels; as the hepatic copper decreases, a synchronous rise in serum ceruloplasmin concentration is observed, suggesting an increasing capacity for ceruloplasmin synthesis with maturity.

Holtzman and Gaumnitz[116] have provided evidence that copper is not the sole regulatory mechanism for ceruloplasmin in rats, in which the rate of synthesis of apo-ceruloplasmin in copper deficiency was found to be the same as that for holo-ceruloplasmin in normocupremic rats, the apo-ceruloplasmin being, however, considerably less stable.

It has been suggested by Evans[108] that ceruloplasmin synthesis is regulated by copper combining with the ceruloplasmin regulatory gene, promoting increased synthesis of ceruloplasmin messenger RNA. The antagonism by cadmium, silver, and zinc to copper metabolism within the hepatocytes has been shown by Whanger and Weswig[110] and Evans et al.,[117] in experiments with bovine liver, to be partially due to the competition of these metals with copper for binding to the storage protein metallothionein in a similar way to their antagonism in absorption at the intestinal mucosa. Alternative suggestions are that the metals prevent copper inducing the apo-ceruloplasmin or that they are incorporated into ceruloplasmin in place of copper. Observations of the deleterious effects of molybdenum on metabolic utilization of copper have been restricted to animals.

B. Extrahepatic Tissue Copper

Most of the tissue copper is present as the proteins already described, particularly superoxide dismutase and metallothionein. It is also found as stable complexes and chelates of organic molecules, such as amino acids, purines, pyrimidines, nucleotides, DNA, RNA, and nonspecific proteins. The characterization of enzymes has revealed copper to play an important role in many enzyme systems,[32] either as an activator or inhibitor, as a prosthetic group or an integral part of the enzyme molecule.

The ferroxidase activity of ceruloplasmin and the activity of superoxide dismutase have already been described.

Cytochrome c oxidase (EC 1.9.3.1), the terminal oxidase in the mitochondrial electron transport chain and dimer of combined cytochrome a and a_3, requires one atom of copper per molecule for its catalytic activity.[118] This enzyme plays a key role in oxidative phosphorylation by oxidizing reduced cytochrome[28] c with molecular oxygen. Studies of the changes in paramagnetism during this process have suggested that reduced cupric radicals play an important part.[119] Severe induced copper deficiency is accompanied by a decreased capacity for oxidative phosphorylation.[120] It has been suggested that a similar effect occurs after tissue damage through irradiation, when organic peroxides maintain the copper atom of the enzyme in an oxidized state and thus unable to react with molecular oxygen. No specific pathology has been associated with this enzyme in copper deficiency states in man.

Monoamine oxidases (EC 1.4.3.4) are known copper-dependent enzymes, and that purified from bovine and porcine plasma has a molecular weight of 195,000 and contains four atoms of copper per molecule. These enzymes catalyze the oxidative deamination of a variety of monoamines forming the corresponding aldehyde, ammonia, and hydrogen peroxide.[121] Lysyl oxidase described by O'Dell[32] catalyzes the oxidative deamination of lysine in the peptide chain of elastin to form peptidyl α-amine adipic acid-semialdehyde, which is the precurser for cross-linking in this compound. This is shown in Figure 2A. A deficiency of copper results in reduced activity of this enzyme

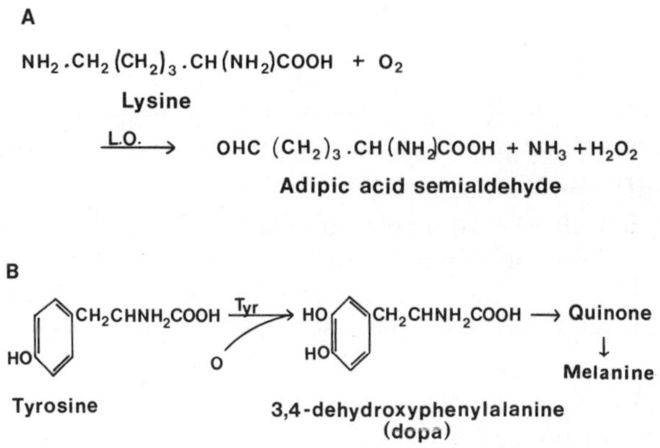

A

$NH_2.CH_2(CH_2)_3.CH(NH_2)COOH + O_2$

Lysine

$\xrightarrow{L.O.}$ $OHC(CH_2)_3.CH(NH_2)COOH + NH_3 + H_2O_2$

Adipic acid semialdehyde

B

HO–[ring]–CH_2CHNH_2COOH $\xrightarrow{Tyr}$ (O) HO, HO–[ring]–CH_2CHNH_2COOH → Quinone → Melanine

Tyrosine

3,4-dehydroxyphenylalanine (dopa)

FIGURE 2. (A) Oxidation of lysine by lysyl oxidase (L.O.) (B) Hydroxylation of tyrosine by tyrosinase (Tyr).

and a subsequent loss of cross-linking in elastin[122] with reduced elasticity and tensile strength of connective tissue protein. The clinical results of this deficiency in pigs and chicks are vascular fragility with concomitant ruptures and aneurysms.

Tyrosinase is another copper-containing enzyme which is responsible for hydroxylation of tyrosine to produce 3,4-dihydro-phenylalanine (DOPA) and further oxidation to produce quinone, which then gives rise to melanine (Figure 2B). The observed depigmentation of wool during copper depletion in sheep strongly suggests that reduced activity of the enzyme tyrosinase to be responsible.

Uricase, laccase, dopamine hydroxylase, and galactoseoxidase are further enzymes in which copper has been shown to be an essential constituent.

VIII. HORMONAL INFLUENCES ON COPPER HOMEOSTASIS

Hormonal influences on copper homeostasis are thought to be mediated largely by changes in copper protein synthesis. Krebs'[123] early observations of high copper levels during pregnancy have been confirmed and are explained in terms of the induction of synthesis of ceruloplasmin by the raised estrogen concentrations in pregnancy. The high levels of copper and ceruloplasmin found in patients taking oral contraceptives containing estrogen[124] is further evidence for this mechanism, as are the increases in copper seen in men receiving estrogen therapy. Raised copper levels during pregnancy[125] have been invoked as a useful diagnostic parameter. The serum concentration of copper normally increases threefold from the third month of pregnancy to term. Therefore, a fall in concentration may indicate placental insufficiency, premature rupture of the membrane and spontaneous abortion, and O'Leary et al.[126] have suggested it as an indicator of fetal death.

The physiological role for the raised copper and ceruloplasmin is not clear, unless increased mobilization of iron for the fetus by ferroxidase activity is considered a worthy contender. Changes in copper during the menstrual cycle have not been found, yet they are seen in estrus in the rat. Testosterone and progesterone administered in rats and humans have also been shown to increase copper[127] and probably, therefore, ceruloplasmin. LH and FSH administered to female rats did not change the plasma copper level, although when copper itself was given intravenously or intracerebrally to rabbits, ovulation was produced and in vitro experiments show that copper incubated with bovine pituitary increases release of GH, LH, TSH, and ACTH.[127] The role of the thyroid gland in copper metabolism remains an enigma. Elevated levels of copper after thyroxine injection and in hyperthyroid patients observed by some workers[89] have been challenged by others[128] who, as well as finding decreased levels in these situations, have also found increased levels after thyroidectomy in rats.

Repeated injections of ACTH have shown a decrease in copper and ceruloplasmin levels in rats.[128] Adrenalectomy or adrenal cortical insufficiency from several causes, including idiopathic Addisons disease and hypopituitarism, is associated with an elevated plasma concentration of copper and reduced urinary excretion. Conversely, elevated corticosteroids occurring in Cushings syndrome and in adrenal cortical carcinomas are associated with decreased plasma levels and increased urinary excretion. Tracer doses of ^{64}Cu in rats with cortisone administration reveal significantly lower concentrations of the isotope in the liver and increased biliary excretion[129] which coincides with an increased bile flow. This evidence suggests that corticosteroids have a regulatory action on copper homeostasis mediated by its choleretic action. Study of the nyctahemoral variations in plasma concentrations has established an inverse relationship to that established for ACTH,[130] a relationship consistent with work described earlier. Nyctahemoral variations observed in urinary copper excretion[91] are more difficult to relate to the activity of the pituitary adrenal axis.

A deficiency of copper in cats has indicated the copper dependency of the enzyme Δ^5-3 ketosteroid isomerase, which converts pregnerolone to progesterone

$$\text{Pregnenolone} \xrightarrow{\Delta^5-3 \text{ Ketosteroid Isomerase}} \text{Progesterone}$$

Copper-deficient cats showed reduced formation of ^3H labeled progesterone, corticosterone, and cortisol from ^3H cholesterol when compared to normals, and additional evidence comes from the observed adrenal cortical hyperplasia and raised ACTH in these animals.

The raised levels of growth hormone occurring in acromegaly are concomitant with raised copper concentrations; hyperphysectomy reduces these levels slightly. The profound effect of growth hormones on protein synthesis is thought to exert its marked effects on copper homeostasis through its action on ceruloplasmin synthesis, although little work has been done to confirm this.

Copper has been shown to play a role in catecholamine metabolism and hence in adrenal medullary function. The oxidase property of ceruloplasmin previously discussed may have a regulatory action upon catecholamine concentration and the copper-dependent enzyme dopamine-β-hydroxylase.

IX. COPPER HOMEOSTASIS IN DISEASE

Disturbances of copper metabolism have been observed in many pathological conditions, although only in two of these is it a cardinal feature. Wilson's disease (hepatolenticular degeneration) and Menkes' kinky hair syndrome are two genetically determined conditions with which copper disturbances are invariably associated.

A. Wilson's Disease

Wilson's disease is a recessive inborn error of metabolism with a prevalence of 1:200,000. It was first described in 1912 by Wilson[6] and is characterized by neurological symptoms, cirrhosis of the liver, and the classical Kayser-Fleischer rings (green-brown deposits) due to copper deposition in the cornea. Characteristically, the disease manifests itself between the ages of 6 and 20 years, although presentation between 30 and 40 years of age occurs rarely. The biochemical abnormalities in this condition include excessive copper accumulation in the liver, brain, kidney, and cornea and in the author's experience, almost invariably a low serum ceruloplasmin with an elevated nonceruloplasmin copper. The biliary and hence fecal excretion of copper is very high. A plethora of theories have been invoked to relate the metabolic and clinical manifestations to each other and to the genetic defect in Wilson's disease. Sass-Kortsak[131] and later, Evans[108] have suggested that the primary lesion is best studied in asymptomatic homozygous patients to enable secondary effects of copper toxicity to be delineated. In the asymptomatic patients, the only biochemical abnormalities observed are an accumulation of hepatic copper, impaired biliary excretion of copper, and, to a lesser extent, reduced ceruloplasmin synthesis.

With progression of the disease, the liver and other tissues become overloaded with nonceruloplasmin-bound copper. Uzman et al.[132] have suggested that the primary defect lies prior to biliary excretion and incorporation of copper into ceruloplasmin and suggests the existence of an abnormal hepatic protein with a high avidity for copper. Evans et al.[133] obtained evidence for an abnormal protein in Wilson's disease, which possesses a greater avidity for copper than metallothionein. Staining techniques, used by Goldfischer and Sternlieb,[134] show a more diffuse distribution of copper in the hepatocytes among some Wilson's patients, and they suggest that a cytoplasmic sulfhydryl-rich protein with a high affinity for copper may exist in Wilson's patients. Further work is required to confirm these findings. This hypothesis would account for the accumulation of copper in tissue, either by the greater affinity for the metal due to this abnormal protein or due to the decreased uptake of the metal by the liver and the concomitant elevation of nonceruloplasmin copper and deposition in the tissues. The cellular damage and hemolysis found in the condition is likely to be due to direct toxic effects of copper. Lindquist[135] has demonstrated that excess copper can initiate lipid peroxidation of lysosomal membranes with leakage of protolytic enzymes. The observed amino aciduria, phosphaturia, and uricosuria in Wilson's disease are probably due to the detrimental action of the lysosomal acid hydrolases released in the renal tubules.[108] More recent work has shown the amino acid excretion to consist mainly of hydroxyproline, a degradation product of collagen.[136]

The relentless inflammation, fatty degeneration, and excessive fibrous changes progress towards postnecrotic cirrhosis during which there may be sudden releases of copper which give rise to severe hemolytic crisis.

The disease has been controlled using a number of chelating agents.[137] BAL (2,3-dimercaptopropanol) was first used and now penicillamine (β-dimethyl cysteine) and triethylene tetramine are used. Penicillamine has numerous side effects, one of which is a loss of taste, commonly accepted to be a consequence of the chelation and loss of copper from the tissue, although some recent work has disputed this.[138] It has been found useful to limit the amount of copper present in the diet and to add an amount of potassium sulfide in order to diminish copper absorption. Early diagnosis and persistent vigorous therapy have reduced, and in some cases halted, the progress of the disease.

B. Menkes' Kinky Hair Syndrome

Menkes' kinky hair syndrome is an X-linked recessive inborn error of metabolism, characterized by slow growth, pili torti (kinky hair), vascular and bone changes, hypothermia, and progressive cerebral degeneration leading ultimately to an early death. The syndrome was observed by Menkes et al.[7] in 1962 and later the abnormality was recognized by Danks et al.[139] to be similar to that manifest by copper-deficient animals. Subsequent examination has demonstrated that these patients are indeed copper deficient, with low hepatic, cerebral, and serum concentrations, depressed copper oxidase activity, and nervous tissue devoid of cytochrome oxidase activity.

The duodenal mucosa from patients with this disease is overloaded with copper and a transport defect appears to be at fault. Recent work with cultured skin fibroblasts from patients with Menkes' syndrome has shown an increased uptake of ^{64}Cu into the cell, with a reduced release compared with that of normal cells.[52] These same workers have shown that much of the copper is combined with a protein of low molecular weight (10^4) and the rest with a protein of molecular weight 10^5. Both of these proteins have a higher copper content than that encountered in normal cells. Other workers,[140] also using skin fibroblasts have shown that the metallothionein in these cells has an abnormal retention of cadmium. This evidence points towards a defect in metallothionein, its metal cleavage, its binding properties, or its concentration in Menkes' syndrome. Clinical improvement is only possible prior to the early irreversible secondary changes and occurs following intravenous administration of copper; the latter has indicated a primary defect of impaired intestinal absorption.

The syndrome indicates the essential nature of copper in human metabolism and suggests physiological roles for some of the copper-dependent enzymes. Surprisingly, neutropenia and anemia are not associated with the condition, an observation which has prompted recent investigations into erythrocyte copper concentration with early reports suggesting this to be normal, and perhaps even high. This may indicate the presence in the erythrocytes of a similar abnormal protein to that which appears to occur in the intestinal mucosa. Further work is required to reconcile this anomaly and fully characterize the lesion.

C. Miscellaneous Diseases

Elevated serum copper has been reported in acute, but not chronic, schizophrenia, in cerebral vascular accident, and anxiety states.[141] Other evidence from studies with lambs suggested that copper, or its metalloprotein, may play critical roles in cerebral functions mediated by its effect on neurotransmitters.[142] Recent in vitro experiments have demonstrated that copper stimulates the rate of autoxidation of the catecholamine neurotoxin 6 hydroxydopamine; they suggest that intraneural copper may have some pathological importance in copper overload (Wilson's disease).[143] The hypnotic drug LSD has been demonstrated to depress catecholamines and increase 5-hydroxytryptamine in the brain. Furthermore, this drug inhibits oxidation of 5-hydroxytryptamine and enhances that of dopamine and noradrenaline, an effect perhaps mediated by the interaction of LSD with ceruloplasmin.[144] Evans[108] has advanced the hypothesis that in acute schizophrenia, stress, with its increased adrenaline release, can account for the elevation in copper. Others suggest that the poor nutritional state of these patients alters copper balance, perhaps through low ascorbic acid concentrations.[144] Neither of these theories has been confirmed or refuted.

Bogden et al.[12] demonstrated abnormally high levels of copper in plasma and cerebral spinal fluid following cerebral infarction as compared with those found following subarachnoid haemorrhage; they suggest the potential diagnostic use for copper determination since these two conditions are difficult to distinguish clinically.

Excessive exercise in rats is reported to increase ceruloplasmin, an observation that is without explanation,[145] unless Evans' stress-mediated response is applicable. Serum and hepatic copper are classically elevated in primary biliary cirrhosis and biliary atresia, two conditions in which copper excretion by the bile is reduced and increases in serum ceruloplasmin and urinary copper excretion are observed. Elevations of the metal in oral contraceptive therapy and during pregnancy have been described earlier and have a hormonal basis. The use of copper-containing intrauterine devices has not been found to increase plasma copper concentrations. LH, FSH, estrogen, and progesterone are not affected by its presence in the uterus in women.[146] The presence of copper in these devices has drawn attention to the effectiveness of metallic copper as a spermicide and to some, as yet unresolved, mechanism which inhibits implantation of the zygote in the uterus; this subject is comprehensively reviewed by Oster and Salgo.[30]

The first evidence that copper might be implicated in cardiac lesions came from a study of copper deficient cattle in Western Australia.[147] The cattle developed a "falling disease" which was believed to be due to acute heart failure, precipitated by mild exercise; autopsy revealed that normal myocardial tissue had been replaced by collagenous tissue. The role that copper-dependent enzymes play in promoting cross-linking of elastin by desmosine and isodesmosine has been described earlier; reduced cross-linking can give rise to aortic rupture in animals and may be the reason for the vascular lesions in Menkes' kinky hair syndrome.

Klevay and Hyg[148] have found evidence that an imbalance of zinc and copper contributes to a risk of coronary heart disease. Rats given a diet with a high ratio of zinc to copper (40:1) produced higher serum cholesterol concentrations than did those receiving a low ratio diet (5:1). In the U.S., human mortality from coronary heart disease has been studied in 47 cities and was shown to correlate significantly ($P < 0.002$) with the zinc:copper ratio of milk used in those cities.[42] Further supportive evidence comes from a study of 109 young people; those who had been fed breast milk, which has a lower zinc:copper ratio than cows' milk, showed significantly ($P < 0.005$) larger arterial luminal diameters than those fed on cows' milk. Contradictory evidence to these observations was found by Walravens and Hambridge,[149] who demonstrated no significant difference in plasma cholesterol concentrations of two groups of infants fed a milk formula with zinc:copper ratios of 5:1 and 17:1. Recent work on copper-deficient rats has demonstrated an increased rate of synthesis of cholesterol, resulting in highly significant increases in plasma cholesterol concentrations.[150] They suggest that copper may influence the rate of cholesterol turnover and clearance from the liver.

A group of conditions remain, in which elevated serum copper and ceruloplasmin have been observed and in some cases reported to be of diagnostic value.

Slightly reduced levels of copper and ceruloplasmin have been shown to occur in anorexia nervosa.[151] This is not, perhaps, surprising, although the same study demonstrated an elevated concentration of copper in hair, an observation which is without explanation. A recent study of obese patients prior to treatment by intestinal bypass surgery[152] showed serum copper levels significantly reduced following surgery. One patient with a plasma copper of 90 $\mu g/100 m\ell$ developed leukopenia, which subsequently responded to replacement therapy using oral copper sulfate.

Plasma copper was shown by Bogden and Troiano[153] to be significantly raised in alcoholics during hospital-based alcohol withdrawal; they also demonstrated that the patients, who developed delerium tremens or prolonged hallucinary states, showed the highest serum copper concentrations. It was concluded that copper has a value in indicating those patients susceptible to these major side effects of this therapy.

A number of trace elements including copper have recently been implicated in multiple sclerosis,[154] from a study in which 20 elements were determined in the hair of 40 multiple sclerosis patients and 42 controls, copper showed a highly significantly lower level among the multiple sclerosis group. Hodgkin's disease is a condition in which consistent elevations of copper are observed. A retrospective study by Thorling and Thorling[10] of 241 patients suffering this condition has shown a significant correlation between the degree of elevation and the histological grading of the tumor. These and other works suggest that normal serum copper concentration is a criterion for complete remission. Many other neoplastic conditions are reported to be accompanied by raised copper levels, including leukemia, osteosarcoma,[155] and breast cancer, and again in these it has been suggested that copper can be a useful indicator of remission.

A recent study of osteogenic sarcoma among 18 patients concluded that although serum copper was significantly increased, its nonspecific nature made it of little value as a marker of tumor activity.[156] Another recent study of 126 patients with malignant tumors of the digestive tract showed the presence of elevated serum copper concentrations compared to a normal population.[157] This elevation was more marked among patients with advanced metastatic carcinoma, particularly those with liver involvement, and it was concluded that a copper:zinc ratio may be of some prognostic value in these patients.

Postoperative infection,[158] salmonella infection,[159] myocardial infarction,[14] rheumatoid arthritis,[78] burns and surgical trauma are other states in which disturbed copper homeostasis has been repeatedly observed.

In all of these conditions, increases in serum copper concentrations are mirrored by those of ceruloplasmin. Closer inspection reveals that a number of other proteins are also present in raised concentrations. They include c-reactive protein, haptoglobin, α_1 acid glycoprotein, α_1 antitrypsin, and fibrinogen. This group of proteins are variously described as acute phase proteins and acute reacting proteins; they are altered in a host of nonspecific infections and inflammatory conditions[104] and the changing concentration of one of them, fibrinogen, is responsible for the popular, nonspecific diagnostic test, erythrocyte sedimentation rate.

It can be inferred that the raised concentration of copper observed in many pathological conditions occurs as a consequence of the tissue injury and inflammatory processes involved, in which some, as yet ill-defined humoral mediator(s) (sometimes called leukocyte endogenous mediators [LEM])[161] stimulate nonspecific protein synthesis in the liver with the observed elevations of serum ceruloplasmin and thus copper.

Support for this inference has been recently obtained in rheumatoid arthritis, when treatment of the condition with anti-inflammatory drug therapy produced a gradual reduction in serum copper which was mirrored by changes to the acute phase protein haptoglobin and α_1 acid glycoprotein.[79] Brown et al.[161] have shown that the reduction in serum copper brought about by anti-inflammatory drugs varies with the particular chemotherapy adopted, an indication perhaps of the anti-inflammatory properties of individual drugs.

Copper has itself been implicated in the chemotherapy of inflammatory diseases,[162] such as rheumatoid arthritis, in which it has been demonstrated that copper acetate has anti-inflammatory activity. Copper chelates of a variety of drugs, including aspirin and penicillamine, are reported to possess up to 20 times the anti-inflammatory activity of the basic drug although this has been disputed.[163] It has been postulated that the anti-inflammatory activity is mediated through copper chelates of the drugs administered.

Further chemotherapeutic effects of copper have been reported for the prevention of cardiac arrhythmia induced by prostaglandins,[164] and it was concluded that copper blocks this effect of the prostaglandins, an effect which may be significant in inflammatory conditions where prostaglandin E_2 is considered a major mediator of the inflammatory response. The most spectacular use for copper chemotherapy in recent literature was that reported for the treatment of secondary amenorrhoea in which 47 patients underwent 25 days of electrophoretic treatment using an anode soaked in copper sulfate and placed on the sacral region.[17] Menstruation occurred in 34 of the women and 4 became pregnant!

REFERENCES

1. **Hart, E. B., Steenbock, H., Waddell, J., and Elvehjem, C. A.,** Iron in nutrition. VII. Copper as a supplement to iron for haemoglobin building in the rat, *J. Biol. Chem.,* 77, 797, 1928.
2. **Curzon, G. and O'Reilly, S.,** A coupled iron-caeruloplasmin oxidation system, *Biochem. Biophys. Res. Commun.,* 2, 284, 1960.
3. **Karpel, J. T. and Peden, V. H.,** Copper deficiency in long-term parenteral nutrition, *J. Pediatr.,* 80, 32, 1972.
4. **Rooney, P. J., Walkins, C., Ahola, S. J., Gray, G. I. L., and Carson, D. W.,** A short-term double-blind controlled trial of prenazone in rheumatoid arthritis, *Curr. Med. Res. Opin.,* 2, 43, 1973.
5. **Glazebrook, A. J.,** Wilson's disease, *Edinburgh Med. J.,* 52, 83, 1945.
6. **Wilson, S. A. K.,** Progressive lenticular degeneration: a familiar nervous disease associated with cirrhosis of the liver, *Brain,* 34, 295, 1912.
7. **Menkes, J. H., Alter, M., Steigleder, G. K., Weakley, D. R., and Sung, J. H.,** A sex-linked recessive disorder with retardation of growth, peculiar hair and focal cerebral and cerebellar degeneration, *Pediatrics,* 29, 764, 1962.
8. **O'Leary, J. A.,** Serum copper levels as a measure of placental function, *Am. J. Obstet. Gynecol.,* 105, 636, 1969.
9. **Schenker, J. G., Jungrois, E., and Pocishuk, W. Z.,** Serum copper levels in normal and pathologic pregnancies, *Am. J. Obstet. Gynecol.,* 105, 933, 1969.
10. **Thorling, E. B. and Thorling, K.,** The clinical usefulness of serum copper determinations in Hodgkin's Disease, *Cancer,* 38, 225, 1976.
11. **Sadamitsu, I. and Sadao, A.,** Plasma copper and zinc levels in patients with malignant tumours of digestive organs, *Cancer,* 42, 626, 1978.
12. **Bogden, J. D., Troiano, R. A., and Joselow, M. M.,** Copper, zinc, magnesium and copper in plasma and cerebrospinal fluid of patients with neurological diseases, *Clin. Chem.,* 23, 485, 1977.
13. **Bogden, J. D. and Raymond, A. T.,** Plasma, calcium, copper, magnesium and zinc concentrations in patients with the alcohol withdrawal syndrome, *Clin. Chem.,* 24, 1553, 1978.
14. **Versieck, J., Barbier, F., Speecke, A., and Hosie, J.,** Influence of myocardial infarction on serum manganese, copper and zinc concentrations, *Clin. Chem.,* 21, 578, 1975.
15. **Underwood, E. J.,** Copper requirements, in *Trace Elements in Human and Animal Nutrition,* 4th ed., Academic Press, New York, 1977, 87.
16. **Delves, H. T.,** The clinical value of trace-metal measurements, in *Essays in Medical Biochemistry,* Vol. 2, Marks, V., Hales, C. N., Eds., The Biochemical Society and The Association of Clinical Biochemists, London, 1976, 37.
17. **Dickova, K., Ninov, V., Toshev, G., and Hristova, N.,** An attempt for treatment of secondary amenorrhoea with copper salts by means of electrophoresis, *Akush. Ginekal. (Soffia),* 17(Part 3), 213, 1978.
18. **Henry, R.,** Copper, in *Clinical Chemistry Principles and Techniques,* 2nd ed., Henry, R., Cannon, D. C., and Winkelman, J. W., Eds., Harper & Row, London, 1974, 695.
19. **Williams, H. L., Johnson, D. J., and Haut, M. J.,** Simultaneous spectrophotometry of Fe^{2+} & Cu^{2+} in serum denatured with guanidine hydrochloride, *Clin. Chem.,* 23, 237, 1977.
20. **Walshe, A.,** The application of atomic absorption spectra to chemical analysis, *Spectrochim. Acta,* 7, 108, 1955.
21. **Ward, A. F., Mitchell, D. G., Kahl, M., and Aldons, K. M.,** Determination of copper in plasma and serum by use of a micro-sampling cup in atomic absorption spectroscopy, *Clin. Chem.,* 20, 1199, 1974.
22. **L'vov, B. V.,** The analytical use of atomic absorption spectra, *Spectrochim. Acta,* 17, 1140, 1961.
23. **Evenson, M. A. and Warren, B. L.,** Determination of serum copper by atomic absorption with the use of the graphite cuvette, *Clin. Chem.,* 21, 619, 1975.
24. **Evenson, M. A. and Anderson, C. A.,** Ultramicro analysis for copper, cadmium and zinc in human liver tissue by use of atomic absorption spectrophotometry and the heated graphite tube atomizer, *Clin. Chem.,* 21, 537, 1975.
25. **Delves, H. T.,** Trace-metal analysis of protein fractions by atomic absorption spectrophotometry, in 3rd Int. Conf. Atomic Spectroscopy, Toronto, 1973.
26. **Kiniseley, R., Fassel, V. A., and Butler, C. C.,** Applications of inductively coupled plasma-excitation sources to the determination of trace metals in biological fluids, *Clin. Chem.,* 19, 807, 1973.
27. **Anand, V. D., White, J. M., and Hipolito, V. N.,** Some aspects of specimen collection and stability in trace element analysis of body fluids, *Clin. Chem.,* 21, 595, 1972.
28. **Thiers, R. E.,** Contamination in trace element analysis and its control, in *Methods of Biochemical Analysis,* Vol. 5, Glick, Ed., Interscience, New York, 1957, 273.

29. **Scheinberg, I. H.**, The effects of heredity and environment on copper metabolism, *Med. Clin. North Am.*, 60, 705, 1974.
30. **Oster, G. and Salgo, M. P.**, Copper in mammalian reproduction, *Adv. Pharmacol. Chemother.*, 14, 327, 1977.
31. **Lahey, M. E. and Schubert, W. G.**, New deficiency syndrome occuring in infancy, *Am. J. Dis. Child.*, 93, 31, 1957.
32. **O'Dell, B. L.**, Biochemistry of copper, *Med. Clin. North Am.*, 60, 687, 1976.
33. **Friis Hansen, B.**, Body composition during growth: in vivo measurements and biochemical data correlated to differential anatomical growth, *Pediatrics*, 47, 264, 1971.
34. **Holliday, M. A.**, Letter to the editor, *Pediatrics*, 54, 524, 1974.
35. **Casper, R. C., Kirschner, B., and Jacob, R. A.**, Zinc and copper status in anorexia nervosa, *Psychopharmacol. Bull.*, 14, 53, 1978.
36. **Sjollema, B.**, Kupfermangel als un sache von krankheiten bei pflanzen und tieren, *Biochem. Z.*, 267, 151, 1933.
37. **Neal, W. M., Becker, R. B., and Shealy, A. L.**, *Science*, 74, 418, 1931.
38. **Stogdale, L.**, Chronic copper poisoning in dairy cows, *Aust. Vet. J.*, 54, 139, 1978.
39. **Chuttani, H. K., Gupta, P. S., Gulati, S., and Gupta, D. N.**, Acute copper sulfate poisoning, *Am. J. Med.*, 39, 849, 1965.
40. **Peters, R. and Walshe, J. M.**, Studies on the toxicity of copper I. The toxic action of copper in-vitro and inv-vivo, *Proc. R. Soc. London Ser. B*, 166, 273, 1966.
41. **Boulard, M., Blume, K. G., and Beutler, E.**, The effect of copper on red cell enzyme activity, *J. Clin. Invest.*, 51, 459, 1972.
42. **Underwood, E. J.**, Copper in Human health and nutrition, in *Trace Elements in Human and Animal Nutrition*, 4th ed., Academic Press, New York, 1977, 91.
43. **Guthrie, B. and Robinson, M. F.**, Daily intakes on manganese, copper zinc and cadmium by New Zealand women, *Br. J. Nutr.*, 38, 55, 1977.
44. National Research Council: Food and Nutrition Board, Recommended Dietary Allowances, 8th ed., National Academy of Science, Washington, D.C., 1974.
45. World Health Organization Expert Committee on Trace Elements in Human Nature, Trace elements in human nature, *Tech. Report Series*, No. 532, 15, 1973.
46. **Chapman, H. L. and Bell, M. C.**, Relative absorption and excretion by beef cattle of copper from various sources, *J. Anim. Sci.*, 22, 82, 1963.
47. **Davies, N. T. and Cambell, J. K.**, The effects of cadmium on intestinal copper absorption and binding in the rat, *Life Sci.*, 20, 955, 1977.
48. **Pulido, P., Kägi, J. H. R., and Vallee, B. L.**, Isolation of some properties of human metallothionein, *Biochemistry*, 5, 1768, 1966.
49. **Bush, J. A., Mahoney, J. P., Markowitz, H., Gubler, C. J., Cartwright, G. E., and Wintrobe, M. M.**, Studies on copper metabolism. XVI. Radioactive copper studies in normal subjects and in patients with hepatolenticular degeneration, *J. Clin. Invest.*, 34, 1766, 1955.
50. **Weber, P. M., O'Reilly, S., Pollycove, M., and Shipley, L.**, Gastrointestinal absorption of copper: studies with ^{64}Cu & ^{95}Zo, a whole body counter and the scintillation camera, *J. Nucl. Med.*, 10, 591, 1969.
51. **Crampton, R. F., Matthews, D. M., and Poisner, R.**, Observations on the mechanism of absorption of copper by the small intestine, *J. Physiol. (London)*, 178, 111, 1965.
52. **Beratis, N. G., Price, P., Labadie, G., and Hirschhorn, K.**, ^{64}Cu metabolism in Menkes' and normal cultured skin erythroblasts, *Pediatr. Res.*, 12, 699, 1978.
53. **Garnica, A. D., Chan, W. Y., and Rennert, O. M.**, Role of metallothionein in copper transport in patients with Menkes' syndrome, *Ann. Clin. Lab. Sci.*, 8, 302, 1978.
54. **Bradshaw, R. A., Shearer, W. T., and Gurd, F. R. N.**, Sites of binding of copper(II) ion by peptide (1-24) of bovin serum albumin, *J. Biol. Chem.*, 243, 3817, 1968.
55. **Lau, S., Kruck, T. P. A., and Sarkar, B.**, A peptide molecule mimicking the copper(II) transport site of human serum albumin, *J. Biol. Chem.*, 249, 5878, 1974.
56. **Appleton, D. W. and Sarkar, B.**, The absence of specific copper(II)-binding sites in dog albumin, *J. Biol. Chem.*, 246, 5040, 1971.
57. **Holmberg, C. G. and Laurell, C. B.**, Investigations in serum copper II. Isolation of the copper containing protein and a description of some of its properties, *Acta Chem. Scand.*, 2, 550, 1948.
58. **Magdoff-Fairchild, B., Lovell, F. M., and Low, B. W.**, An X-ray crystallographic study of caeruloplasmin, *J. Biol. Chem.*, 244, 3497, 1969.
59. **Ryden, L.**, Single chain structure of human caeruloplasmin, *Eur. J. Biochem.*, 26, 380, 1972.
60. **Fee, J. A.**, Copper protein systems containing the "Blue" copper centre, *Struct. Bonding*, 23, 1, 1975.

61. **Veldsema, A. and Van Gelder, B. F.**, The ratio of type-1 and type-2 Cu(II) in human caeruloplasmin, *Biochim. Biophys. Acta,* 293, 322, 1973.
62. **Marriott, J. and Perkins, D. J.**, Relationships between the copper atoms of caeruloplasmins.* Studies of the exchange of ^{64}Cu with caeruloplasmin, *Biochim. Biophys. Acta,* 117, 387, 1966.
63. **Carrico, R. J., Malmström, B. G., and Vänngord, T.**, A study of the reduction and oxidation of human caeruloplasmin. Evidence that a diamagnetic chromophore in the enzyme participates in the oxidase mechanism, *Eur. J. Biochem.,* 22, 127, 1971.
64. **Owen, C. A.**, Metabolism of radio copper(Cu^{64}) in the rat, *Am. J. Physiol.,* 209, 900, 1965.
65. **Hsieh, H. S. and Frieden, E.**, Evidence for caeruloplasmin as a copper transport protein, *Biochem. Biophys. Res. Commun.,* 67, 1326, 1975.
66. **Sternlieb, I., Morell, A. G., Tucker, W. D., Greene, M. W., and Scheinberg, I. H.**, The incorporation of copper into caeruloplasmin in-vivo. Studies with copper^{64} and copper,^{67} *J. Clin. Invest.,* 40, 1834, 1961.
67. **Matsuda, I., Pearson, T., and Holtzman, N. A.**, Determination of apocaeruloplasmin by radioimmunoassay in nutritional copper deficiency, Menkes' kinky hair syndrome, Wilson's disease and umbilical cord blood, *Pediatr. Res.,* 8, 821, 1974.
68. **Osaki, S., McDermott, J. A., and Frieden, E.**, Proof for the ascorbate oxidase activity of caeruloplasmin, *J. Biol. Chem.,* 239, 3570, 1964.
69. **Rice, E. W.**, Correlation between serum copper, caeruloplasmin activity and c-reactive protein, *Clin. Chim. Acta,* 5, 632, 1960.
70. **Roeser, H. P., Lee, G. R., Nacht, S., and Cartwright, G. E.**, The role of caeruloplasmin in iron metabolism, *J. Clin. Invest.,* 49, 2408, 1970.
71. **Osaki, S., Johnson, D. A., and Frieden, E.**, The possible significance of the ferrous oxidase activity of caeruloplasmin in normal human serum, *J. Biol. Chem.,* 241, 2746, 1966.
72. **Ragen, H. A., Nocht, S., Lee, G. R., Bishop, C. R., and Cartwright, G. E.**, Effect of caeruloplasmin on plasma iron in copper-deficient swine, *Am. J. Physiol.,* 217, 1320, 1969.
73. **Gutteridge, J. M. C.**, Caeruloplasmin: a plasma protein, enzymes and antiocidant, *Ann. Clin. Biochem.,* 15, 293, 1978.
74. **Topham, R. W. and Frieden, E.**, Identification and purification of a non-caeruloplasmin ferroxidase of human serum, *J. Biol. Chem.,* 245, 6698, 1970.
75. **Al-Timimi, D. J. and Dormandy, T. L.**, The inhibition of lipid autoxidation by human caeruloplasmin, *Biochem. J.,* 168, 283, 1977.
76. **Stocks, J., Gutteridge, J. M. C., Sharp, R. J., and Dormandy, T. L.**, Assay using brain homogenate for measuring the antioxidant activity of biological fluids, *Clin. Sci. Mol. Med.,* 47, 215, 1974.
77. **Cartwright, G. E., Markowitz, H., Shields, G. S., and Wintrobe, M. M.**, Studies on copper metabolism. XXIX. A critical analysis of serum copper and caeruloplasmin concentrations in normal subjects, patients with Wilson's disease and relatives of patients with Wilson's disease, *Ann. J. Med.,* 28, 555, 1960.
78. **Scudder, P. R.**, Copper Metabolism in Rheumatoid Arthritis and Related Disorder, Master's thesis, University of London, 1977.
79. **Halloran, S. P**, Aspects of Copper Homeostasis in Rheumatoid Arthritis and Related Disorders, Master's thesis, University of Surrey, 1978.
80. **Sinha, S. N. and Gabrieli, E. R.**, Serum copper and zinc levels in various pathological conditions, *Am. J. Clin. Pathol.,* 54, 570, 1970.
81. **Schenker, J. G., Hellerskein, S., Jungries, E., and Polishuk, W. Z.**, Serum copper and zinc levels in patients taking oral contraceptives, *Fertil. Steril.,* 22, 229, 1971.
82. **Zachheim, H. S. and Wolf, P.**, Serum copper in psoriasis and other dermatoses, *J. Invest. Dermatol.,* 58, 28, 1972.
83. **Lahey, M. E., Guhler, C. J., Cartwright, G. E., and Wintrobe, M. M.**, Studies on copper metabolism. VII. Blood copper in normal human subjects, *J. Clin. Invest.,* 32, 322, 1953.
84. **Turner, W. E., Carter, R. J., Bailey, G. G., Smith, B. W., Stroud, P. E., and Bayse, D. D.**, Serum Zinc and Copper levels in the Natonal Health and Nutrition Examination Survey, *Clin. Chem.,* 24(6), 1028, 1978.
85. **Yunice, A. A., Lindeman, R. D., Czerwinski, A. W., and Clark, M.**, Influence of age and sex on serum copper and caeruloplasmin levels, *J. Gerontol.,* 29, 277, 1974.
86. **Halstead, J. A., Hackley, B. M., and Smith, J. C.**, Plasma, zinc and copper in pregnancy and after oral contraceptives, *Lancet,* 2, 278, 1968.
87. **Clemetson, A.**, Chronic copper poisoning in sheep, *Aust. Vet. J.,* 42, 34, 1966.
88. **Hambridge, K. M. and Droegemueller, W.**, Changes in plasma and hair concentrations of zinc, copper, chromium and manganese during pregnancy, *Obstet. Gynecol.,* 44, 666, 1974.
89. **Nielsen, A. L.**, Serum copper and thyrotoxicosis and myoedema, *Acta Med. Scand.,* 118, 431, 1944.

90. **Delves, H. T., Clayton, B. E., and Bicknell, J.,** Concentrations of trace metals in the blood of children, *Br. J. Prev. Soc. Med.,* 27, 100, 1973.
91. **Lifschitz, M. D. and Henkin, R. I.,** Circadian variation in copper and zinc in man, *J. Appl. Physiol.,* 31, 88, 1971.
92. **Mann, T. and Keilin, D.,** Haemocuprein and hepatocuprein, copper-protein compounds of blood and liver in mammals, *Proc. R. Soc. London Ser. B,* 126, 303, 1939.
93. **Carrico, R. J. and Dutsch, H. F.,** Isolation of human hepatocuprein and cerebrocuprein. Their identity with erythrocuprein, *J. Biol. Chem.,* 244, 6087, 1969.
94. **McCord, J. M. and Fridovich, I.,** Superoxide dismutase. An enzymic function for erythrocuprein (hemocuprein), *J. Biol. Chem.,* 244, 6049, 1969.
95. **Gutteridge, J. M. C.,** Superoxide dismutase (erythrocuprein) and free radicals in clinical chemistry, *Ann. Clin. Biochem.,* 13, 393, 1976.
96. **Marberger, H., Huber, W., Bartsch, G., Schulte, T., and Swoboda, P.,** Orgotein: a new anti-inflammatory metalloprotein drug evaluation of clinical efficacy and safety in inflammatory conditions of the urinary tract, *Int. Urol. Nephrol.,* 6, 61, 1974.
97. **Babior, B. M., Kipnes, R. S., and Carnutee, J. T.,** Biological defense mechanisms. The production by leucocytes of superoxide, a potential bacteriocidal agent, *J. Clin. Invest.,* 52, 741, 1973.
98. **Bohenkamp, W. and Weser, U.,** Copper deficiency and erythrocuprein (2 Cu, 2 Zn-superoxide dismutase), *Biochem., Biophys. Acta,* 444, 396, 1976.
99. **Williams, D. M., Loukopoulos, D., Lee, G. R., and Cartwright, G. E.,** Role of copper in mitochondrial iron metabolism, *Blood,* 48, 77, 1976.
100. **Klaude, D. S.,** Anaemia of lead intoxication: a role for copper, *J. Nutr.,* 107, 1779, 1977.
101. **Neumann, P. Z., and Silverberg, M.,** Metabolic pathways of red blood cell copper in normal humans and in Wilsons disease, *Nature (London),* 213, 775, 1967.
102. **Vuori, E., Seppälä, A. H., and Kilpiö, J. O.,** The effect of age and sex on the concentration of copper in aorta, heart, kidney, liver, lung, pancreas and skeletal muscle, *Scand. J. Work Environ. Health,* 4, 167, 1978.
103. **Owen, C. A. and Hazelrig, J. B.,** Metabolism of Cu^{64}-labelled copper by the isolated rat liver, *Am. J. Physiol.,* 210, 1059, 1966.
104. **McConkey, B., Crockson, R. A. and Crockson, A. P.,** The assessment of rheumatoid arthritis. A study based on measurement of the acute-phase reactants, *Q. J. Med. N.S.,* 41, 115, 1972.
105. **Gollan, J. K., Davis, P. S., and Deller, D. J.,** Binding of copper by human alimentary secretions, *Am. J. Clin. Nutr.,* 24, 1025, 1971.
106. **Gregoriadis, G., Morell, A. G., Sternlieb, I., and Scheinberg, L. I.,** Catabolism of desialylated caeruloplasmin in the liver, *J. Biol. Chem.,* 245, 5833, 1970.
107. **Van Berge-Henegouwen, G. P., Tangedahl, T. N., Hofman, A. F., Northfield, T. C., La Rueso, N. F., and McCall, J. T.,** Biliary secretions of copper in healthy man. Quantitation by an intestinal profusion technique, *Gastroenterology,* 72, 1228, 1977.
108. **Evans, G. W.,** Copper homeostasis in the mammalian system, *Physiol. Rev.,* 53, 535, 1973.
109. **Bremner, I. and Young, B. W.,** Isolation of (copper, zinc)-thioneins from the livers of copper-injected rats, *Biochem. J.,* 157, 517, 1976.
110. **Whanger, P. D. and Weswig, P. H.,** Effect of some copper antagonists on induction of caeruloplasmin in the rat, *J. Nutr.,* 100, 341, 1970.
111. **Berger, N. A. and Eichborn, G. L.,** Interaction of metal ions with polynuclestides and related compounds. XV. Nuclear magnetic resonance studies of the binding of copper(II) to nucleotides and polynucleotides, *Biochemistry,* 10, 1857, 1971.
112. **Goldfischer, S.,** Liver cell lysosomes in Wilson's disease: acid phosphatase activity by light and electron microscope, *Am. J. Pathol.,* 43, 511, 1963.
113. **Porter, H.,** Neonatal hepatic mitochondrocuprein. IV. Sulphitolysis of the cystine-rich crude copper protein and isolation of a peptide containing more than 35% half — cystine, *Biochim. Biophys. Acta,* 229, 143, 1971.
114. **Neifakh, S. A., Monakhov, N. K., Shaposhnikov, A. M., and Zubzhitski, Y. N.,** Localisation of caeruloplasmin biosynthesis in human and monkey liver cells and its copper regulation, *Experientia,* 25, 337, 1969.
115. **Evans, G. W., Majors, P. F., and Cornatzer, W. E.,** Induction of caeruloplasmin synthesis by copper, *Biochem. Biophys. Res. Commun.,* 41, 1120, 1970.
116. **Holtzman, N. A. and Gaunitz, B. M.,** Studies on the rate of release and turnover of caeruloplasmin and apo-caeruloplasmin in rat plasma, *J. Biol. Chem.,* 245, 2354, 1970.
117. **Evans, G. W., Majors, P. F., and Cornatzer, W. E.,** Mechanisms for cadmium and zinc antagonism of copper metabolism, *Biochem. Biophys. Res. Commun.,* 40, 1142, 1970.
118. **Griffiths, D. E. and Wharton, D. C.,** Studies of the electron transport system. XXXV. Purification and properties of cytochrome oxidase, *J. Biol. Chem.,* 236, 1850, 1961.

119. **Beinert, H., Griffiths, D. E., Wharton, D. C., and Sands, R. H.,** Properties of the copper associated with cytochrome oxidase as studied by paramagnetic resonance spectroscopy, *J. Biol. Chem.*, 237, 2337, 1962.
120. **Gallagher, G. H., Judah, J. D., and Rees, K. R.,** The biochemistry of copper deficiency. I. Enzymological disturbances, blood chemistry and excretion of amino acids, *Proc. R. Soc. London, Ser. B,* 145, 134, 1956.
121. **Tabor, C. W., Tabor, H., and Rosenthal, S. M.,** Purification of amino oxidase from beef plasma, *J. Biol. Chem.*, 208, 645, 1954.
122. **Hill, C. H. and Starcher, B.,** Effect of reducing agents on copper deficiency in the chick, *J. Nutr.*, 85, 271, 1965.
123. **Krebs, H. A.,** Uber das Kupfer in Merschlichen Blutserum, *Klin. Wochenschr.*, 7, 584, 1928.
124. **Carruthers, M. E., Hobbs, C. B., and Warren, R. L.,** Raised serum copper and caeruloplasmin levels in subjects taking oral contraceptives, *J. Clin. Pathol.*, 19, 498, 1966.
125. **O'Leary, J. A. and Spellacy, W. N.,** Zinc and copper levels in pregnant women and those taking oral contraceptives, *Am. J. Obstet. Gynecol.*, 103, 131, 1969.
126. **O'Leary, J. A., Novalis, G. S., and Vosburgh, G. J.,** Maternal serum copper concentrations in normal and abnormal gestations, *Obstet. Gynecol.*, 28, 112, 1966.
127. **Henkin, R. J.,** Trace metals in endocrinology, *Med. Clin. North Am.*, 60, 779, 1976.
128. **Evans, G. W. and Wiederanders, R. E.,** Effect of hormones on caeruloplasmin and copper concentrations in the plasma of the rat, *Am. J. Physiol.*, 214, 1152, 1968.
129. **Mearrick, P. T. and Mistilis, S. P.,** Excretion of radio copper by the neo-natal rat, *J. Lab. Clin. Med.*, 74, 421, 1969.
130. **Munch-Petersen, S.,** The variations in serum copper in the course of 24 hours, *Scand. J. Clin. Lab. Invest.*, 2, 48, 1950.
131. **Sass-Kortsak, A.,** Copper Metabolism, *Adv. Clin. Chem.*, 8, 1, 1965.
132. **Uzman, L. L., Iber, F. L., Chalmers, T. C., and Knowlton, M.,** The mechanism of copper depositions in the liver in hepatolenticular degeneration (Wilson's disease), *Am. J. Med. Sci.*, 231, 511, 1956.
133. **Evans, G. W., Cornatzer, W. E., Dubois, R. S., and Hambridge, K. M.,** Characterisation of hepatic copper proteins from mammalian species and a human with Wilson's disease, *Fed. Proc. Fed. Am. Soc. Exp. Biol.*, 30, 461, 1971.
134. **Goldfischer, S. and Sternlieb, I.,** Changes in the distributions of hepatic copper in relation to the progression of Wilson's disease, (hepatolenticular degeneration), *Am. J. Pathol.*, 53, 883, 1968.
135. **Lindquist, R. R.,** Studies on the pathogenesis of hepatolenticular degeneration. III. The effect of copper on rat liver lysosomes, *Am. J. Pathol.*, 53, 903, 1968.
136. **Asatoor, A. M., Milne, M. D., and Walshe, J. M.,** Urinary excretion of peptides of hydroxyproline in Wilson's disease, *Clin. Sci. Mol. Med.*, 51, 369, 1976.
137. **Anon.,** Don't forget Wilson's disease, *Br. Med. J.*, 1384, Nov 1978.
138. **Knudsen, L. and Weismann, K.,** Taste dysfunction and changes in zinc and copper metabolism during penicillamine therapy for generalized scleroderma, *Acta Med. Scand.*, 204, 75, 1978.
139. **Danks, D. M., Campbell, P. E., Walfer-Smith, J., Stevens, B. J., Gillespie, J. M., Blomfield, J., and Turner, B.,** Menkes' kinky hair syndrome, *Lancet,* 1, 1100, 1972.
140. **Garnica, A. D., Chan, W. Y., and Rennert, O. M.,** Role of metallothioneins in copper transport in patients with Menkes' syndrome, *Ann. Clin. Lab. Sci.*, 8, 302, 1978.
141. **Orsini, M., Jibril, A. O., and Voliai, F. I.,** Trace metals in cerebrospinal fluid, *Clin. Chem.*, 22, 1210, 1976.
142. **O'Dell, B. L., Smith, R. M., and King, R. A.,** Effect of copper status on brain neurotransmitter metabolism in the lamb, *J. Neurochem.*, 26, 451, 1976.
143. **Heikkila, R. E. and Cabbat, F. S.,** The stimulation of hydroxydopamine autoxidation by bivalent copper: potential importance in the neurotoxic process, *Life Sci.*, 23, 33, 1978.
144. **Putnam, F. W.,** *The Plasma Proteins, Structure, Function and Genetic Control,* Vol. 2., Academic Press, New York, 1975, 51.
145. **Dowdy, H. P. and Dohm, G. L.,** Effect of training and exercise on serum caeruloplasmin in rats, *Proc. Soc. Exp. Biol. Med.*, 139, 489, 1972.
146. **Antipova, N. B., Zagorskaya, E. A., Anashkina, G. A., Manuilova, I. A., and Sokolova, Z. P.,** The content of gonadotrophic and ovarian hormones in the blood of women using intrauterine agents with copper, *Sov. Med.*, 8, 80, 1978.
147. **Underwood, E. J.,** Copper deficiency and functions, in *Trace Elements in Human and Animal Nutrition,* 4th ed., Academic Press, New York, 75, 1977.
148. **Klevay, L. M. and Hyg, S. D.,** Hypercholesterolemia in rats produced by an increase in the ratio of zinc to copper ingested, *Am. J. Clin. Nutr.*, 26, 1060, 1973.

149. **Walravens, P. A. and Hambridge, K. M.**, Growth of infants fed a zinc supplement formula, *Am. J. Clin. Nutr.*, 29, 1114, 1976.
150. **Allen, K. G. D. and Klevay, L. M.**, Cholesterol metabolism in copper deficient rats, *Life Sci.*, 22, 1691, 1978.
151. **Casper, R. C., Kirschner, B., and Jacob, R. A.**, Zinc and copper status in anorexia nervosa, *Psychopharmacol. Bull.*, 14, 53, 1978.
152. **Atkinson, R. L., Dahms, W. T., Bray, G. A., Jacob, R., and Sandstead, H. H.**, Plasma zinc and copper in obesity and after intestinal bypass, *Ann. Intern. Med.*, 89, 491, 1978.
153. **Bogden, J. D. and Troiano, R. A.**, Plasma calcium, copper, magnesium and zinc concentrations in patients with the alcohol withdrawal syndrome, *Clin. Chem.*, 24, 1553, 1978.
154. **Ryan, D. E., Holzbecher, J., and Stuart, D. C.**, Trace elements in scale-hair of persons with multiple sclerosis and of normal individuals, *Clin. Chem.*, 24, 1996, 1978.
155. **Shifrine, M. and Fisher, G. L.**, Caeruloplasmin levels in sera from human patients with osteosarcoma, *Cancer*, 38, 244, 1976.
156. **Breiter, D. N., Diasio, R. B., Neifeld, J. P., Roush, M. L., and Rosehberg, S. A.**, Serum copper and zinc measurements in patients with osteogenic sarcoma, *Cancer*, 42, 598, 1978.
157. **Inutsuka, S. and Araki, S.**, Plasma copper and zinc levels in patients with malignant tumours of digestive organs, *Cancer*, 42, 626, 1978.
158. **Beisel, W. R.**, Trace elements in infectious processes, *Med. Clin. North Am.*, 60, 831, 1976.
159. **Pekarek, R. S., Kluge, R. M., Dupont, H. L., Wannemacher, R. W., Hornick, R. B., Bostian, K. A., and Belsel, W. R.**, Serum zinc, iron and copper concentrations during typhoid fever in man: Effect of chloramphenicol therapy, *Clin. Chem.*, 21, 528, 1975.
160. **Bendstrup, P.**, Serum copper, serum iron and total iron-binding capacity of serum in patients with chronic arthritis, *Acta Med. Scand.*, 146, 384, 1953.
161. **Brown, D. H., El Gnobarey, A., Smith, W. E., and Teape, J.**, Serum copper levels in rheumatoid arthritis, *Ann. Rheum. Dis.*, 37, 391, 1978.
162. **Elmes, M. E.**, Anti-inflammatory drugs and tissue copper, *Lancet*, 2, 1329, 1974.
163. **Lewis, A. J.**, An appraisal of the anti-inflammatory activity of copper salts, *Br. J. Pharmacol.*, 63, 413, 1978.
164. **Swift, A., Karmazyn, M., Horrobin, D. F., Manku, M. S., Karmli, R. A., Morgan, R. O., and Ally, A. I.**, Low prostaglandin concentrations cause cardiac rhythm disturbances. Effect reversed by low levels of copper or chloroquine, *Prostaglandins*, 15, 651, 1978.

Chapter 15

MERCURY

D. F. Williams

TABLE OF CONTENTS

I. INTRODUCTION

Toxicologically, mercury is one of the more interesting and dangerous metals. Along with lead and cadmium, it is generally regarded as being among the most toxic and its widespread industrial use has emphasized the hazards that this metal poses to man. It has two characteristics which are virtually unique in metals and which combine to give very powerful toxicological properties. First, it is a volatile liquid which leads to high localized atmospheric concentrations. Second, although metallic mercury and inorganic mercury compounds are themselves toxic, biotransformation into short-chain alkyl mercury compounds, by bacteria, occurs readily in aqueous environments, yielding far more toxic substances.

While the toxicological properties of mercury are of great concern to the environmentalist and public health scientists because of industrial pollution, the significance of mercury from the biomaterials point of view is that metallic mercury is used in the preparation of the most common dental filling material, amalgam. The controversy over the industrial exposure of dental personnel to mercury has raged for a long time, but it is certainly accepted now that great care has to be taken in its use and a strict mercury hygiene regime adopted. It may be that amalgam will become the first biomaterial to be barred from use because of a concern for the safety of the clinical staff rather than the patient.

Although the clinical problems of mercury in dentistry are discussed at some length in *Biocompatibility of Dental Materials* of this series, the general toxicology of the metal and its compounds is included in this volume for completeness and comparison with that of the other metallic biomaterials.

II. OCCURRENCE, PROPERTIES, AND USES OF MERCURY

Occurrence — The most important ore of mercury is cinnabar, HgS, from which it is extracted either by precipitation with aluminum in alkaline solution or by roasting in air. It occasionally occurs as the natural metallic element in small droplets in rock formations. The element is not particularly abundant, and a really high concentration in soil would be 10 ppm, the usual levels being nearer 100 ppb.

Properties — As noted earlier, mercury is a liquid, being the only metal to exist in this state at room temperature. The melting point is −38.9°C and the boiling point is 357.3°C. It has a silvery appearance which is highly reflective when the metal is pure, but dulls when there are traces of impurities. It is heavy with a specific gravity of 13.55 and atomic weight 200.61. It is extremely volatile.

Uses — The metal itself has widespread industrial use in scientific apparatus, mercury vapor lamps, control instrumentation, electrolytic cells, and so on. Uses of inorganic mercury are largely dependent on its antifouling and antibacterial action, including examples in the paint, agricultural, cosmetic, and pharmaceutical industries; at one time mercury compounds were used in the treatment of syphilis and skin diseases. Organic mercury compounds similarly depend on their effects on microorganisms for their applications, a number of different compounds being used as seed dressings for example.

III. MERCURY IN ANIMAL TISSUES

There is not a great deal of evidence concerning levels of mercury in normal subjects who have not been exposed to the element. Kosta et al. reported the levels in humans to be 0.03 ppm fresh weight of thyroid, 0.04 ppm pituitary, 0.03 ppm liver, 0.004 ppm brain, and 0.14 ppm kidney.[1] Joselow et al.[2] found mercury levels ranging from 0.05 to 0.3 ppm wet weight in most tissue, but also reported the highest levels, at 2.7 ppm, in the kidneys. A similar range of 0.5 to 2.5 ppm mercury in dry tissue was observed by Howie and Smith.[3] Parts of the body exposed to the atmosphere such as skin, hair, and nails tend to give higher values because of exogenous contamination. A head hair level of 5.5 ppm was recorded by Rodger and Smith, for example.[4]

The range of mercury levels in whole blood was reported to be 2 to 9 ng/g by Kellershohn et al.[5] and Goldwater states that 74% of the normal population have less than 5 ng/g mercury in their blood.[6] According to Kellershohn,[5] the red cells contain twice the mercury concentration found in the plasma.

IV. METABOLISM OF MERCURY AND ITS COMPOUNDS

A. Absorption

1. Intestinal Absorption

Dietary levels of mercury — Browning[7] estimates that traces of the order of 0.005 to 0.25 ppm mercury occur in nearly all foods, very much in agreement with the mean level of 0.03 ppm quoted by Beliles.[8] Magos[9] states that the background level in fruit, vegetables, eggs, and meat is less than 0.05 ppm. Thus, the daily intake in a subject on a normal diet would be of the order of 20 μg. Fish often contain more mercury than other foods because of the presence of methyl mercury in some contaminated waters. Magos[9] reports values of 0.5 ppm in certain species with a maximum of 14 ppm in the Pacific blue marlin.

Rate of absorption — Absorption in the gastrointestinal tract is very much dependent on the form of the mercury. According to Nordberg,[10] absorption of elemental mercury is extremely limited, probably amounting to less than 0.01% of the administered dose. Absorption of inorganic mercury compounds has been variously estimated to be around 7%,[10] 6%,[11] and 15% while that of organic mercury compounds is much higher. Nordberg[10] states that the absorption of methylmercuric compounds is about 95% of the administered dose in humans, a figure close in agreement to that of Miettinen.[12] Taguchi[11] reported a lower-figure of 73% for the absorption of methylmercury chloride in rats.

2. Inhalation

Mercury vapor — Metallic mercury, as noted earlier, is extremely volatile. According to Beliles,[8] a concentration of 200 $\mu g/m^3$ may be found in the atmosphere near areas of high soil mercury levels, while Magos[9] reports that the concentration in an unventilated space arising from the volatility of elemental mercury may reach 13 mg/m^3. The maximum recommended concentration in industrial atmospheres[13] is 0.05 mg/m^3, although this is often exceeded in dental and other laboratories.[14]

Absorption in the lung — Vostal and Clarkson[15] have demonstrated that mercury vapor is almost completely absorbed through the alveolar membrane. Nordberg[10] estimates that 80% of mercury vapor reaching the alveoli is absorbed. Hayes and Rothstein[16] report a range of 75 to 100% for the absorption of mercury in the rat following inhalation, and Magos[17] quotes similar results. The amount absorbed in humans has been given as 50%[18] and 75%[19] and the very rapid diffusion of mercury through the alveolar membrane has been demonstrated on several occasions.[20,21]

3. Other Routes

Both mercury vapor and methylmercury are readily absorbed by the skin and, indeed, this route may represent a significant portal of entry on industrial exposure and in the use of cosmetic and pharmaceutical preparations.

B. Transport and Distribution

1. Metallic Mercury

While 60 μg mercury dissolve in 1 ℓ of air at 40°C, 1.5 mg dissolves in body lipoids at the same temperature,[23] this partition coefficient favoring considerable diffusion through the alveolar membrane and dissolution in the blood lipoids. In the blood there is a tendency for the mercury vapor to be oxidized[22] even though it is normally very resistant to oxidation in air. However, this is a relatively slow process so that unoxidized vapor (Hg^o) remains dissolved in the blood stream long enough for it to reach the blood brain barrier.[17,20] The brain, therefore, takes up more mercury in this way than when equivalent doses of inorganic mercury salts are injected.[20,24]

The biological oxidation of Hg^o to Hg^{2+} is largely brought about by the enzyme catalase by mechanisms which have been discussed by Kudsk.[22] The oxidation process can be delayed by any compound which interferes with catalase activity,[25] such as ethanol.[26]

2. Inorganic Mercury

As discussed earlier, inorganic mercury salts are poorly absorbed. Friberg[27] gave subcutaneous injections of mercuric chloride to rats and compared the retention, distribution, and excretion of mercury with rats given equivalent amounts of methylmercury dicyandiamide. Approximately 100 times less mercury was found in the blood of the mercuric chloride treated rats, 10 times less in the brain, and half as much in the liver.

The biological half-life of orally derived mercuric salts is between 30 and 60 days,[10] which is about half that for organic mercury compounds.[12] Inorganic mercury accumulates in the kidneys where it is largely bound to sulfhydryl groups. Retention in the kidneys is longer than in the rest of the body.[28] In the kidneys mercury is taken up from both the tubular urine and the peritubular capillaries.[29] At low dose levels much of the inorganic mercury is eliminated via the bile, but at higher doses elimination is largely via the urine.[30]

3. Organic Mercury Compounds

There are two classes of organic mercury compounds, often referred to as organomercurials, in this context. First, there are the short-chain alkylmercury compounds, the most important of which is methylmercury, which are highly toxic. Second, there are other organomercurials, including the aryl and alkoxyalkyl mercurials which are rapidly metabolized to inorganic mercury. The most common examples here are phenyl- and methoxyethyl-mercury compounds which have been used as fungicides and diuretics. In these compounds the mercury-carbon bond is rapidly cleaved by enzymes in the soluble fraction of liver[31] so that the distribution of mercury after administration of such organomercurials is much the same as with inorganic mercury. Although the absorption of these organomercurials from the gastrointestinal tract is greater than for inorganic mercury, the uptake of mercury by the tissues following conversion is slower so the toxicity, and particularly the extent of renal damage, is less.[32]

Methyl mercury — This is probably the most lethal and widely investigated compound of mercury. According to Magos,[9] and confirming the figure given by Miettinen,[12] the biological half-life of methylmercury in man is about 70 days, which means that a daily intake of 100 μg mercury in this form results in a body burden of 10 mg at steady state in a 70-kg man. Although methylmercury is also converted into inorganic mercury, the rate of decomposition is much slower.[33] Furthermore, there is complete reabsorption when excreted in the bile[34] and urinary excretion is low.[13] All these factors contribute to the long biological half-life. In contrast to inorganic mercury which preferentially accumulates in the kidneys, methylmercury is located in the blood, especially the red blood cells, and also concentrates in certain organs. In particular, methylmercury is able to cross the blood-brain barrier very readily,[35] Nordberg quoting a steady state concentration of nearly 2 μg mercury per gram brain tissue in man.[10] This aspect is covered in more detail in the later section on toxicology.

C. Excretion

The pattern of excretion of mercury is naturally dependent upon both absorption and distribution processes. It is, however, generally agreed that excretion is greater in the feces than the urine and that lesser routes such as exhalation, sweat, hair, and milk also play a part. Although urine levels of groups of workers tend to reflect general body burdens,[36] it is difficult to correlate levels with exposure in individual subjects since the mercury content of urine is highly variable. A range of 0.1 to 13.3 μg/100 mℓ has been reported,[3] although these levels may be increased by several orders of magnitude in industrially exposed workers.[37] Browning[7] quotes an average value of 0.5 μg excreted daily in the urine and 10 μg in the feces.

V. TOXICITY OF MERCURY

A. General

The toxicity of mercury and mercurials is largely due to the high affinity that mercury cations, such as Hg^{2+}, $C_6H_5Hg^+$, and CH_3Hg^+, have to sulfhydryl (−SH) groups, as reviewed by Rothstein.[38] These −SH groups are found in some diffusable low molecular weight substances such as cysteine, CoA, lipoate, and thioglycolate, but more importantly are located in most proteins. Thus, although the mercury compounds are highly specific for the sulfhydryl group, they are highly nonspecific in their targets because of the wide distribution of this group. They can disturb almost all functions in which proteins are involved and especially can inhibit most enzymes if present in sufficient concentration.[39]

In spite of this large range of potential sites of action, the toxicological effects of mercury are, in fact, highly specific, with well-defined symptoms and progress for the different form of the metal. The specific target for a particular type of mercurial is controlled both by the chemical form, as discussed in detail later, and the distribution of the mercurial in the body.[40] Here the actual location of the target protein and the ability of the mercurial to reach the target are clearly important. It is still not entirely clear what the target proteins are, although it is known that mercury is taken up by lysosomes, with subsequent alterations to lysosomal enzymes.[41,42] The ability to reach the target is dependent on factors such as circulatory dynamics, the distribution of mercury in the blood between cells and plasma, the ability of the mercury to penetrate cell membranes, and so on.[40] These factors and the clinical toxicology are now discussed in relation to elemental, inorganic, and organic mercury.

B. Metallic Mercury

Ingested liquid mercury is reputedly nontoxic,[43] although there is not a great deal of evidence available on the subject. As noted above, the toxicological problems with metallic mercury arise because of the high vapor pressure of the mercury. Acute poisoning by metallic mercury is relatively rare and toxicity is more often seen as a chronic effect, usually in workers exposed to mercury vapor over a long period of time.

The effects of inhaled mercury vapor do, in fact, vary as a function of the time of exposure.[44] With acute exposure to high levels of vapor, the lung is the critical organ.

In some early experiments, Christensen et al.[45] observed pulmonary edema in mice subjected to lethal doses of mercury vapor, death being preceded by cyanosis, extreme weakness, vomiting, and diarrhea. With subacute exposure of short duration, the kidney becomes the critical organ while prolonged exposure to lower levels results in the brain becoming the critical organ. This change in critical target organ arises because of the distribution network outlined earlier; only a small amount of the absorbed mercury accumulates in the brain so that it takes prolonged exposure for a critical concentration to be established there, but once this has been reached, elimination is very slow.

The effects on the brain and central nervous system of chronic exposure to mercury vapor are well documented. Ashe et al.[46] reported experiments in which rabbits were exposed to mercury vapor at concentrations between 0.1 and 6 mg Hg/m^3 where concentrations above 3 $\mu g/g$ produced moderate histopathological changes in the brain. Levels between 1 and 2 $\mu g/g$ produced definite but mild changes, while there were no observable effects at levels below 0.5 $\mu g/g$. It is quite possible for behavioral effects to occur at concentrations lower than those which produce distinct morphological changes.

Symptoms of chronic mercury poisoning in dogs arising from the inhalation of vapor, described in some detail by Fraser et al.[47] include gingivitis, diarrhea, anorexia, loss of weight, and disturbance to the central nervous system. Gingivitis, excessive salivation, and dermatitis may also occur in industrially exposed workers. Anorexia, weight loss, anemia, and muscular weakness may also occur, and Smith et al.[36] have shown that loss of weight correlates with exposure to mercury vapor.

Details of the effects of mercury on the central nervous system have been discussed by Norton.[48] Discussing elemental mercury and inorganic mercury, he describes psychological changes (erethismus mercurialis), deposition of mercurials in the anterior chamber of the eye (mercurialentis), tremor, and signs of autonomic dysfunction, including excessive salivation. The emotional changes and autonomic nervous system effects observed with this type of mercury are not seen with organic mercury.

C. Inorganic Mercury

The target organ for inorganic mercurials is very clearly the kidney. Clinical signs of poisoning from ingested mercuric salts include gastroenteritis, with abdominal pain, nausea, and so on, leading to renal dysfunction with anuria and uremia. Fortunately, exposure to high levels of inorganic mercury is extremely rare. The most common causes of acute inorganic mercury poisoning are suicide and homicide, although chromic effects arising from prolonged occupational exposure can also occur.

The effects on the kidney are seen from the work of Berlin[44] and others. In cases of known poisoning due to mercuric salts, values of between 10 and 70 ppm have been found for the mercury in the kidneys, a conclusion in agreement with the observations of Kazantis et al.[49] who found 10 and 25 ppm in two kidney samples obtained from renal biopsy.

Mercuric chloride is the most widely investigated salt in this context; it is soluble and fairly readily absorbed. Ingestion leads to an ashen-grey appearance of the mouth due to precipitation of protoplasm of the mucous membranes,[8] followed by the gastrointestinal disturbances and renal dysfunction noted above. Death is usually due to these renal lesions, especially of the tubular epithelium. The extent of the renal changes is dose-related. In animals, at low doses of mercuric chloride, necrosis of the terminal portions of the renal tubules occurs, the initial and middle portions of the proximal tubules being also affected at higher doses. Among other reported effects are ultrastructural changes in the mitochondria of the proximal tubules and ischemia in the renal cortex at high doses.[10]

Ellis and Fang[50] found that about two thirds of mercury, derived from inorganic mercury, deposited in rat kidneys is present in the soluble postmicrosomal fraction, about 30% in the cellular nuclei, and the rest divided between the mitochondria and microsomes. Komsta-Szumska et al.[51] gave intravenous injections of mercuric chloride solution to rats and also found that most (54%) accumulated in the soluble fraction and a large amount (30%) in the nuclear fraction. Mitochondrial and microsomal fractions initially accounted for 11 and 6% of the total mercury, but this increased with repeated injections. They found that in the soluble fraction low molecular weight, metallothioein-like proteins were responsible for the mercury accumulation.

Mercuric nitrate is also of significance in inorganic mercury poisoning, for it was widely used in the felt hat industry when a solution of mercuric nitrate was applied to rabbits fur to improve the felting qualities. In this case however, it appears that volatile components were released during this "carrotting" process which, after being inhaled, led to disturbances of the central nervous system, as with inhaled elemental mercury vapor.[52] The effects on these workers were very pronounced, with severe psychological and emotional problems and this syndrome is thought to have given rise to the common expression "mad as a hatter".

Mercurous chloride is far less soluble than mercuric chloride and leads to quite different effects. This substance at one stage was used as a cathartic agent and teething powder and was the cause of acrodyria, or pink appearance, with hypokeratosis and edema.

D. Organic Mercury

As discussed earlier, organic mercurials can be divided into two types from the toxicological point of view, the short-chain alkylmercury compounds such as methylmercury and those unstable organomercurials that are rapidly metabolized to inorganic mercury.

1. Unstable Organomercurials

Unstable organomercurials are organomercurials with a covalent bond between mercury and carbon,[9] the other ionic valency being taken up by an anion. The common and relevant examples here are phenyl- and methoxyethylmercury salts. These compounds display a low order of toxicity and, although still quite widely used as fungicides and diuretics, give rise to few occupational or accidental poisoning episodes. On those occasions that have been reported, the symptoms tend to be loss of appetite, diarrhea, weight loss, and fatigue, with occasional renal effects such as albuminuria. The lack of toxicity appears to be due to the rapid biotransformation of these substances, described earlier, which leads to a conversion to inorganic mercury compounds. Gage[53] showed that after a subcutaneous application of methoxyethyl mercury, about 60%, appeared in the air as ethylene, the compound being broken down nonenzymatically to release ethylene and mercury, the latter migrating to the kidneys. A similar type of process occurs with phenylmercury yielding a phenol conjugate that appears in the urine. Occasionally irritant effects of phenylmercury on the skin are noted, but more serious effects are rare. Magos[9] states that the LD_{50} values of phenyl and methoxyethyl-mercury compounds given parenterally are 3 to 8 times higher than the LD_{50} of inorganic mercury salts. However, the toxicities of the two are about the same when ingested because of the higher rate of absorption with these organomercurials. There is no evidence that these unstable organomercurials are able to pass the blood-brain barrier.

2. Alkyl Mercury Compounds

In these substances the strong carbon-mercury bond is not readily dissociated, and toxic effects are attributed to the action of the intact molecule.[54] Consequently, compounds with the shortest carbon chain are the most toxic since they can pass across membranes more readily. Methylmercury compounds are, in fact, the most toxic of all mercurials with ethylmercury compounds a little less severe. An important factor in the causation of alkylmercury and especially methylmercury poisoning is the ability of certain bacteria to induce biotransformation of inorganic mercury into methylmercury. The most notorious situation in which this happened on a large scale was the so-called Minamata epidemic.[56] Between 1953 and 1960 a factory close to the Minamata River in Japan, producing acrylonitrile and other chemicals, released effluent into the river that contained substantial amounts of mercuric chloride. This was transformed into methylmercury, which then accumulated in the fish that populated those waters. Some polluted fish and shellfish from the river contained up to 20 ppm methylmercury[55] and the people who consumed local fish suffered organomercury poisoning, 46 deaths being reported. Other epidemics include those of Niigata (1964 to 1965) and Iraq (1971 to 1972),[57] in all cases there being significant mortality and morbidity. In the latter case the outbreak was caused by the consumption of homemade bread prepared from wheat treated with methylmercury fungicide.

The toxicity and target organs for methylmercury vary with animal species. In man and other primates the target is the central nervous sytem, while in lower mammals the peripheral nervous system may be affected. As discussed below, the fetal brain is also a critical organ in humans if methylmercury is ingested during pregnancy.

The symptoms of neurological disturbances due to methylmercury poisoning in man may occur weeks or months after exposure to a high dose, indicative of the long biological half-life. Symptoms include numbness of lips, mouth, hands, and feet, ataxia, visual disturbances, and difficulty in speaking. In severe cases, there are mental changes, involuntary movements, and possibly death. Infants exposed prenatally are severely affected even though the mothers may show no or minimal signs of methylmercury poisoning. The babies show severe damage to the central nervous system with symptoms similar to those of cerebral palsy.

As first described by Hunter et al.,[58] methylmercury is readily absorbed by the body, either in the gastrointestinal system, the respiratory tract, or through the skin. As the neurological symptoms imply, it readily passes the blood brain barrier after being rapidly taken up by erythrocytes. The brain concentrations of mercury derived from this source are much higher than those from any other mercurial. In the brain, the methylmercury decreases the number of neurones in the cerebellum. The damage is permanent because of the strong affinity that the mercury has for the sulfur in the sulfhydryl groups in the cell membrane proteins.

VI. MERCURY HAZARDS IN DENTISTRY

Amalgam has been used in dentistry for over 100 years and for much of this time has been the subject of controversy in relation to its potential for causing mercury poisoning. The cause for concern is not so much from the point of view of safety of the patient, who is exposed to very little mercury, but of the dental personnel who are often exposed to a relatively high level of mercury vapor in the atmosphere for much of their working life.

A. The Use of Mercury in Dentistry

Mercury is used in the preparation of dental amalgam, the resorative material that is used for virtually all fillings in the posterior teeth and for many in unobtrusive aspects of anterior teeth. The amalgam is prepared by mixing together metallic mercury with a silver-tin alloy. In conventional amalgams, used virtually unchanged in composition since the beginning of the century, this alloy corresponds to the intermetallic compound Ag_3Sn (known as the γ phase) and contains up to 6% copper to modify certain properties. The reaction with mercury produces two new phases, Ag_2Hg_3 (γ_1) and Sn_7Hg (γ_2), but does not go to completion and a considerable amount of Ag_3Sn remains in the amalgam. The material is placed in the prepared cavity of the tooth while this reaction is taking place, in other words, while it is still in a relatively fluid state. Although the trend in recent years has been to use the minimum amount of mercury, there will still be some free mercury in the amalgam, the majority of which is removed as the dentist forces the amalgam into the cavity, the liquid mercury rising to the surface, where it is scraped away.

Modifications have been made to both the composition of the alloy, the physical form of the alloy, and the method of mixing it with mercury in recent years. The γ_2 phase is both the weakest and most corrosion resistant and so some formulations contain much higher amounts of copper, which tends to form a complex with tin, preventing the formation of the γ_2 phase. More importantly, the usual practice for mixing the mercury with the amalgam alloy involved proportioning liquid mercury in a suitable container and transferring it to a pestle and mortar for the mixing. Most commercially available amalgams are now supplied as preproportioned mercury and alloy contained within separate chambers of a capsule. The dividing membrane is punctured immediately before use and mixing is achieved with the aid of a vibrator which holds the capsule and shakes it at high speed. Thus handling of the mercury is avoided and the risk of spillage considerably reduced.

B. The Hazard to the Patient

It is clear that there is no risk of systemic mercury poisoning to the patient from the use of dental amalgam. Nixon et al.[59] have shown that the enamel of teeth containing amalgam fillings will contain far more mercury than control teeth. The enamel of unerupted teeth has less than 0.1 ppm mercury, that of erupted, unfilled teeth has a mean of 2.6 ppm, while the enamel in contact with a filling may have 150 to 1200 ppm mercury. There is also an increased excretion of mercury for several days after the insertion of a filling, but there is no evidence for systemic distribution of mercury derived in this way or of any associated abnormal effects. The increased excretion is probably due to ingested fragments of amalgam that find their way into the mouth during the insertion of the filling. Although amalgam does corrode, where one of the products is mercury itself, it is unlikely that such mercury will find its way into the systemic circulation. Much of the mercury that is produced reacts with the retained γ phase, while any that is released into the saliva will be poorly absorbed from the gastrointestinal tract. The release of mercury from fillings has been discussed by Frykholm.[60]

Occasionally fragments of amalgam are accidentally inserted into oral soft tissues during the placement of a filling.[61,62] Such pieces become engulfed by macrophages and, in the case of large particles, by foreign body giant cells. Electron probe microanalysis of the amalgam has shown that corrosion takes place releasing mercury into the surrounding tissue,[63,64] this reaction apparently being responsible for the tissue reaction that is observed.

C. The Hazard to Dental Personnel

Undoubtedly there is a far greater risk to dental personnel. There is in the literature one reported case of fatal intoxication by mercury arising in dentistry.[65] This involved a 42-year-old woman who worked for 20 years as a dental surgery assistant. Although hitherto in good health, she suddenly developed symptoms of vomiting, pain in the lumbar region of the abdomen, and edema of the legs and face. Within a few days hospital treatment was necessary and a nephrotic syndrome diagnosed. Serum sodium and chlorine were depressed and severe renal failure followed, death occurring in a short time. Although at autopsy the kidneys appeared normal macroscopically, mercury was found widely as fine particles on microscopic examination within the epithelial and endothelial cells of the glomerulus and Bowmans capsule, and the proximal and distal convoluted tubules. The mercury level was 520 ppm, compared to control values of less than 10 ppm. It was concluded that death was due to acute renal failure arising from mercury intoxication, the mercury accumulating in the kidneys over a long period of time.

Other symptomatic but nonfatal cases of mercury poisoning have also been recorded.[66] Four dental workers developed symptoms after a serious but unreported spillage had occurred, with the development of severe headaches, diplopia, loss of muscular coordination, hallucinations, and loss of memory. The amount spilled was very large and acute or subacute symptoms were soon identified. Urine levels were found to be very high, but all four responded well to treatment by oral N-acetyl D-penicillamine.

A disturbing aspect is the observations by many investigators of raised mercury levels in apparently asymptomatic dental workers. Battistone et al.,[67] for example, found the mean urine mercury level in dentists and their assistants to be in the range 5 to 124 $\mu g/\ell$, which was significantly higher than the 3- to 17-$\mu g/\ell$ range in the controls. Similarly, Marks and Taylor[68] showed that while the normal early-morning urine mercury level is usually less than 5.5 nmol/mmol creatinine, 62.6% of dental workers have a level greater than 5 nmol/mmol, with a maximum at 74 nmol/mmol.

It is of course, very difficult to determine the significance of slightly raised mercury levels. Battistone et al.[69] performed further studies that involved measuring blood mercury levels. While a range of 0 to 10 ng Hg/mℓ blood should include 85% of the U.S. population,[70] 77% of the dentists studied had levels in this range and even the higher levels were well below clinically toxic levels. The difficulty arises in trying to correlate the rather ill-defined and almost ubiquitous symptoms associated with mild mercury poisoning, such as headaches and irritability, with exposure to mercury.

The minimization of the hazards to dental workers clearly lies in reducing the amount of mercury released into the atmosphere in the dental clinics. So-called mercury hygiene has been discussed at length in the literature[71-86] and codes of practice have been published.[87] The reader is referred to this literature and to the chapter by Hefferen in *Biocompatibility of Dental Materials*[88] of this series for further details.

REFERENCES

1. **Kosta, L., Bryne, A. R., and Zelenko, V.,** Correlation between selenium and mercury in man following exposure to inorganic mercury, *Nature (London)*, 254, 238, 1975.
2. **Joselow, M. M., Goldwater, L. J., and Weinburg, S. B.,** Absorption and excretion of mercury in man. XI. Mercury content of normal human tissues, *Arch. Environ. Health*, 15, 64, 1967.
3. **Howie, R. A. and Smith, H.,** Mercury in human tissue, *J. Forensic Sci. Soc.*, 7, 90, 1967.
4. **Rodger, W. J. and Smith, H.,** Mercury absorption by fingerpring officers using grey powder, *J. Forensic Sci. Soc.*, 7, 86, 1967.

5. **Kellershohn, C., Comar, D., and Le Poec, C.**, Determination of the mercury content of human blood by activation analysis, *J. Lab. Clin. Med.*, 66, 168, 1965.
6. **Goldwater, L. J.**, Occupational exposure to mercury, *J. R. Inst. Public Health*, 27, 279, 1964.
7. **Browning, E.**, *Toxicity of Industrial Metals*, 2nd ed., Butterworths, London, 1969, 226.
8. **Beliles, R. P.**, Metals, in *Toxicity: The Basic Science of Poisons*, Casarett, L. J. and Doull, J., Eds., MacMillan, New York, 1975, 484.
9. **Magos, L.**, Mercury and mercurials, *Br. Med. Bull.*, 31, 241, 1975.
10. **Norberg, G. F., Ed.**, *Effects and Dose-Response Relationships of Toxic Metals*, Elsevier, Amsterdam, 1976, 21.
11. **Taguchi, Y.**, Studies on microdetermination of total mercury and the dynamic aspects of methyl mercury compounds in vivo, *Jpn. J. Hyg.*, 25, 553, 1971.
12. **Miettinen, J. K.**, Absorption and elimination of dietary mercury and methylmercury in man, in *Mercury, Mercurials and Mercaptans*, Miller, M. W. and Clarkson, T. W., Eds., Charles C Thomas, Springfield, Ill., 1973, 233.
13. **Berlin, M. H., et al.**, Maximum allowable concentrations of mercury compounds, *Arch. Environ. Health*, 19, 891, 1969.
14. **Mayz, E., Corn, M., and Barry, G.**, Determination of mercury in air at University facilities, *Am. Ind. Hyg. Assoc. J.*, 32, 373, 1971.
15. **Vostal, J. J. and Clarkson, T. W.**, Mercury as an environmental hazard, *J. Occup. Med.*, 15, 649, 1973.
16. **Hayes, A. and Rothstein, A.**, The metabolism of inhaled mercury vapour in the rat, *J. Pharmacol.*, 138, 1, 1962.
17. **Magos, L.**, Mercury-blood interactions and mercury uptake by the brain after vapour exposure, *Environ. Res.*, 1, 323, 1967.
18. **Teisinger, J. and Fiserova-Bergerova, V.**, Pulmonary retention and excretion of mercury vapour in man, *Ind. Med. Surg.*, 34, 580, 1965.
19. **Neilsen-Kudsk, F.**, Absorption of mercury vapour from the respiratory tract in man, *Acta Pharmacol.*, 23, 250, 1965.
20. **Magos, L.**, Uptake of mercury by the brain, *Br. J. Ind. Med.*, 25, 315, 1968.
21. **Berlin, M., Norberg, G., and Sereniun, F.**, On the site and mechanism of mercury vapour absorption in the lung, *Arch. Environ. Health*, 18, 42, 1969.
22. **Kudsk, F. N.**, Biological oxidation of elemental mercury, in *Mercury, Mercurials and Mercaptans*, Muller, M. W. and Clarkson, T. W., Eds., Charles C Thomas, Springfield, Ill., 1973, 355.
23. **Clarkson, T. W., Gatzy, J., and Dalton, C.**, *Studies on the Equilibrium of Mercury Vapour and Blood*, Division of Radiation Chemistry and Toxicology, Report UR-582, University of Rochester, Rochester, N.Y., 1961, 64.
24. **Berlin, M., Jerksell, L. G., and von Ubisch, H.**, Uptake and retention of mercury in the mouse brain, *Arch. Environ. Health*, 12, 33, 1966.
25. **Magos, L., Sugata, Y., and Clarkson, T. W.**, Effects of 3 amino — 1,2,4, — triazole on mercury uptake by in vitro human blood samples and by whole rats, *Toxicol. Appl. Pharmacol.*, 28, 367, 1974.
26. **Magos, L., Clarkson, T. W., and Greenwood, M. R.**, The depression of pulmonary retention of mercury vapour by ethanol: identification of the site of action, *Toxicol. Appl. Pharmacol.*, 26, 180, 1973.
27. **Friberg, L.**, Studies on the metabolism of mercuric chloride and methylmercury dicyandiamide, *Arch. Ind. Health*, 20, 42, 1959.
28. **Norberg, G. F. and Skerfving, S.**, in *Mercury in the Environment*, Friberg, L. and Vostal, J., Eds., CRC Press, Cleveland, 1972, 29.
29. **Clarkson, T. W. and Magos, L.**, Effects of 2,4-dinitrophenol and other metabolic inhibitors on the renal deposition and excretion of mercury, *Biochem. Pharmacol.*, 19, 3029, 1970.
30. **Rothstein, A. and Hayes, A. D.**, The metabolism of mercury in the rat studied by isotope techniques, *J. Pharmacol. Exp. Ther.*, 130, 166, 1960.
31. **Daniel, J. W.**, The biotransformation of organomercury compounds, *Biochem. J.*, 130, 64, 1972.
32. **Clarkson, T. W.**, The pharmacology of mercury compounds, *Annu. Rev. Pharmacol.*, 12, 375, 1972.
33. Task Group on Metal Accumulation, *Environ. Physiol. Biochem.*, 3, 65, 1973.
34. **Norseth, T. and Clarkson, T. W.**, Intestinal transport of ^{203}Hg-labelled methyl mercury chloride, *Arch. Environ. Health*, 22, 568, 1971.
35. **Swensson, A.**, Investigations on toxicity of some organic mercury compounds, *Acta Med. Scand.*, 143, 365, 1952.
36. **Smith, R. G., Vorwald, A. J., Patril, L. S., and Mooney, T. F.**, Effects of exposure to mercury in the manufacture of chlorine, *Am. Ind. Hyg. Assoc. J.*, 31, 687, 1970.

37. **Joselow, M. M. and Goldwater, L. J.,** Absorption and excretion of mercury in man. XII. Relationship between urinary mercury and proteinuria, *Arch. Environ. Health,* 15, 155, 1967.
38. **Rothstein, A.,** Mercaptans, the biological targets for mercurials, in *Mercury, Mercurials and Mercaptans,* Miller, M. W. and Clarkson, T. W., Eds., Charles C Thomas, Springfield, Ill., 1973, 68.
39. **Webb, J. L.,** *Enzymes and Metabolic Inhibitors,* Vol. 2, Academic Press, New York, 1966, 729.
40. **Rothstein, A.,** Sulphydryl groups in membrane structure and function, in *Current Topics in Membranes and Transport,* Bonner, G. and Kleinzeller, A., Eds., Academic Press, New York, 1970.
41. **Fowler, B. A.,** The morphologic effects of dieldrin and methyl mercuric on pars recta segments of rat kidney proximal tubules, *Am. J. Pathol.,* 69, 163, 1972.
42. **Fowler, B. A., Brown, H. W., Lucier, G. W., and Beard, M. E.,** Mercury uptake by renal lysosomes of rats ingesting methyl mercury hydroxide, *Arch. Pathol.,* 98, 297, 1974.
43. **Williams, D. R.,** *An Introduction to Bio-inorganic Chemistry,* Charles C. Thomas, Springfield, Ill., 1976, 367.
44. **Berlin, M.,** Dose-response relations and diagnostic indices of mercury concentrations in critical organs upon exposure to mercury and mercurials, in *Effects and Dose-Response Relationships of Toxic Metals,* Norberg, G. F., Ed., Elsevier, Amsterdam, 1976, 325.
45. **Christensen, H., Krogh, M., and Nielsen, M.,** Acute mercury poisoning in a respiration chamber, *Nature (London),* 139, 626, 1937.
46. **Ashe, W. F., Largent, E. J., Dutra, F. R., Hubbard, D. M., and Blackstone, M.,** Behaviour of mercury in the animal organism following inhalation, *Arch. Ind. Hyg. Occup. Med.,* 7, 19, 1957.
47. **Fraser, A. M., Melville, K. I., and Stehle, R. C.,** Mercury-laden air, *J. Ind. Hyg.,* 22, 297, 1934.
48. **Norton, S.,** Toxicology of the central nervous system, in *Toxicology, the Basic Science of Poisons,* Casarett, L. J. and Doull, J., Eds., MacMillan, New York, 1975, 161.
49. **Kazantis, G., Sciller, F. R., Asscher, A. W., and Drew, R. G.,** Albuminuria and the nephrotic syndrome following exposure to mercury and its compounds, *Q. J. Med.,* 31, 403, 1962.
50. **Ellis, R. N. and Fang, S. C.,** Elimination, tissue accumulation and cellular incorporation of mercury in rats receiving an oral dose of ^{203}Hg-labelled phenylmercuric acetate and mercuric acetate, *Toxicol. Appl. Pharmacol.,* 11, 104, 1953.
51. **Komsta-Szumska, E., Chmielnicka, J., and Piotrowski, J. K.,** Binding of inorganic mercury by subcellular fractions and proteins of rat kidneys, *Arch. Toxicol.,* 37, 57, 1976.
52. **Neal, P. A., Flinn, R. H., Edwards, T. I., and Reinhart, W. H.,** Mercurialism and its control in the felt-hat industry, *U.S. Publ. Health Serv. Bull.,* No. 263, 1941.
53. **Gage, J. C.,** The metabolism of methoxyethyl mercury and phenyl mercury in the rat, in *Mercury, Mercurials and Mercaptans,* Miller, M. W. and Clarkson, T. W., Eds., Charles C Thomas, Springfield, Ill., 1973, 346.
54. **Kurkland, L. T.,** An appraisal of the epidemiology and toxicology of alkylmercury compounds, in *Mercury, Mercurials and Mercaptans,* Miller, M. W. and Clarkson, T. W., Eds., Charles C Thomas, Springfield, Ill., 1973, 23.
55. **Kitamurg, S., Sumino, K., Hayakawaa, K., and Shibata, T.,** Dose-response relationship of methyl mercury, in *Effects and Dose-Response Relationships of Toxic Metals,* Norberg, G. F., Ed., Elsevier, Amsterdam, 1976, 262.
56. **Kutsuna, M., Ed.,** *Minamata Disease,* Kumamoto University Press, Kumamoto, Japan, 1968.
57. **Clarkson, T. W. and Marsh, D. O.,** The toxicity of methylmercury in man: dose-response relationships in adult populations, in *Effects and Dose-Response Relationships in Adult Populations,* Norberg, G. F., Ed., Elsevier, Amsterdam, 1976, 246.
58. **Hunter, D., Bomford, R. R., and Russell, D. S.,** Poisoning by methylmercury compounds, *Q. J. Med.,* 9, 193, 1940.
59. **Nixson, G. S., Paxton, G. D., and Smith, H.,** Estimation of mercury in human enamel by activation analysis, *J. Dent. Res.,* 44, 654, 1965.
60. **Frykholm, K. O.,** Mercury from dental amalgam, *Acta Odont. Scand. Suppl.,* 22, 15, 1957.
61. **Orban, B.,** Discolorations of the oral mucous membranes by metallic foreign bodies, *J. Periodontol.,* 17, 55, 1946.
62. **Weathers, D. R. and Fine, R. M.,** Amalgam tattoo of oral mucosa, *Arch. Dermatol.,* 110, 727, 1974.
63. **Eley, B. M., Garrett, J. R., and Harrison, J. D.,** Analytical ultrastructural studies on implanted dental amalgam in guinea pigs, *Histochem. J.,* 8, 647, 1976.
64. **Harrison, J. D., Rowley, P. S. A., and Peters, P. D.,** Amalgam tattoos: light and electron microscopy and electron probe microanalysis, *J. Pathol.,* 121, 83, 1977.
65. **Cook, T. A. and Yates, P. O.,** Fatal mercury intoxication in a dental surgery assistant, *Br. Dent. J.,* 127, 553, 1969.
66. **Merfield, D. P., Taylor, A., Gemmell, D. M., and Parrish, J. A.,** Mercury intoxication in a dental surgery following unreported spillage, *Br. Dent. J.,* 141, 179, 1976.

67. **Battistone, G. C., Summons, D. W., and Miller, R. A.,** Mercury excretion in military dental personnel, *Oral Surg.*, 35, 47, 1973.
68. **Marks, V. and Taylor, A.,** Urinary mercury excretion in dental workers, *Br. Dent. J.*, 146, 269, 1979.
69. **Battistone, G. C., Hefferren, J. J., Miller, R. A., and Cutright, D. E.,** Mercury; its relation to the dentists health and dental practice characteristics, *J. Am. Dent. Assoc.*, 92, 1182, 1976.
70. **World Health Organization,** International study of normal values for toxic substances, *WHO Occupa. Health*, 66, 39, 1966.
71. **Stevens, J. T., Box, J. M., and Pelleu, G. B.,** Mercury vapor levels in dental spaces, *Mil. Med.*, 140, 114, 1975.
72. **Hefferen, J. J.,** Mercury surveys of the dental officers, *J. Am. Dent. Assoc.*, 89, 902, 1974.
73. **Miller, S. L., Domey, R. G., and Elston, S. F.,** Mercury vapour levels in the dental office, *J. Am. Dent. Assoc.*, 89, 1084, 1974.
74. **Schneider, M.,** An environmental study of mercury contamination in dental offices, *J. Am. Dent. Assoc.*, 89, 1092, 1974.
75. **Cutright, D. E., Miller, R. A., and Battistone, G. C.,** Systemic mercury levels caused by inhaling mist during high-speed amalgam grinding, *J. Oral. Med.*, 28, 100, 1973.
76. **Lenihan, J. M. A., Smith, H., and Harvey, W.,** Mercury hazards in dental practice, *Br. Dent. J.*, 135, 365, 1973.
77. **Buckwald, H.,** Exposure to dental workers to mercury, *Am. Ind. Hyg. Assoc. J.*, 33, 492, 1972.
78. **Nixson, G. S. and Rowbotham, T. C.,** Mercury hazards associated with high speed mechanical amalgamators, *Br. Dent. J.*, 131, 308, 1971.
79. **Stewart, F. H. and Stradling, G. N.,** Monitoring techniques for mercury vapour in dental surgeries, *Br. Dent. J.*, 131, 299, 1971.
80. **Rupp, N. W. and Paffenbarger, G. C.,** Significance to health of mercury used in dental practice in a review, *J. Am. Dent. Assoc.*, 82, 1401, 1971.
81. **Chandler, H. H., Rupp, N. W., and Paffenbager, G. C.,** Poor mercury hygiene from ultrasonic amalgam condensation, *J. Am. Dent. Assoc.*, 82, 553, 1971.
82. **Gronka, P. A., Bonkoskie, R. L., Tomchek, G. J., Bach, F., and Rakow, A. B.,** Mercury vapour exposure in dental offices, *J. Am. Dent. Assoc.*, 81, 923, 1970.
83. **Grossman, L. I.,** Amount of mercury vapour in the air of dental offices and laboratories, *J. Dent. Res.*, 28, 435, 1947.
84. **Frykholm, K. O.,** Exposure of dental personnel to mercury during work, *Sven. Tandlaek. Tidskr.*, 63, 763, 1970.
85. **Joselow, M. M., Goldwater, L. J., Alvarez, A., and Herndon, J.,** Absorption and excretion of mercury in man. XV. Occupational exposure during dentistry, *Arch. Environ. Health*, 17, 39, 1968.
86. **Eames, W. B., Gasper, J. D., and Mohler, H. C.,** The mercury enigma in dentistry, *J. Am. Dent. Assoc.*, 92, 1199, 1976.
87. Council on Dental Materials and Devices, American Dental Association, Recommendations in dental mercury hygiene, *J. Am. Dent. Assoc.*, 96, 487, 1978.
88. **Hefferren, J. J. and Rao, G. S.,** Toxicity of mercury, in *Biocompatibility of Dental Materials*, Smith, D. C. and Williams, D. F., Eds., CRC Press, Boca Raton, Fla., in press.

Chapter 16

ARGYRIA: SILVER IN BIOLOGICAL TISSUES

R. J. Pariser

TABLE OF CONTENTS

I. DEFINITION

Argyria (argyrism) is the deposition of silver in biological tissues. This deposition can occur at the site of contact with the silver-containing substance (localized argyria) or at sites distant from that of the original contact (generalized argyria). The term argyrosis has been applied to the deposition of silver in ocular tissues, either as localized argyria or as part of systemic argyria.

II. HISTORY

Silver was used in antiquity, possibly for medicinal purposes, but the first such recorded use appears in eighth century Moslem writings.[1,2] The first case of what was probably generalized argyria was described in the tenth century by Avicenna, who noted bluish discoloration of the eyes in a person who had ingested silver.[3] In 14th and 15th century alchemy the supposed relation between silver, the moon, and the brain was the basis for the medicinal use of silver in nervous disorders. The term "lunar caustic" (silver nitrate) derives from this association. Angelo Sala was said to have described darkening of the skin due to ingestion of silver in 1614.[2] The account of Fourcroy in 1791, as translated by Stillians,[4] presents a clear description of a case of generalized argyria and some surprisingly accurate speculations on the pathogenesis of the cutaneous discoloration:

> A protestant minister in the suburbs of Hamburg, suffering from an obstruction of the liver, took, on the advice of a quack, a solution of silver nitrate. After several months of this treatment, his skin gradually changed in color and finally became almost black. Remaining so for several years, it then began to get lighter in color.
>
> This observation, which was given me by M. Swediaur, is interesting, for it seems that the silver solution taken internally must act first on the stomach and then through all of the organs before being carried into the skin. It seems that the salt passed rapidly into the absorbent system. One would gladly follow further the course of this peculiar change in the skin.

The incidence of argyria apparently paralleled the popularity of medicinal silver. Increases in both occurred until the middle of the 19th century, after which was a lull lasting until the early 20th century, when various colloidal silver preparations were developed for topical use.[4] In the 1930s the use of silver arsphenamine in syphilotherapy gave rise to many new cases of generalized argyria.[5] In recent years argyria has become increasingly rare, but a recent report emphasizes the continuing potential for abuse of silver-containing products and the subsequent development of argyria.[6]

III. CLINICAL MANIFESTATIONS

A. Localized Argyria

Deposition of silver directly into tissue has been divided into the "tatoo" form, in which tiny particles of the metal enter the skin as foreign bodies, and the "smooth" form, in which soluble or colloidal silver enters and is deposited in the tissue. On rare occasion metallic silver objects implanted in tissue have been reported to discolor nearby tissues.[4]

The "smooth" form of localized argyria has been observed following topical application of silver-containing liquids to mucous membranes of the nose, pharynx, and bladder, the conjunctiva, or to skin wounds and burns.[2,7] Silver-amalgam tattooing of oral mucous membranes from dental fillings is probably analagous to the implantation type mentioned above. In all types the usual clinical finding is blue-black discoloration of the involved surface.

Silver deposition in the eye (argyrosis) is a special form of localized argyria in which the conjunctiva, cornea, and lacrimal apparatus are sites of deposition.[2] The term argyrosis has also been used to describe the ocular findings in generalized argyria, such as the limbic ring resembling the Kayser-Fleischer ring.[8]

Buckley reported a case of localized argyria of the fingers in a duplicating machine operator, whose hands were repeatedly immersed in solutions containing sodium argentothiosulfates.[9] Silver-containing granules were located around the sweat gland pores, the route through which percutaneous penetration presumably occurred.

B. Generalized Argyria

1. Pathogenesis

Generalized argyria results when silver is disseminated and deposited throughout the body, usually after absorption through the gastrointestinal tract or by parenteral administration. In numerous cases local application of silver-containing medicaments to the mucous membranes of the nose and throat resulted in generalized argyria. It is likely that in most, if not all, such cases the silver was inadvertently ingested and absorbed from the gastrointestinal tract. Likewise, cases in which absorption through the respiratory tract has been suggested could certainly have been due to swallowing of silver particles suspended in pulmonary secretions.[2] Occasional cases of generalized argyria following instillation of silver-containing liquids into the urethra, bladder, vagina, and rectum have been reported. Rare cases have been ascribed to percutaneous absorption.[2] The dissemination of silver presumably occurs via the hematogenous route, as persistent argyremia has been detected after silver ingestion.[10]

The quantity of silver intake required to produce clinically evident systemic argyria is apparently quite variable. The average is stated to be around 25 to 30 g,[11] but cases have been reported in which 6 g of oral silver nitrate or 6.3 g of intravenous silver arsphenamine (corresponding doses of elemental silver, 3.8 and 0.9 g, respectively) produced generalized argyria.[2] The capacity of human tissue to bind silver without apparent ill effects is attested to by the case of the "blue man" of Barnum and Bailey fame, whose total body content of silver was estimated at around 100 g.[11]

Traces of silver have been detected in various tissues in newborns and adults who had no history of silver intake.[12] Its presence apparently results from the ubiquitous nature of the element in trace quantities in the environment. Gaul and Staud found that the amount of silver in human tissue was proportional to age and concluded that silver gradually accumulated through low level exposure to the metal.[13] In normals as well as argyrics, this slow accumulation of silver proceeds without apparent harm; in fact, the only well-documented toxic effects of silver compounds result from the locally corrosive or irritative effects of large single doses.[14]

2. Clinical Findings

The clinical findings of generalized argyria are limited to discoloration of the skin, mucous membranes, and nails. Although occasional disorders have been described in argyrics,[2,15] the bulk of evidence indicates that silver deposition causes no physiologic derangements in any organ. The cutaneous discoloration may be a subtle bluish tint, a slatey-gray color, or a deep blue-black. The cadaveric hue has been described as "resembling a corpse suddenly come to life."[16]

The relative importance of melanin and silver in the production of argyric color is unsettled. The deposition of silver is probably uniform throughout the skin, but the pigmentation is most pronounced in sun-exposed areas.[6] It has been suggested that silver directly stimulates the production of melanin, especially in areas subjected to light exposure.[17,18] With rare exceptions, the pigment of argyria is stable and persists indefinitely even after silver intake has stopped. Argyria in children has been reported to improve with age, presumably by dilution of the pigment as the surface area of the skin increases.[4]

The mucous membranes, like the skin, are discolored in generalized argyria. If the silver-containing product directly contacts a mucous membrane, an element of localized argyria may accentuate this finding. The nail finding in generalized argyria is an azure-blue discoloration of the proximal nail bed. The abnormal color extends slightly beyond the lunula of the fingernails. Toenail changes occur less commonly.[19,20]

IV. PATHOLOGY

A. Localized Argyria

In localized argyria of the "smooth" type, the histopathology consists of deposition of brownish-black silver granules in the papillary dermis and along the dermoepidermal junction.[2] Buckley described the deposition as occurring preferentially around sweat glands.[9] Silver granules have occasionally been seen in the epithelium of the skin and mucous membranes, a finding absent in generalized argyria.[2]

B. Generalized Argyria

Because it deposits in connective tissue, silver has been found in virtually all organs in generalized argyria. In all tissues, the microscopic picture is the presence of dark granules in otherwise normal structures without degenerative or inflammatory changes. Due to its superiority in detecting small particles in thin sections, darkfield microscopy has proven valuable in locating areas of silver deposition in tissue.[6]

In the skin, the silver granules are preferentially deposited in the upper dermis, arranged along the elastic fibers in a pattern which has been likened to chains of streptococci. Granules are also scattered throughout the dermal collagen. Most characteristic is the localization of granules in the basal lamina of the secretory portion of the eccrine sweat glands, forming characteristic rings around the cross sections of the glands. The blood vessel walls, perineural tissue, arrector pili muscles, and dermoepidermal junction are occasional sites of deposition. With rare exceptions, granules are not found in the epidermis or sebaceous glands.[6,11,17,21,22]

Details of the gross and microscopic pathology of generalized argyria are listed in Table 1. In general the granules deposit in the connective tissue and vessels of most organs, but tend to be excluded from their parenchyma.

Two recent studies reached different conclusions as to the chemical nature of the granules in the skin or argyrics. In a case due to silver nitrate ingestion, Buckley and Terhaar[17] concluded that the deposits were metallic silver, based on differential solubility methods. Pariser,[6] using the same method, concluded that they were silver sulfide in a study of three cases arising from the use of silver protein solutions. Silver sulfide has also been detected in localized argyria.[21]

V. DIFFERENTIAL DIAGNOSIS

The differential diagnosis of generalized argyria includes a variety of diseases in which cutaneous discoloration occurs (Table 2). In those conditions due to abnormal color of the blood (cyanosis, polycythemia, and methhemoglobinemia), the color is blanched by pressure. Cyanosis in particular has been a common cause of confusion. Many argyric patients have been subjected to fruitless searches for cardiopulmonary disease.[22,23] The other conditions listed in Table 2 can generally be distinguished by their associated clinical findings. Deposition of other heavy metals, particularly bismuth, may be more difficult to differentiate from argyria.[24,25]

Table 1
PATHOLOGY OF GENERALIZED ARGYRIA[2,11]

Organ	Gross color changes	Pattern of silver granule deposition
Skin	Slatey blue-gray discoloration, especially in sun-exposed areas; azure lunules	Elastic fibers in dermis, dermal blood vessel walls, basal lamina of secretory coils of eccrine glands
Heart	Slatey-blue discoloration of pericardium, myocardium, epicardium	Pericardium, subendocardial space, between (not in) myocardial cells
Vascular tree	Blue-black vessel walls, expecially in atheromatous plaques; veins less involved	Throughout all layers of vessel walls
Gastrointestinal tract	Blue-gray discoloration of mucous membranes and peritoneum	None in epithelia or glandular cells; granules in connective tissue of mucosa, submucosa, and muscularis, especially in elastic fibers
Liver	Capsule and branches of portal vein bluish	Connective tissue of capsule and periportal areas; hepatic and portal vein walls; none in hepatocytes or Kupfer cells
Pancreas	Slight bluish discoloration	Connective tissue around and in the pancreas; none in acinar or islet cells
Spleen	Brown to black surface; blue cut surface	Capsule and connective tissue fibers; some granules in macrophages
Lymph nodes	Dark, especially mesenteric nodes	In reticular spaces; some in macrophages
Kidney	Grayish surface; dark color at corticomedullary junction and papillae of pyramids	Glomerular basement membrane, membrane of Henle's loop, tubules, and collecting ducts; none in epithelial cells
Respiratory system	Trachea, vocal cords, epiglottis brownish colored; slight discoloration of lung parenchyma	Elastic fibers in pleura and parenchyma
Nervous system	Meninges and choroid plexus dark	None in neural parenchyma; granules in connective tissue and choroid plexus

Table 2
DIFFERENTIAL DIAGNOSIS OF ARGYRIA

Abnormal color of blood
- Cyanosis
- Polycythemia
- Methhemoglobinemia

Abnormal skin color due to disease
- Addison's disease
- Hemochromatosis
- Melanosis due to melanoma
- Wilson's disease

Abnormal skin color due to drugs
- Antimalarials[26,27]
- Chlorpromazine[28]
- Gold[21]
- Mercury[29]
- Bismuth[24,25]

Abnormal nail color
- Wilson's disease[30]
- Pseudomonas onychia[31,32]
- Quinacrine administration[33]
- Mercury inunction[34,35]

Eye findings
- Kayser-Fleischer ring of Wilson's disease[8]

VI. TREATMENT

It is generally agreed that no effective treatment for argyria exists. Reduction of skin pigment following intradermal injection of potassium ferricyanide and sodium thiosulfate has been reported, but it is not clear whether silver or melanin was removed.[36-38] In one case where it was measured, this treatment did not alter the silver content of the skin.[4] Although melanin may contribute to the discoloration of argyria, efforts directed at reducing melanin pigment in argyrics have apparently not been made.

Generalized argyria is now a decidely rare disorder, but the potential for the production of new cases still exists, since a variety of silver-containing medications are still in use.[6] The use of colloidal silver solutions such as silver protein (Argyrol®), silver iodide (Neo-Silvol®), or silver nitrate on mucous membranes of the mouth, nose, or throat probably represents the greatest threat, especially if these preparations can be obtained without proper medical supervision. Silver sulfadiazene, in widespread use topically to treat burns, has not yet been associated with localized or generalized argyria. Since treatment of argyria is ineffective, close supervision of patients using silver medicinally or avoidance of such products altogether is essential if permanent deposition is to be avoided.

REFERENCES

1. **Gager, L. T. and Ellison, E. M.**, Generalized (therapeutic) argyria, *Int. Clin.*, 4, 118, 1935.
2. **Hill, W. R. and Pillsbury, D. M.**, *Argyria — The Pharmacology of Silver*, Williams & Wilkins, Baltimore, 1939.
3. **Aryres, S., Jr.**, Localized argyria following treatment of burn with tannic acid and silver nitrate, *Arch. Dermatol.*, 38, 645, 1938.
4. **Stillians, A. W.**, Argyria, *Arch. Dermatol.*, 35, 67, 1937.
5. **Gaul, L. E. and Staud, A. H.**, Clinical spectroscopy: quantitative distribution of silver in the body or its physiopathologic retention as a reciprocal of the capillary system, *Arch. Dermatol.*, 3, 433, 1935.
6. **Pariser, R. J.**, Generalized argyria, *Arch. Dermatol.*, 114, 373, 1978.
7. **Plewig, G., Lincke, H., and Wolff, H. H.**, Silver-blue nails, *Acta Derm. Venereol.*, 57, 413, 1977.
8. **Whelton, M. J. and Pope, F. M.**, Azure lunules in argyria, *Arch. Int. Med.*, 121, 267, 1968.
9. **Buckley, W. R.**, Localized argyria, *Arch. Dermatol.*, 88, 531, 1963.
10. **Blumberg, H. and Carey, T. N.**, Argyremia: detection of unsuspected and obscure argyria by the spectrographic demonstration of high blood silver, *JAMA*, 103, 1521, 1934.
11. **Gettler, A. O., Rhoads, C. P., and Weiss, S.**, A contribution to the pathology of generalized argyria with a discussion of the fate of silver in the human body, *Am. J. Pathol.*, 3, 631, 1927.
12. **Czitober, H., Frishauf, H., and Leodolter, I.**, Quantitative untersuchungen bei universeller argyrose mittels neurtonenaktivierungsanalyze, *Virchows Arch. A*, 350, 44, 1970.
13. **Gaul, L. E. and Staud, A. H.**, Clinical spectroscopy. Seventy cases of generalized argyrosis following organic and colloidal silver medication, including a biospectrometric analysis of ten cases, *JAMA*, 104, 1387, 1935.
14. **Goodman, L. C. and Gilman, A., Eds.**, *The Pharmacological Basis of Therapeutics*, 5th ed., MacMillan, New York, 1975.
15. **Zech, P., Colon, S., Labeeuw, R., Blanc-Brunat, N., Richard, P., and Perol, M.**, Syndrome néphrotique avec dépôt d'argent, *Nouv. Presse Med.*, 2, 161, 1973.
16. **Bryant, B. L.**, Argyria resulting from intranasal medication, *Arch. Otolaryngol.*, 31, 127, 1940.
17. **Buckley, W. R. and Terhaar, C. J.**, The skin as an excretory organ in argyria, *Trans. St. John's Hosp.*, 59, 34, 1973.
18. **Mehta, A. C. and Butterworth, K. D.**, Argyria: electron microscopic study of a case, *Br. J. Dermatol.*, 78, 175, 1966.
19. **Hill, W. R. and Montgomery, H.**, Argyria: with special reference to the cutaneous histopathology, *Arch. Dermatol.*, 44, 588, 1941.
20. **Prose, P. H.**, An electron microscopy study of human generalized argyria, *Am. J. Pathol.*, 42, 293, 1963.
21. **Buckley, W. R., Oster, C. F., and Fassett, D. W.**, Localized argyria. II. Chemical nature of the silver containing particles, *Arch. Dermatol.*, 92, 697, 1965.
22. **Rich, L. L., Epinette, W. W., and Nasser, W. K.**, Argyria presenting as cyanotic heart disease, *Am. J. Cardiol.*, 30, 290, 1972.
23. **Levine, S. A. and Smith, J. A.**, Argyria confused with heart disease, *N. Engl. J. Med.*, 226, 682, 1947.
24. **Spiegel, L.**, A discoloration of the skin and mucous membranes resembling argyria, following the use of bismuth and silver arsphenamine, *Arch. Dermatol.*, 23, 266, 1931.
25. **Lueth, H. C., Sutton, D. C., McMullen, C. J., and Muehlberger, C. W.**, Generalized discoloration of skin resembling argyria following prolonged oral use of bismuth: a case of "bismuthia," *Arch. Int. Med.*, 57, 1115, 1936.
26. **Zachariae, H.**, Pigmentation of skin and oral mucosa after prolonged treatment with chloroquin, *Acta Derm. Venereol.*, 43, 149, 1963.
27. **Tuffanelli, D., Abraham, R. K., and Dubois, E. I.**, Pigmentation from antimalarial therapy, *Arch. Dermatol.*, 88, 419, 1963.
28. **Hays, G. B., Lyle, C. B., Jr., and Wheeler, C. E.**, Slate-gray color in patients recieving chlorpromazine, *Arch. Dermatol.*, 90, 1, 1964.
29. **Burge, K. M. and Winkleman, R. K.**, Mercury pigmentation — an electron microscopic study, *Arch. Dermatol.*, 102, 51, 1970.
30. **Bearn, A. G. and McKusick, V. A.**, An unusual change in the finger nails in two patients with hepatolenticular degeneration (Wilson's disease), *JAMA*, 166, 904, 1958.
31. **Bauer, M. F. and Cohen, H.**, The role of *Pseudomonas aeruginosa* in infections about the nails, *Arch. Dermatol.*, 75, 394, 1957.

32. **Chernosky, M. F. and Dukes, C. D.,** Green nails: the importance of *Pseudomonas aeruginosa* in onychia, *Arch. Dermatol.,* 88, 548, 1963.
33. **Lutterlock, C. H. and Shallenberger, D. L.,** Unusual pigmentation developing after prolonged suppressive therapy with quinacrine HCl, *Arch. Dermatol.,* 53, 349, 1946.
34. **Calloway, J. L.,** Transient discoloration of the nails due to mercury bichloride, *Arch. Dermatol.,* 36, 62, 1937.
35. **Butterworth, T. and Stream, L. P.,** Mercurial pigmentation of the nails, *Arch. Dermatol.,* 88, 55, 1963.
36. **Stillians, A. W. and Lawless, T. K.,** An intracutaneous method in treating argyria, *Arch. Dermatol.,* 17, 153, 1928.
37. **Stillians, A. W. and Lawless, T. K.,** The intradermal treatment of argyria, *JAMA,* 92, 20, 1929.
38. **Stillians, A. W.,** Argyria: the Practical Value of its Endermic Treatment, Rep. 8th Int. Cong. Dermatol. Syphilis, 1930, 714.

INDEX

A

C

Coronary atherosclerosis, aluminum levels in, I: 192

E

F

G

H

I

K

L

M

N

O

P

Q

R

S

T

Z